This book outlines the basic science underlying the prediction of stress and velocity distributions in granular materials. It takes the form of a textbook suitable for post-graduate courses, research workers and for use in design offices. The nature of a rigid-plastic material is discussed and a comparison is made between the Coulomb and Conical (extended von Mises) models. The methods of measuring material properties are described and an interpretation of the experimental results is considered in the context of the Critical State Theory.

The early chapters consider the traditional methods for predicting the forces on planar retaining walls and the walls of bunkers and hoppers. These approximate methods are described and their accuracy discussed. Later chapters give details of the exact methods of stress and velocity prediction, covering both the radial stress and velocity fields and the method of characteristics. The analysis of stress and velocity discontinuities is also considered as is the prediction of the mass/core flow transition. The final chapter covers the discharge rate of materials through orifices, dealing with both the correlations of experimental results and theoretical prediction. The influence of interstitial pressure gradients is also considered leading to an analysis of the flow of fine materials and the effects of air-augmentation. The book ends with an assessment of Jenike's method for predicting the circumstances under which cohesive arching prevents flow.

The book will be an invaluable text for all those working with or doing research into granular materials. Exercises and solutions are provided which will be particularly useful for the student.

STATICS AND KINEMATICS
OF GRANULAR MATERIALS

STATICS AND KINEMATICS
OF GRANULAR MATERIALS

R. M. NEDDERMAN

Department of Chemical Engineering
University of Cambridge
and Ely Fellow of
Trinity College, Cambridge

CAMBRIDGE
UNIVERSITY PRESS

Published by the Press Syndicate of the University of Cambridge
The Pitt Building, Trumpington Street, Cambridge CB2 1RP
40 West 20th Street, New York, NY 10011-4211, USA
10 Stamford Road, Oakleigh, Victoria 3166, Australia

First published 1992

Printed and bound in Great Britain by
Woolnough Bookbinding Ltd, Irthlingborough, Northamptonshire

A catalogue record for this book is available from the British Library

Library of Congress cataloguing in publication data
Nedderman, R. M.
Statics and kinematics of granular materials / R. M. Nedderman.
p. cm.
Includes bibliographical references
ISBN 0-521-40435-5 (hardback)
1. Granular materials. 2. Strains and stresses. 3. Kinematics.
I. Title.
TA418.78.N44 1992
620.1'9–dc20 91-39970 CIP

ISBN 0 521 40435 5 hardback

Contents

Notation

Note on subscripts

The co-ordinate variables, x, y, z, r, θ, χ are used as subscripts to denote the appropriate component of the parameter. Stresses have two subscripts, the first denoting the plane on which the stress acts and the second, the direction in which the associated force acts. The subscript w denotes conditions at the wall.

The subscripts A and P denote the active and passive cases.

The overbar denotes an average quantity.

Superscripts $^+$ and $^-$ denote conditions on either side of a discontinuity.

The list of symbols below gives only those parameters which appear on several occasions.

Symbols that appear only transiently and are defined in the text are not included.

a	Direction cosine (appendix 1); distance
b	Horizontal distance; breadth
c	Cohesion; parameter defined by equation (5.4.19)
c_e	Equivalent cohesion defined by equation (6.5.9)
d	Particle diameter
e	Centre of Mohr's strain rate circle; voids ratio as defined by equation (6.2.3)
$f(\ )$	Function
f_c	Unconfined yield strength
ff	Flow factor, defined by equation (10.8.6)
g	Acceleration due to gravity
h	Height; effective height of cone (§5.7)
h_c	Depth of tension cracking

k	Ratio μ_w/μ (chapter 4); parameter in Rosin–Rammler distribution (§6.3); permeability defined by equation (6.4.1); parameter in the Beverloo correlation, equation (10.2.4)
l	Length
m	Mass
m	Parameter defined by equation (5.8.10)
m^*	Parameter defined by equation (5.8.21)
m^{**}	Parameter defined by equation (5.8.28)
n	Parameter in Warren Spring equation (6.5.8); co-ordinate normal to a surface
p	Pressure; mean of the major and minor principal stresses
p^*	Distance of the centre of Mohr's circle from the point of intersection of the Coulomb line with the σ axis ($p^* = p + c \cot \phi$)
p_c	Value of p during consolidation
p_i	Isotropic pressure
p_m	Size distribution function by mass (§6.3)
P_n	Size distribution function by number (§6.3)
p_0	Reference pressure; value of p on the centre-line
q	Radius of Mohr's stress circle (§6.2); radial stress field parameter, $p/\gamma r$ (chapters 7 & 8)
q_0	Dimensionless surcharge, $Q_0/\gamma b$
r	Radial co-ordinate; radius
r^*	Dimensionless radial co-ordinate, $2r/D$
r_0	Radius of the free-fall arch
s	Co-ordinate along a surface; deviatoric stress (chapter 9)
t	Tan α (chapter 4)
u	Velocity in the x-direction; Cartesian co-ordinate (appendix 1)
u^*	Change in normal velocity across a discontinuity
v	Velocity in the y-direction; Cartesian co-ordinate (appendix 1)
v^*	Change in shear velocity across a discontinuity
v_i	Interstitial velocity
v_r, v_θ etc.	Velocities in appropriate co-ordinate directions
v_R	Relative velocity
v_s	Specific volume (§6.2)
v_0	Reference specific volume (§6.2)
w	Velocity in the z-direction
x	Cartesian co-ordinate; parameter in log-normal distribution (§6.3)
y	Cartesian co-ordinate; parameter in log-normal distri-

bution (§6.3)

z	Cartesian co-ordinate; distance down far wall (§4.8); parameter in Rosin–Rammler distribution (§6.3)
A	Area; parameter defined by equation (7.4.6) or equation (9.4.4); angle defined by equation (7.9.5)
B	Parameter defined by equation (5.8.8); parameter defined by equation (7.4.7) or equation (9.4.5)
C	Parameter in Warren Spring equation (6.5.8); parameter defined by equation (7.4.8) or equation (7.4.28) or equation (9.4.6); parameter in the Beverloo correlation, equation (10.2.2)
D	Diameter; parameter defined by equation (7.4.9) or equation (9.4.7)
D_0	Orifice diameter
D_a	Arithmetic mean diameter, defined by equation (6.3.31)
D_{max}	Parameter in Avrami distribution, equation (6.3.16)
D_g	Geometric mean diameter
D_{gm}	Geometric mean diameter by mass
D_H	Hydraulic mean diameter defined by equation (5.5.6)
D_s	Surface mean diameter, defined by equation (6.3.32)
D_v	Volume mean diameter, defined by equation (6.3.33)
D_{vs}	Volume/surface mean diameter, defined by equation (6.3.34)
E	Parameter defined by equation (7.4.10) or equation (9.4.8)
E_1	Parameter in Ergun equation, defined by equation (6.4.7)
E_2	Parameter in Ergun equation, defined by equation (6.4.8)
E_α	Parameter defined by equation (7.11.9)
E_β	Parameter defined by equation (7.11.10)
F	Force; parameter defined by equation (7.4.11) or equation (7.4.29) or equation (9.4.9)
FF	Flow function (§6.6)
$F(x,\phi_d)$	Function in the Rose and Tanaka correlation, equation (10.2.5)
$F(\theta)$	Velocity function defined by equation (8.4.12)
G	Shear modulus; plastic potential (§8.3); mass flow rate of gas (chapter 10)
H	Dimensionless depth, h/b
I_1, I_2, I_3	Stress invariants (§6.2)
J	Force on far wall
K	Rankine's coefficient of earth pressure (Janssen's constant) (chapters 3 & 5)
K_n, K_n'	Normalisation factors (§6.3)

L	Rotation matrix
M	Mass; parameter defined by equation (9.2.6); momentum flux (chapter 10)
M_c	Value of M in a compression test
M_e	Value of M in an extension test
N	Normal force; number of particles (§6.3)
P	Force; perimeter (§5.5); dimensionless form of p, $p/\gamma a$
P_0	Value of P on the axis of symmetry
P_m	Cumulative size distribution by mass
$P^{1/8}_n$	Cumulative size distribution by number
Q	Surcharge; volumetric flow rate
Q^*	Volumetric flow rate per unit width
Q_0	Uniform surcharge
R	Radius of Mohr's circle; radius; radius of Enstad element (§5.10); dimensionless radial co-ordinate, r/a
R'	Radius of Mohr's strain rate circle
Re	Reynolds number, defined by equation (6.4.10)
R_H	Hydraulic mean radius, defined by equation (5.5.12)
T	Ultimate tensile strength, defined by equation (6.5.8); total energy content, (§10.5)
U	Velocity; superficial velocity
U_i	Interstitial velocity
V	Velocity; volume; specific volume (§6.2); parameter in equation (10.4.3)
V_p	Velocity in the plug flow region
W	Weight; mass flow rate
W_B	Mass flow rate predicted
W_0	Mass flow rate in the absence of air-augmentation
X	Force; Enstad parameter, defined by equation (5.10.13); dimensionless distance x/a
Y	Enstad parameter, defined by equation (5.10.14); dimensionless distance y/a; yield function (§8.3)
Z	Dimensionless depth, z/b
α	Angle; hopper half-angle; parameter in equation (6.6.9)
β	Angle of inclination of the top surface; angle defined by equation (5.10.1); compressibility factor (§6.2)
γ	Shear strain (§3.2); weight density $\rho_b g$
δ	Effective angle of internal friction (§6.6); small quantity; angle defined in figure 7.28
δ_{ij}	Kroenecke delta function
ε	Void fraction; angle defined by equation (3.3.4)
ε_V	Volumetric strain

$\dot{\varepsilon}$	Direct strain rate
$\dot{\varepsilon}_V$	Volumetric strain rate
η	Angle between the wall and the vertical
θ	Angle; angular co-ordinate
ζ	Angle measured anticlockwise from the x-direction to the characteristic direction; angle measured anticlockwise from the x-plane to the plane of the discontinuity
κ	Parameter taking the value -1 in active failure and $+1$ in passive failure; parameter defined by equation (10.5.23)
λ	Angle defined by equation (2.3.10); compressibility parameter (chapters 6 & 10); plasticity parameter (chapters 8 & 9)
λ_S	Shape factor (§6.3)
μ	Coefficient of internal friction
μ_e	Equivalent coefficient of internal friction, defined by equation (6.5.9)
μ_f	Viscosity of fluid
μ_g	Viscosity of gas
ν	Angle of dilation
ξ	Angle measured anticlockwise from the plane of the discontinuity to the major principal plane (chapter 7); angle defined by equation (9.3.10)
ρ	Density
ρ_b	Bulk density
ρ_f	Fluid density
ρ_g	Gas density
ρ_s	Solid density
σ	Normal stress; standard deviation (§6.3)
$\sigma_1, \sigma_2, \sigma_3$	Principal stresses
σ_c	Normal stress under consolidation conditions (§6.5)
σ_{oct}	Octahedral stress, defined by equation (9.2.3)
σ_R	Reduced normal stress, σ/σ_c
σ_y	Yield stress
τ	Shear stress
τ_c	Shear stress under consolidation conditions
τ_R	Reduced shear stress, τ/σ_c
ϕ	Angle of internal friction
ϕ_c	Angle of friction measured in a compression test
ϕ_e	Equivalent angle of internal friction defined by equation (6.5.9)
ϕ_e	Angle of friction measured in an extension test (chapter 8)
ϕ_w	Angle of wall friction

ψ	Angle measured anticlockwise from the x-direction (or the r-direction of cylindrical co-ordinates) to the major principal stress direction; stream function (§8.2)
ψ^*	Angle measured anticlockwise from the r-direction of polar or spherical co-ordinates to the major principal stress direction
χ	Angular co-ordinate
ω	Angle defined by equation (3.7.8)
Δ	Increment; angle of end point locus (§6.6); angle defined by equation (7.12.4)
ΔP	Pressure difference
Λ	Angle defined by equation (2.5.8); compressibility parameter (§6.2)

1

Introduction

A granular material can be defined as any material composed of many individual solid particles, irrespective of the particle size. Thus the term granular material embraces a wide variety of materials from the coarsest colliery rubble to the finest icing sugar. The handling of granular materials is of the greatest importance in the chemical industry, it being estimated that on a weight basis, roughly one-half of the products and at least three-quarters of the raw materials are in the form of granular solids. When one adds to this the vast tonnages of wheat, sugar, iron ore, cement, sand and gravel that have to be stored and transported, the importance of granular materials becomes self-evident. The early dominance of oil in the history of the chemical industry has tended to emphasize the importance of fluid mechanics as one of the main constituents of a chemical engineer's education. It is only relatively recently that the economic advantage of a study of the behaviour of granular materials has been appreciated. One reason for this is that solids handling equipment has long been designed empirically and it is only over the last thirty years that the subject has been placed on a firm numerical basis. By contrast, the quantitative nature of fluid mechanics has been well-known since the work of Darcy in the 1840's.

Granular materials, or bulk solids as they are sometimes called, are usually stored in hoppers or bunkers. These can vary in size from a conventional salt-cellar, containing a few grams, to large installations holding several thousands, or even tens of thousands, of tonnes. The large bunkers storing agricultural products such as grain or sugar can be up to 20 m in diameter and 60 m in height and are a prominent feature of many rural landscapes. The words bunker, hopper, silo and bin tend to be used indiscriminately but it is recommended practice to

1

use the word 'bunker' if the walls are parallel, forming a container of constant cross-section and to use 'hopper' when the walls converge towards a relatively small opening at the base. The words 'bin' and 'silo' are general, covering both bunkers and hoppers, and also covering the frequent case of a cylindrical bunker surmounting a conical hopper as in figure 1.1. The British Code of Practice for silo design, issued by the British Materials Handling Board (BMHB, 1987), at present in draft form, recommends the use of the ugly compound 'silo–bin', but this will not be used here.

As implied in the title, this book is mainly concerned with two aspects of the behaviour of granular materials, both of direct relevance to the problems of storage and transportation of granular materials. The first part of the book, chapters 2 to 7, is concerned with the statics of granular materials. Here the objective is to predict the stress distributions within a granular material with particular emphasis on predicting the forces on the walls of the silo in which the material is stored. The second part deals principally with kinematics, the study of the motion of flowing granular materials. This topic has implications for the prediction of stress distributions, as it is found that frequently there are flowing and stagnant zones within a discharging bunker and that the forces on the walls depend on whether or not the material adjacent to the wall is moving. Finally, in the last chapter we pay attention to the dynamics of granular materials in order to predict the discharge rate from silos.

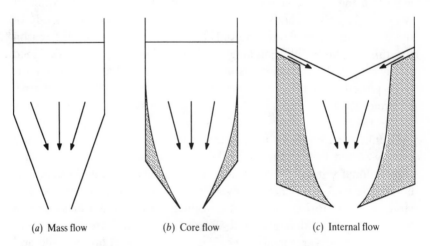

(*a*) Mass flow   (*b*) Core flow   (*c*) Internal flow

Figure 1.1 Flow patterns in discharging bins.

The analysis of stress and velocity distributions in granular materials is based on principles laid down in the eighteenth century by Coulomb and which have been developed by the soil mechanicists. The emphasis in our case is however different. Whereas in soil mechanics the main objective is to prevent movement of the soil, the converse is true in bulk solids handling especially when considering the discharge rates from silos. Credit must be given to Jenike and his co-workers who in the 1960's first applied the principles of soil mechanics to silo design and developed design procedures that are still in common use. His three famous bulletins and in particular, his third bulletin, take the form of design manuals containing a mixture of rigorous analyses and empirically derived recommended procedures. This style has been followed some 20 years later by Arnold and co-workers (1980) and in the various national codes of practice for hopper design.

The objective of this book is somewhat different. It is intended primarily as an undergraduate or postgraduate textbook in which the basic principles are treated rigorously. The emphasis is on the fundamental science and a careful distinction between established fact and rule-of-thumb is attempted. The subject matter of chapters 1 to 6 and chapter 10 is based on lectures given as part of a course leading to the Master of Engineering (M. Eng.) degree in the Department of Chemical Engineering at Cambridge. The remaining chapters contain material required by research students working on the flow of granular materials. Some of the examples given in the book are based on questions set in Part II of the Chemical Engineering Tripos at Cambridge with kind permission of the Council of the Senate.

The writers of codes of practice and design manuals are obliged to propose some recommendation for all pertinent aspects of silo construction and where no established scientific facts are available, they have to resort to empiricism and propose rules-of-thumb. Unfortunately the temptation to borrow an empirical result from a previous code is often irresistible and in the successive uncritical copying, what started as an admittedly empirical rule-of-thumb, often acquires the status of established fact. One of the objectives of this book is to examine these aspects of traditional wisdom, in an attempt to see which have some scientific foundation. The writer of a textbook does have at least one advantage over the writer of a design manual. When he reaches the limit of his competence he can resort to Aristotle's stratagem and freely admit that he does not know. The silo designer will therefore find this work incomplete but it is hoped that it will enable him to understand the basic principles on which the usual design procedures are based. Though primarily intended for undergraduate and postgraduate use, an attempt has been made to make the book comprehensible on its own so that it can be used by those who do not have the advantage of readily available advice.

In chapter 2 we will establish various fundamental concepts about the analysis of stress and strain. Many readers will already be familiar with these ideas and may wish to pass on directly to chapter 3. However their attention is drawn particularly to §2.3 and §2.5 in which the sign conventions for stress and strain rate are defined, as these differ from those commonly used in fluid mechanics and elasticity. In chapter 3 the nature of the idealised granular material, the so-called Coulomb material, is described and various approximate methods of predicting stress distributions in such materials are outlined in chapters 4 and 5. In chapter 6 we investigate the methods available for measuring the physical properties of a granular material and consider how accurately these conform to the idealised model. Chapter 7 considers the mathematically exact methods of stress analysis and we compare the results with the approximate results of chapters 4 and 5. The analysis of velocity distributions is considered in chapter 8 and this leads on to the consideration of an alternative to the Coulomb model of a granular material which is discussed in chapter 9. Chapters 7 to 9 are, however, mathematically complex and the less mathematically inclined reader might prefer to omit them and pass directly to chapter 10 which deals with discharge rates through orifices.

First, however, we need to define certain terms which will be used frequently throughout the book and discuss some of the more fundamental properties of granular materials.

Perhaps the most important single property of a granular material is the particle size and consequently many authors have sought to classify materials according to their mean particle size. Richards (1966), for example, proposes the classification given in table 1.1. Whilst this is helpful in a qualitative way, it has not been followed slavishly in this book where we will use the phrases 'granular material' and 'bulk solids' indiscriminately to denote all materials composed of many solid particles, irrespective of size.

The measure of particle size must be chosen with care. If the particles are spherical, the diameter is clearly the most convenient dimension but for non-spherical particles some characteristic dimension has to be selected. Frequently this is the equivalent spherical diameter, that is the diameter of the sphere of the same volume, but this is not the

Table 1.1

Particle size range	Name of material	Name of individual component
0.1 μm–1.0 μm	Ultra-fine powder	Ultra-fine particle
1.0 μm–10 μm	Superfine powder	Superfine particle
10 μm–100 μm	Granular powder	Granular particle
100 μm–3.0 mm	Granular solid	Granule
3.0 mm–10 mm	Broken solid	Grain

only possibility and other measures of particle size are discussed in chapter 6. Furthermore, most materials contain a range of particle sizes so that we must also consider the particle size distribution. It is found that the presence of a few fines can have a marked effect on the behaviour of the material, so that the simple classification based on mean particle size alone is an over-simplification. Particle size distributions are also considered in chapter 6.

Though ultimately one expects that we will be able to predict the majority of material properties from the particle size distribution, the subject has not yet developed that degree of understanding. Some considerable progress has been made, notably by Molerus (1982), and Briscoe and Adams (1987) but it is still convenient in many cases to treat a granular material as a continuum and to measure its bulk properties without enquiring in detail about their causation. In particular, both the density and frictional properties are more commonly measured than predicted and in this respect granular materials do not differ from conventional fluids. No doubt in time, we will be able to predict the density and viscosity of a liquid from its molecular properties, but these quantities are, for the present, always obtained experimentally.

There is, however, a particular problem with regard to density, since there are two densities of interest, the density of the particles themselves, which we will call the *solid density* and denote by ρ_s, and the density of the mixture of solid and interstitial gas which is known as the *bulk density* ρ_b. Provided the particles are not porous, the solid density can be measured by the usual techniques of liquid displacement and the bulk density can be obtained from the ratio of the mass and volume of a sample.

These two densities are related by

$$\rho_b = \rho_s(1 - \varepsilon) \tag{1.1}$$

where ε is the void fraction defined as the volumetric fraction of the material occupied by the interstitial gas. Strictly equation (1.1) should be written

$$\rho_b = \rho_s(1 - \varepsilon) + \rho_g\varepsilon \qquad (1.2)$$

where ρ_g is the gas density. However, since the gas density is typically one-thousandth of that of the solid, equation (1.1) is sufficiently accurate.

Whilst the particles themselves may be compressible, the change in solid density over the range of stresses normally encountered is usually small, so that ρ_s is effectively a constant for a given material. On the other hand, the bulk density is found to vary significantly with applied stress, mainly as a result of rearrangement of the particles. Unfortunately on reduction of the stress, the material does not necessarily expand and as a result the bulk density depends not only on the current stress in the material but also on its stress history. Thus for a given material ρ_s may be treated as a constant but the value of ρ_b will depend on the present *and* past treatment of the material.

When considering the flow pattern within a discharging bunker, it is usual to distinguish between mass and core flow. In a mass flow hopper, all the material is in motion as illustrated in figure 1.1(*a*). In such a hopper the first material to be loaded is the first to be discharged, giving the 'first in, first out' flow pattern. However, mass flow can only occur in comparatively narrow hoppers. If the hopper half-angle α is large the flow will be confined to a narrow core surrounded by stagnant material as illustrated in figures 1.1(*b*) and 1.1(*c*). If the core is narrower than the width of the silo, as in figure 1.1(*c*), the material near the top will cascade down the top surface into the flowing core and will be discharged before material at a lower level, giving the 'first in, last out' pattern. However, the width of the flowing core normally increases with height and for a tall, narrow silo the flowing core will reach the upper parts of the walls. The Draft British Code of Practice (BMHB, 1987) subdivides what is usually known as core flow, i.e. the patterns illustrated by figures 1.1(*b*) and 1.1(*c*), into core flow in the strict sense in which the core reaches the upper parts of the walls, as in figure 1.1(*b*), and internal flow in which the flowing core never reaches the wall, as in figure 1.1(*c*). In this work we will use the older definitions of mass flow, in which all the material is moving, and core flow, in which some of the material is stagnant.

2

The analysis of stress and strain rate

2.1 Introduction

In this chapter we will develop relationships for the analysis of stress and rates of strain which will be familiar to many readers from their knowledge of fluid mechanics or elasticity. Such readers may wish to proceed directly to chapter 3, but their attention is drawn to §2.3 and §2.5 in which the sign conventions used in this book are defined, since these differ from those commonly used in fluid mechanics.

The nature of forces and stresses is discussed in §2.2 and in particular we note that force is a vector but that stress is a somewhat more complicated quantity and cannot therefore be resolved by the familiar techniques of vector resolution. The simplest method for determining the stress components on a particular plane is known as Mohr's circle and this is derived in §2.3 and compared with alternative methods in appendix 1.

Forces are generated as a result of stress gradients and these are related to the acceleration of the material by Euler's equation which is derived in §2.4. Finally in §2.5, we define the strain rate in terms of the velocity gradients and note that strain rates, like stresses, can be analysed by means of a Mohr's circle.

In an attempt to reduce the tedium of this chapter, most of the derivations are presented only for Cartesian co-ordinates and the results for other co-ordinate systems are given, without derivation, in the appendices.

2.2 Force, stress and pressure

It is assumed that the reader is fully familiar with the concept of force and with the fact that force is a vector. As a consequence of its vectorial nature, a force F can be expressed in terms of its components F_x, F_y and F_z parallel to the three co-ordinate directions and by convention these components are taken to be positive when acting in the direction of the co-ordinate increasing. Forces can be resolved in any chosen direction by the techniques of vector algebra but it is more usual to rely on a graphical construction known as the triangle of forces. This construction is, however, so simple that it is often possible to write down the answer by inspection without the necessity of drawing the diagram itself. In particular the component of a force F in a direction inclined to it by the angle θ is $F \cos \theta$, a result sometimes known as the cosine law of vector resolution.

The concept of stress is less familiar and is best illustrated by considering an elementary cuboid with edges parallel to the co-ordinate directions as shown in figure 2.1. It is usual to name the faces of such a cuboid according to the directions of their normals and there are therefore two x-faces as shown in the figure. On each face there may be a force and we will denote that acting on one of the x-faces by F_x. Since the cuboid is of infinitesimal size the force on the other x-face will not differ significantly. The force F_x will not necessarily be normal to the x-face and we can resolve it into its components in the three co-ordinate directions, F_{xx}, F_{xy} and F_{xz}. Dividing by the area of the x-face, A_x, we obtain the stresses on that face and it is usual to distinguish between the normal stress σ_{xx}, obtained from F_{xx}, and the other two stresses which are called shear stresses and denoted by τ_{xy} and τ_{xz}.

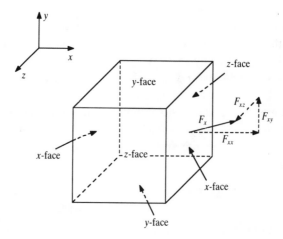

Figure 2.1 Components of force acting on the face of an elementary cuboid.

There are similarly three stress components on each of the two remaining pairs of faces, so that in total we have nine stress components which may be written in the form,

$$
\begin{array}{ccc}
\sigma_{xx} & \tau_{xy} & \tau_{xz} \\
\tau_{yx} & \sigma_{yy} & \tau_{yz} \\
\tau_{zx} & \tau_{zy} & \sigma_{zz}
\end{array}
$$

It should be noted that in this formulation, the first subscript refers to the face on which the stress acts and the second subscript to the direction in which the associated force acts. A vector has three components in three-dimensional space and it is therefore clear that a stress, having nine components, cannot be a vector.

The components of a stress in any other set of co-ordinate directions can be obtained by matrix manipulations or by the techniques of tensor analysis. Fortunately most of the problems with which we are concerned are essentially two-dimensional; for such systems the very much simpler device known as Mohr's circle can be used and this has particular advantages as it fits conveniently with the basic relationships governing the behaviour of granular materials. Mohr's circle is considered in the next section and the matrix methods of co-ordinate transformation are given in appendix 1.

Circumstances can occur in which all three normal stresses are equal and all the shear stresses are zero. This is more common in the field of fluid mechanics and is known as a state of isotropic pressure. Pressure, which is a scalar since it acts equally in all directions, is therefore a particular case of a stress. Unfortunately, the words 'pressure' and 'stress' tend to be used indiscriminately. For example the Draft British Code of Practice (BMHB, 1987) recommends the use of 'stress' within the material but denotes the stresses exerted on the containing walls as 'pressures'. This usage is contrary to that commonly found in mechanics and will be avoided in this book. We will use the word 'stress' to apply to both internal and external stresses and reserve the word 'pressure' to the cases when the stress state is isotropic or when we need to consider the motion of the interstitial medium, which is usually air.

2.3 Two-dimensional stress analysis – Mohr's circle

Many of the problems of industrial importance have sufficient symmetry, either planar or cylindrical, to make a two-dimensional analysis realistic.

This is a great convenience as the manipulation of stresses in two-dimensional systems is very much easier than in three dimensions. The rather more complicated analysis of stress in three-dimensional systems is outlined in appendix 1.

The method we will use is known as Mohr's circle. This method does, however, have one disadvantage in that it requires a different sign convention from that required for matrix or tensorial manipulation of stresses. Some authors have attempted to combine the sign conventions by unsatisfactory devices such as reversing the signs of shear stresses if the subscripts are in alphabetical order. It is the opinion of the present author that it should be acknowledged that different sign conventions are necessary and that one should keep to the one appropriate for the technique in use.

Since granular materials can only rarely take tension, it is convenient to take compressive stresses as positive and, having selected this convention, the use of Mohr's circle requires that shear stresses should be taken as positive when acting on the element in an anticlockwise direction. Recalling that Mohr's circle is applicable only to two-dimensional situations, we can illustrate our sign convention by means of figure 2.2. The directions in which the stresses acting on the element are numerically positive are shown by the arrows in this figure.

If we take moments about an axis normal to the paper we find that, for stability,

$$\tau_{xy} = -\tau_{yx} \qquad (2.3.1)$$

and thus the shear stresses occur as a complementary pair.

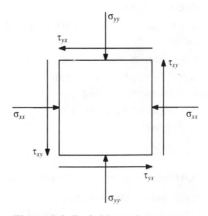

Figure 2.2 Definition of stresses.

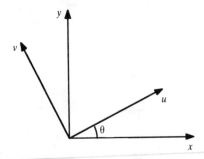

Figure 2.3 Definition of co-ordinate axes.

Let us consider a set of Cartesian axes (x,y) and a second set (u,v) where the u-axis is inclined at angle θ anticlockwise from the x-axis as shown in figure 2.3. Our objective is to predict the stress components σ_{uu}, τ_{uv}, σ_{vv} and τ_{vu} from the known values of σ_{xx}, τ_{xy}, σ_{yy} and τ_{yx}. We can do this by considering a wedge-shaped element of unit depth normal to the paper having faces parallel to the x and y co-ordinates and a face inclined at angle θ anticlockwise from the x-face as shown in figure 2.4. The stresses on the x- and y-faces are (σ_{xx}, τ_{xy}) and (σ_{yy}, τ_{yx}) and are positive in the directions shown. The stresses on the remaining face are strictly σ_{uu} and τ_{uv} but they will be denoted simply by σ and τ for convenience.

If we take the area of the hypotenuse plane to be unity, the area of the x-face will be $\cos\theta$ and that of the y-face will be $\sin\theta$. Thus the forces on the x-face are $\sigma_{xx}\cos\theta$ and $\tau_{xy}\cos\theta$ and those on the y-face are $\sigma_{yy}\sin\theta$ and $\tau_{yx}\sin\theta$.

Resolving the forces in the direction of σ we have

$$\sigma = \sigma_{xx}\cos\theta\cos\theta - \tau_{xy}\cos\theta\sin\theta + \sigma_{yy}\sin\theta\sin\theta + \tau_{yx}\sin\theta\cos\theta$$
$$(2.3.2)$$

and resolving in the direction of τ gives

$$\tau = \sigma_{xx}\cos\theta\sin\theta + \tau_{xy}\cos\theta\cos\theta - \sigma_{yy}\sin\theta\cos\theta + \tau_{yx}\sin\theta\sin\theta$$
$$(2.3.3)$$

Substituting from equation (2.3.1) and recalling that

$$\cos 2\theta = 1 - 2\sin^2\theta = 2\cos^2\theta - 1 \qquad (2.3.4)$$

and that

$$\sin 2\theta = 2\sin\theta\cos\theta \qquad (2.3.5)$$

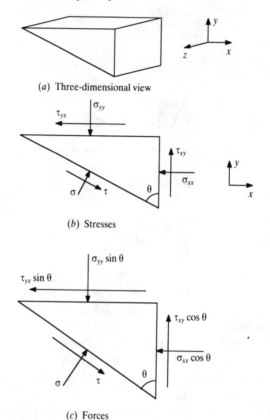

(a) Three-dimensional view

(b) Stresses

(c) Forces

Figure 2.4 Stresses and forces on a wedge-shaped element.

we have

$$\sigma = \tfrac{1}{2}(\sigma_{xx} + \sigma_{yy}) + \tfrac{1}{2}(\sigma_{xx} - \sigma_{yy})\cos 2\theta - \tau_{xy}\sin 2\theta \qquad (2.3.6)$$

and

$$\tau = \tfrac{1}{2}(\sigma_{xx} - \sigma_{yy})\sin 2\theta + \tau_{xy}\cos 2\theta \qquad (2.3.7)$$

Defining the symbols p, R and λ by

$$p = \tfrac{1}{2}(\sigma_{xx} + \sigma_{yy}) \qquad (2.3.8)$$

$$R^2 = \left(\frac{\sigma_{xx} - \sigma_{yy}}{2}\right)^2 + \tau_{xy}{}^2 \qquad (2.3.9)$$

$$\tan 2\lambda = \frac{2\tau_{xy}}{(\sigma_{xx} - \sigma_{yy})} \qquad (2.3.10)$$

equations (2.3.6) and (2.3.7) take the form

$$\sigma = p + R\cos(2\theta + 2\lambda) \qquad (2.3.11)$$

$$\tau = R\sin(2\theta + 2\lambda) \qquad (2.3.12)$$

It can be seen that these two equations define a circle on (σ,τ) axes which is known as Mohr's circle. The circle has its centre at the point $(p,0)$ and its radius is R as shown in figure 2.5. Every point on the circle represents the combination of σ and τ on some plane and in particular the stresses on the x- and y-planes, (σ_{xx},τ_{xy}) and (σ_{yy},τ_{yx}) are marked by the points X and Y respectively. From equations (2.3.11) and (2.3.12) we see that the stresses on a plane inclined at θ anticlockwise from the x-plane are given by the end of the radius inclined at 2θ anticlockwise from the radius to the point X. In particular the stresses on the y-plane, for which θ is 90° are therefore given by the other end of the diameter from the point X. Inevitably therefore, $\tau_{xy} = -\tau_{yx}$ and it is this result that necessitates the use of a sign convention in which complementary shear stresses are equal and opposite and which prohibits the use of the sign convention required for matrix or tensor manipulation.

The stresses on the u-plane (σ,τ), or more correctly (σ_{uu},τ_{uv}), are given by the point U on figure 2.5. We see from equation (2.3.11) and (2.3.12) that the radius to the point U is inclined at an angle 2θ to the radius to point X. Thus, we move round Mohr's circle in the same direction as we rotate our axes but through *twice* the angle. If we had

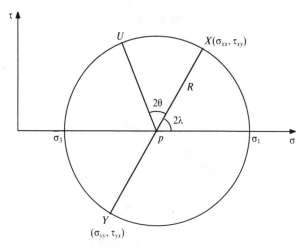

Figure 2.5 Mohr's circle for stresses.

taken as a sign convention that normal stresses were positive when compressive and that shear stresses were positive when clockwise, we would have found that rotation in Mohr's circle was in the opposite direction to that in physical space. This is clearly less convenient than the sign convention we have adopted.

There are two planes of particular interest, namely those on which the shear stress is zero. These are indicated in figure 2.5 and are known as the principal planes. The corresponding stresses σ_1 and σ_3 are called the major and minor principal stresses. From the figure it can be seen that the major principal plane lies at an angle λ clockwise from the x-plane and that the minor principal plane lies at an angle $90 - \lambda$ anticlockwise from the x-plane. Since the principal planes lie at opposite ends of a diameter, they are inevitably at right-angles to each other.

We can illustrate this analysis with a simple example. Let us consider a situation in which $\sigma_{xx} = 12$ kN m^{-2}, $\sigma_{yy} = 4$ kN m^{-2} and $\tau_{xy} = 3$ kN m^{-2}. We can identify the points X and Y since they have co-ordinates $(12,3)$ and $(4,-3)$ and plot them on a Mohr's diagram as in figure 2.6. These two points lie on opposite ends of the diameter, which can therefore be drawn. The centre of the circle can be found by inspection or from equation (2.3.8) as the point $(8,0)$. Consideration of the triangle OXA or equation (2.3.9) gives $R = 5$ kN m^{-2}. Thus the principal stresses are 8 ± 5 kN m^{-2} i.e. $\sigma_1 = 13$ kN m^{-2} and $\sigma_3 = 3$ kN m^{-2}. Also by inspection tan $2\lambda = 3/4$ or $\lambda = 18.43°$. Thus the major principal stress acts in a direction inclined at $18.43°$ clockwise from the x-axis. We can find the stresses on a plane at $30°$ anticlockwise from the x-plane by constructing the radius at $2 \times 30° = 60°$ anticlock-

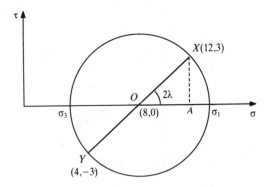

Figure 2.6 Example of the use of Mohr's circle.

wise from the radius to point X. The co-ordinates of the end of this radius, (σ_{uu}, τ_{uv}) are clearly

$$\sigma_{uu} = p + R\cos(2\theta + 2\lambda) = 8 + 5\cos(60 + 36.86) = 7.40\,\text{kN m}^{-2}$$

$$\tau_{uv} = R\sin(2\theta + 2\lambda) = 4.96\,\text{kN m}^{-2} .$$

The stresses on the perpendicular plane are

$$\sigma_{vv} = 8 + 5\cos(36.86 + 60 + 180) = 8.60\,\text{kN m}^{-2}$$

$$\tau_{vu} = 5\sin(36.86 + 60 + 180) = -4.96\,\text{kN m}^{-2} .$$

It should be noted that Mohr's circle can be used as a graphical construction, in which case the results are inevitably approximate, or can be used as the basis for a geometrical analysis, as above, in which case the results are exact.

It can be seen that this two-dimensional analysis will be valid in real, three-dimensional, space if the force on the z-plane of figure 2.4 has no component in the x- or the y-direction. Thus the shear stresses τ_{zx} and τ_{zy} must be zero and consequently the stress σ_{zz} must be a principal stress. Therefore, in general we can draw a set of three nesting Mohr's circles as shown in figure 2.7. Each circle represents rotation about one of the three principal axes. The three principal stresses are given by the intersections of the circles with the σ axis and are known (despite the rules of English grammar) as the major, the intermediate and the minor principal stresses and are conventionally denoted by σ_1, σ_2 and σ_3 where $\sigma_1 > \sigma_2 > \sigma_3$. In almost all the analyses in this book we will be concerned only with the Mohr's circle containing both the major and minor principal stresses, as it is found

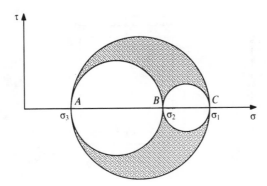

Figure 2.7 Mohr's circles for the rotation about the three principal axes.

experimentally that the value of the intermediate principal stress is irrelevant in a great many of the situations of interest. Thus, as a result of the nature of granular materials we can normally work in two dimensions and ignore the intermediate principal stress and its associated Mohr's circles.

The sign convention defined above cannot, in general, be used for three-dimensional systems, but can be adapted for systems of axial symmetry. Here we may use the convention as defined above in the positive quadrant, but in the negative quadrant we must work in mirror image as this quadrant becomes the positive quadrant when viewed from behind the paper.

2.4 The stress gradient and Euler's equation

If we consider the two-dimensional infinitesimal element shown in figure 2.8, we can note that the stresses on opposite sides of the element will differ if there is a stress gradient. We will denote these stress differences by $\delta\sigma_{xx}$ etc.

Recalling that the lengths of the sides are δx and δy and taking unit distance normal to the paper, we can evaluate the force per unit volume in the x-direction, P_x, by

$$P_x\delta x\delta y = \sigma_{xx}\delta y + \tau_{yx}\delta x - (\sigma_{xx} + \delta\sigma_{xx})\delta y - (\tau_{yx} + \delta\tau_{yx})\delta x \quad (2.4.1)$$

or

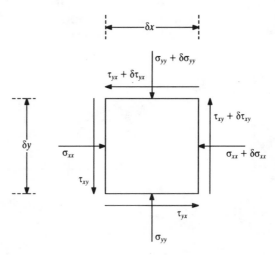

Figure 2.8 Stresses on an infinitesimal element.

$$P_x = -\frac{\partial \sigma_{xx}}{\partial x} - \frac{\partial \tau_{yx}}{\partial y} \qquad (2.4.2)$$

Similarly there is a force in the y-direction given by

$$P_y = -\frac{\partial \sigma_{yy}}{\partial y} + \frac{\partial \tau_{xy}}{\partial x} \qquad (2.4.3)$$

It should be noted that in these equations all stresses appear as their derivatives with respect to their first subscript, and that a particular component of P is obtained by summing all the terms with the appropriate second subscript. The difference in sign in equations (2.4.2) and (2.4.3) results solely from the sign convention imposed upon us by the use of Mohr's circle.

In most of the analyses of chapters 3 to 7, we will be considering the statics of a mass of material and we will take a set of Cartesian axes with x measured horizontally to the left and y measured vertically downwards. Thus the forces P_x and P_y are respectively 0 and $-\rho_b g$ where ρ_b is the bulk density. Thus, from (2.4.2) we have

$$\frac{\partial \sigma_{xx}}{\partial x} + \frac{\partial \tau_{yx}}{\partial y} = 0 \qquad (2.4.4)$$

and from (2.4.3)

$$\frac{\partial \sigma_{yy}}{\partial y} + \frac{\partial \tau_{yx}}{\partial x} = \rho_b g \qquad (2.4.5)$$

where we have made the substitution, $\tau_{xy} = -\tau_{yx}$ for subsequent convenience.

If the material is in motion, the force per unit volume, resulting from both the stress gradients and the gravitational effects, will equal the product of the mass per unit volume and the acceleration. Thus,

$$P_x = \rho_b \frac{Du}{Dt} = \rho_b \left(\frac{\partial u}{\partial t} + u\frac{\partial u}{\partial x} + v\frac{\partial u}{\partial y} \right) \qquad (2.4.6)$$

where u and v are the velocities in the x- and y-directions, t is time and Du/Dt denotes the total derivative, defined by the second part of this equation. Thus we have that

$$\rho_b \left(\frac{\partial u}{\partial t} + u\frac{\partial u}{\partial x} + v\frac{\partial u}{\partial y} \right) + \frac{\partial \sigma_{xx}}{\partial x} + \frac{\partial \tau_{yx}}{\partial y} = 0 \qquad (2.4.7)$$

and similarly in the y-direction

$$\rho_b\left(\frac{\partial v}{\partial t} + u\,\frac{\partial v}{\partial x} + v\,\frac{\partial v}{\partial y}\right) + \frac{\partial \sigma_{yy}}{\partial y} + \frac{\partial \tau_{yx}}{\partial x} - \rho_b g = 0 \qquad (2.4.8)$$

where as before we have assumed that y is measured vertically downwards.

Similar expressions in other co-ordinate systems are given in appendix 2.

Equations (2.4.7) and (2.4.8), and their equivalents in other co-ordinate systems are variously known as *Euler's Equation*, the *Momentum Equation* or the *Equation of Motion*. They are of perfectly general utility, being equally applicable to the motion of both fluids and granular materials.

2.5 Mohr's circle for rate of strain

Mohr's circles, similar to the stress circle described in the previous section, can be used to resolve several quantities including both strain and rate of strain. The latter is of importance when we come to consider velocity distributions and the full derivation of Mohr's circle for rate of strain is given in appendix 3. In this section we will merely summarise the results.

The rate of elongational, or direct, strain in the x-direction is usually defined by

$$\dot{\varepsilon}_{xx} = \frac{\partial u}{\partial x} \qquad (2.5.1)$$

where u is the x-component of the velocity. This is the rate of extensional strain, but since we have chosen to treat compressive stress as positive, it is convenient to take compressive strain rate as positive also. Thus, we will define the rate of direct strain in the x-direction by

$$\dot{\varepsilon}_{xx} = -\frac{\partial u}{\partial x} \qquad (2.5.2)$$

Similarly, we will define the rate of direct strain in the y-direction by

$$\dot{\varepsilon}_{yy} = -\frac{\partial v}{\partial y} \qquad (2.5.3)$$

and we will define the rate of shear strain by

$$\dot{\gamma}_{xy} = \frac{\partial u}{\partial y} + \frac{\partial v}{\partial x} \tag{2.5.4}$$

As usual with Mohr's circle we must define the complementary rate of shear strain by

$$\dot{\gamma}_{yx} = -\dot{\gamma}_{xy} \tag{2.5.5}$$

It is shown in appendix 3 that we can construct a Mohr's circle on axes of $\dot{\varepsilon}$ and $\dot{\gamma}/2$ as shown in figure 2.9 and this has properties similar to those of Mohr's stress circle. The rates of strain in the x-direction are given by the point X which has co-ordinates $(\dot{\varepsilon}_{xx}, \dot{\gamma}_{xy}/2)$. At the opposite end of the diameter is the point Y with co-ordinates $(\dot{\varepsilon}_{yy}, \dot{\gamma}_{yx}/2)$, and the strain rates in a direction inclined at θ to the x-direction can be found from the co-ordinates of the end of the radius inclined at 2θ to the radius to point X.

It can be seen that the co-ordinate of the centre of the circle, e, is given by

$$e = \tfrac{1}{2}(\dot{\varepsilon}_{xx} + \dot{\varepsilon}_{yy}) \tag{2.5.6}$$

The radius of the circle is given by

$$R^2 = \frac{(\dot{\varepsilon}_{xx} - \dot{\varepsilon}_{yy})^2}{4} + \frac{\dot{\gamma}_{xy}^2}{4} \tag{2.5.7}$$

and the angle Λ is given by

$$\tan 2\Lambda = \frac{\dot{\gamma}_{xy}}{\dot{\varepsilon}_{xx} - \dot{\varepsilon}_{yy}} \tag{2.5.8}$$

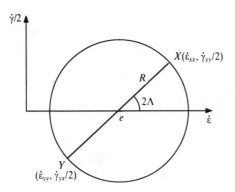

Figure 2.9 Mohr's circle for strain rates.

The quantity e represents the rate of compressive volumetric strain and is zero for an incompressible material and Λ is the angle between the x-direction and the major principal rate of strain direction. Under some circumstances the angle Λ of Mohr's strain rate circle equals the angle λ of the stress circle. When this is so the principal axes of stress and strain rate are coincident and this is sometimes known as the *Principle of Co-axiality*. It is closely related to St. Venant's Principle which is well-known in the field of elasticity and is discussed in greater detail in chapter 8.

3

The ideal Coulomb material

3.1 Introduction

In this chapter we will consider the simplest model of a granular material, known as the *ideal Coulomb material*. It must be appreciated from the outset that this model gives an over-simplified picture of the behaviour of real granular materials and more accurate models will be discussed in chapters 6 and 9. None-the-less, the concept of the ideal Coulomb material forms the basis of a great many analyses of commercial importance and furthermore provides a firm foundation on which to develop important ideas of more general validity. The Coulomb material fulfils the same role in the study of granular materials as the Newtonian fluid does in viscous flow.

We will define the ideal Coulomb material in §3.2 and combine this with Mohr's stress circle in §3.3 to derive the Mohr–Coulomb failure analysis. This analysis is the corner-stone of most studies of stress distribution in granular materials. Following this we will discuss the *Rankine states* and their implication for the prediction of stresses in §3.4. The concept of an angle of repose and its relationship to the parameters in the Coulomb model will be discussed in §3.5 and §3.6 and finally in §3.7 we will consider the interaction of a Coulomb material with a retaining wall.

3.2 The Coulomb yield criterion

The behaviour of a granular material is best understood by comparison with the more familiar behaviour of solids or fluids. Let us consider a sample of material such as that shown schematically in figure 3.1 and observe what happens when it is subjected to an applied force. If the sample is an elastic solid the application of the force will result in a

21

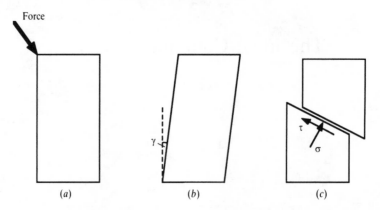

Figure 3.1 Distortion of an element due to the application of a force.

particular deformation. In general for an elastic solid, we can relate the resulting shear strain γ to the applied shear stress τ by a constitutive relationship of the form

$$\tau = f(\gamma) \tag{3.2.1}$$

However, for many, but by no means all, elastic materials the relationship between stress and strain is linear and equation (3.2.1) takes the simple form

$$\tau = G\gamma \tag{3.2.2}$$

where G is the shear modulus. This is one of the forms of Hooke's Law and such elastic materials are known as Hookean solids.

If, however, the sample is a viscous fluid, we observe that the *rate of deformation* is a function of the applied stress and we can say that for the general viscous fluid

$$\tau = f(\dot{\gamma}) \tag{3.2.3}$$

where $\dot{\gamma}$ is the rate of shear strain. As with solids there are many fluids for which the simplest form of this relationship i.e.

$$\tau = \mu_f \dot{\gamma} \tag{3.2.4}$$

is sufficiently accurate. Here μ_f is the viscosity and such fluids are known as Newtonian fluids.

If we conduct the same experiment on a granular material, we find that for small values of the applied force there is a small elastic deformation. This is most important for the understanding of the behaviour of granular materials and will be considered in greater detail in later sections but need not concern us at this stage. However, when the force reaches a critical value, the material divides itself into two blocks which slide past each other as shown in figure 3.1(*c*). It is a matter of observation that for many materials the shear stress on the dividing plane is independent of either the extent or rate of the deformation. Such behaviour is characteristic of plastics (though it must be pointed out that most polymers are not plastics in the rheological sense). Thus, if we ignore the elastic deformations in the two blocks on either side of the dividing plane, we can introduce the concept of the *rigid–plastic failure mode*. That is to say, we are assuming that the material divides itself into two *rigid* blocks separated by a narrow *plastic* zone. The plastic zone is commonly about ten particle diameters in width which is often narrow compared with the typical dimensions of the system. It is, therefore, usually sufficient to assume that the plastic zone is a plane of negligible width and it is variously referred to as the yield, or slip, or failure plane.

As mentioned above, the shear stress on the slip plane is independent of either the extent or rate of the deformation but it is found to depend on the normal stress σ acting on the plane, i.e.

$$\tau = f(\sigma) \tag{3.2.5}$$

As with elastic solids and viscous fluids there exist many materials for which this relationship takes a linear form. Such materials are known as *ideal Coulomb materials* and represent a special case of rigid–plastic materials in the same way that Hookean solids and Newtonian fluids are special cases of elastic solids and viscous fluids. The *Coulomb yield criterion* takes the form

$$\tau = \mu\sigma + c \tag{3.2.6}$$

where μ and c are called the coefficient of friction and the cohesion respectively and are functions of the nature of the material. Care must be taken not to confuse the coefficient of friction with the Newtonian viscosity. Both are commonly given the symbol μ but in this book we are only rarely concerned with viscous fluids and will use the symbol μ_f for the viscosity of a fluid and reserve μ for the coefficient of friction.

For many coarse materials c is small and such materials are known as cohesionless materials. In these cases multiplication by the area of the failure plane gives

$$F = \mu N \tag{3.2.7}$$

which is the well-known frictional relationship between large bodies. Most readers will be all too familiar with this relationship since it forms one of the mainstays of the A-level mechanics course.

The physical origins of the frictional properties of granular materials are still obscure though considerable progress has been achieved in relating the frictional behaviour of the shear plane to the frictional behaviour at inter-particle contacts, e.g. by Briscoe and Adams (1987), and Molerus (1982). It is, however, best to regard equation (3.2.7) as an empirical relationship involving the experimentally determined constants μ and c. The situation is no different from that for liquids. Some progress has been made in relating viscosity to molecular properties but, none-the-less, viscosities are always determined experimentally.

The coefficients of friction normally encountered vary from about 0.3 for smooth spherical particles to about 1.5 for angular particles. Values of the cohesion vary from the undetectably small for coarse materials to around 50 kN m^{-2} for a stiff clay. On the whole, the smaller the particle size, the greater the cohesion and this has prompted many workers to look for an explanation in terms of Van der Waals forces. This may well be correct but this explanation has not yet reached a state of development that has predictive value. Another mechanism that undoubtedly gives rise to cohesion in partially damp materials is surface tension. In such materials, liquid bridges are formed between adjacent particles and the resulting surface tension forces contribute to the strength of the material. Experienced builders of sand castles will be aware that the strongest sand is part way up the beach and is far superior to either the dry sand at the top of the beach or the saturated sand at the water's-edge. This illustrates that the value of the cohesion can be very sensitive to the moisture content and is likely to vary markedly from day to day depending on the weather conditions.

As well as this, the value of the cohesion depends on the degree of compaction of the material, another result familiar to sand castle builders, and this topic is considered in detail in chapter 6 together with a description of the techniques available for measuring material properties. In the meanwhile we will assume that the coefficient of friction and the cohesion of the material under consideration have been measured and can be assumed to be constant.

We can therefore summarise the behaviour of the ideal Coulomb material by saying that, if on any plane $\tau < \mu\sigma + c$, no motion will occur. If, on the other hand, $\tau = \mu\sigma + c$, a slip plane will be formed. The extent of slip may be large during steady flow or infinitesimal in a static material. In the latter case the material is said to be in a state of incipient slip, or incipient failure. Values of $\tau > \mu\sigma + c$ cannot occur, except perhaps transiently, since if an attempt were made to increase τ beyond the value given by equation (3.2.6), the material would not be in a state of static equilibrium and the material on one side of the slip plane would accelerate away.

Finally, for reasons that will become obvious in the next section, it is convenient to define an angle of friction ϕ by

$$\tan \phi = \mu \tag{3.2.8}$$

Thus the criterion for incipient yield, equation (3.2.6), is more commonly written

$$\tau = \sigma \tan \phi + c \tag{3.2.9}$$

Angles of friction depend on the nature of the material, ranging from about 20° for smooth spheres to about 50° for angular materials. Table 3.1, which is taken from the Draft British Code of Practice (BMHB, 1987), gives some typical values.

3.3 The Mohr–Coulomb failure analysis

We have seen in the previous section that the Coulomb failure criterion imposes a limit on the magnitude of the shear stress that can occur within a granular material. This criterion can be used in conjunction with Mohr's circle to provide a convenient diagram which is the basis of the so-called *Mohr–Coulomb failure analysis*.

Table 3.1

Material	Weight Density/kN m^{-3}	Angle of internal friction/°
Wheat	6.5–9.0	15–30
Wheat flour	6.0–7.5	28–42
Dry granulated sugar	8.0–10.0	30–40
Dry quartz sand	14.0–17.0	33–40
Ground limestone	11.0–13.5	27–44
Alumina	10.0–12.5	27–44
Fly ash	8.0–11.5	25–35
Hydrated lime	5.0–7.0	25–45

It will be recalled that the upper limit of the shear stress on any plane is given by equation (3.2.6)

$$\tau = \mu\sigma + c \tag{3.3.1}$$

and we have shown in §2.3 that the combinations of σ and τ that exist on the planes through any particular point are given by Mohr's circle. If, therefore, we plot the Coulomb yield criterion (3.3.1) on a Mohr's circle diagram as in figure 3.2, we see that there exist three possibilities:

(i) The Coulomb line may lie entirely above Mohr's circle as shown by the line marked (i). If this is the case, $\tau < \mu\sigma + c$ on all planes passing through the point of interest and no slip plane will be formed. The material is therefore in a state of stable static equilibrium.

(ii) The Coulomb line may touch the circle as shown by line (ii). In this case there exists a plane s (represented on Mohr's circle by the point S) on which $\tau = \mu\sigma + c$ and on all other planes $\tau < \mu\sigma + c$. Thus, slip is about to occur on plane S and the material is said to be in a state of incipient slip, or incipient failure. The extent of slip is controlled by the motion of the boundaries of the material and need not concern us at this stage. All we need to know is that when the Coulomb line touches the circle, slip may occur and that if it does occur, it will do so along plane S.

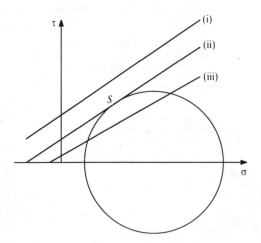

Figure 3.2 Mohr's circle and the Coulomb line.

(iii) If the Coulomb line cuts the circle, as shown by line (iii), there exist planes on which $\tau > \mu\sigma + c$ and this is forbidden in a Coulomb material. We therefore conclude that the Coulomb line cannot cut Mohr's circle.

Thus, the Mohr–Coulomb analysis is based on the concept that slip will only occur if the Coulomb line touches Mohr's circle and that under no circumstances can the Coulomb line cut the circle.

In fact, Coulomb's yield criterion is not concerned with the direction of slip but merely imposes an upper limit on the magnitude of the shear stress that can occur on a plane. Thus strictly equation (3.3.1) should be written

$$|\tau| = \mu\sigma + c \qquad (3.3.2)$$

and the failure criterion should be represented by the pair of lines

$$\tau = \pm(\mu\sigma + c) \qquad (3.3.3)$$

on the Mohr–Coulomb diagram as in figure 3.3. We can therefore see that at critical stability there are two incipient slip planes represented on the diagram by the points S and S' and therefore that incipient slip planes always occur in pairs. If, however, slip does occur, there is no requirement that the extent of slip shall be the same along both planes. This follows from the observation presented in §3.2 that the stresses are independent of the extent or rate of slip. Thus in any particular case, substantial slip may occur along only one of these planes. It must, however, be appreciated that the stress combination on the other plane would permit slip if this was required by the motion of the boundaries of the system.

Recalling that the slope of the Coulomb line, μ, is equal to $\tan\phi$, we see that this line is inclined at angle ϕ to the σ-axis. Thus from the geometry of figure 3.3 we see that the angles SBO and $S'BO$ are both $90 - \phi$ and that the angle SBS' is $180 - 2\phi$. Recalling that angles in Mohr's circle represent twice the rotation in real space, we can deduce that the slip planes are inclined at $90 - \phi$ to each other and are inclined at $\pm(45 - \phi/2)$ to the minor principal plane. It is convenient to denote this latter angle by ε, which is defined formally by

$$\varepsilon = \tfrac{1}{2}(90 - \phi) \qquad (3.3.4)$$

We will see in §3.7 that a criterion similar to the Coulomb yield criterion exists for the slip between a granular material and a wall. To

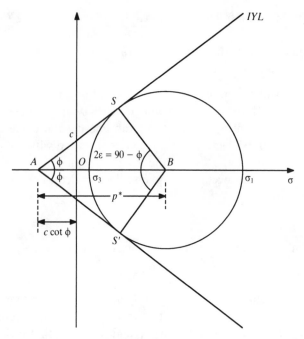

Figure 3.3 The Mohr–Coulomb failure criterion.

distinguish between the two, the Coulomb line discussed above is commonly called the *internal yield locus*, and will be denoted by the abbreviation *IYL* in the relevant figures.

It can be seen from figure 3.3 that the distance AO is $c \cot \phi$, and considering the right-angled triangle ASB, we see that the radius of the circle R is given by

$$R = p^* \sin \phi \qquad (3.3.5)$$

where p^* is the distance between the centre of the circle (point B) and the point where the internal yield locus cuts the σ-axis (point A).

The principal stresses are given by $p^* \pm R - c \cot \phi$, i.e.

$$\sigma_1 = p^*(1 + \sin \phi) - c \cot \phi \qquad (3.3.6)$$

and

$$\sigma_3 = p^*(1 - \sin \phi) - c \cot \phi \qquad (3.3.7)$$

In later sections of this book, we will frequently need to relate the stresses on the x- and y-planes to the parameters p^* and ψ where ψ is the angle measured anticlockwise from the x-plane to the major principal plane. The corresponding Mohr–Coulomb diagram is shown in figure 3.4 and by analogy with equations (3.3.6) and (3.3.7) or by inspection of the figure, we see that

$$\sigma_{xx} = p^* + R\cos 2\psi - c\cot\phi$$
$$= p^*(1 + \sin\phi\cos 2\psi) - c\cot\phi \tag{3.3.8}$$

$$\tau_{yx} = -\tau_{xy} = R\sin 2\psi = p^*\sin\phi\sin 2\psi \tag{3.3.9}$$

$$\sigma_{yy} = p^* - R\cos 2\psi - c\cot\phi$$
$$= p^*(1 - \sin\phi\cos 2\psi) - c\cot\phi \tag{3.3.10}$$

For a cohesionless material, the internal yield locus passes through the origin and the parameter p^* becomes identical to p as defined by equation (2.3.8). Thus for cohesionless materials

$$\sigma_{xx} = p(1 + \sin\phi\cos 2\psi) \tag{3.3.11}$$

$$\tau_{yx} = -\tau_{xy} = p\sin\phi\sin 2\psi \tag{3.3.12}$$

$$\sigma_{yy} = p(1 - \sin\phi\cos 2\psi) \tag{3.3.13}$$

$$\sigma_1 = p(1 + \sin\phi) \tag{3.3.14}$$

$$\sigma_3 = p(1 - \sin\phi) \tag{3.3.15}$$

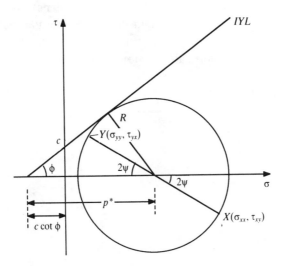

Figure 3.4 Definition of the stress parameters p^* and ψ.

In this case the ratio of the major and minor principal stresses is a constant given by

$$\frac{\sigma_1}{\sigma_3} = \frac{1 + \sin \phi}{1 - \sin \phi} \tag{3.3.16}$$

3.4 The Rankine states

The concept of the *Rankine states* was first developed for an infinite extent of a soil with a horizontal upper surface such as that shown in figure 3.5. Though a geometrical case of minor direct importance for chemical engineers, this situation is still the simplest way of illustrating this very important concept.

Let us consider a vertical plane through the material such as that indicated by the line AA. Considerations of symmetry show that the shear stress on this plane must be zero. (The material to the left of the plane is no more likely to be exerting a downward force on the material to the right than the material to the right is likely to be exerting a downward force on the material to the left.) Similarly, the shear stress on the section $A'A'$ will be zero and a vertical force balance on the material in the control surface $AAA'A'$ gives

$$\sigma_{yy} = \rho_b g y = \gamma y \tag{3.4.1}$$

where ρ_b is the bulk density of the material and g is the acceleration due to gravity. It is common in this subject to denote the product $\rho_b g$ by γ which is known as the weight density of the material and should not be confused with the shear strain rate $\dot{\gamma}_{xy}$. Thus the compressive normal stress in the vertical direction increases linearly with depth. If,

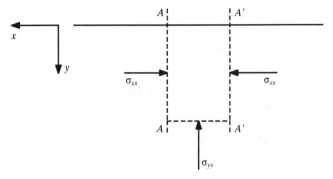

Figure 3.5 Element beneath an infinite horizontal surface.

however, we try to evaluate the horizontal normal stress by resolving horizontally, we find that σ_{xx} is indeterminate. Thus the value of σ_{xx} cannot be determined by statics alone and consideration must be given to the nature of the material.

We have already determined the shear stress τ_{xy} is zero and hence τ_{yx} must also be zero. Thus the stresses on the y-plane at depth y are $\sigma_{yy} = \gamma y$, $\tau_{yx} = 0$ and we can represent these as the point Y on the Mohr–Coulomb diagram of figure 3.6. The point X which represents the stresses on the x-plane, will lie on the σ-axis since $\tau_{xy} = 0$. If we assume for the moment that the point X lies somewhat to the left of Y, as shown by the point X^* for example, we can draw in the Mohr's circle passing through the points X^* and Y. For this selected value of X^*, the circle lies entirely below the internal yield locus, which is marked on the figure as *IYL*, and the material is therefore in a state of static equilibrium. Thus the value of σ_{xx} given by the point X^* is a permissible value.

If we now proceed to assume successively lower values of σ_{xx} we will eventually come to the value σ_A represented by the point X_A. The Mohr's circle now touches the internal yield locus and the material will be in a state of incipient failure. Any further lowering of σ_{xx} will result in slip within the material. Thus the stress σ_A represents the lower limit of the permissible values of σ_{xx}. Similarly, we can increase

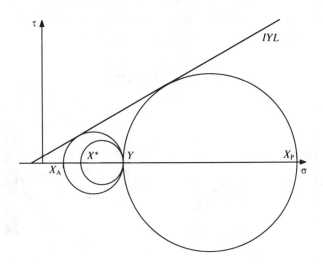

Figure 3.6 Mohr's circles for the Rankine states.

σ_{xx} up to the value σ_P represented by point X_P which is the upper limit of σ_{xx} for stability.

Thus we see that we cannot evaluate σ_{xx} uniquely but can only identify a range of values within which σ_{xx} must lie. The lower extreme value can be obtained physically by confining the material between two vertical smooth walls and moving the walls slowly apart until slip occurs. Thus σ_A is the stress on a wall that is being pushed outwards by the failing material and was termed by Rankine the 'Active Stress'. Similarly, if the walls are pushed slowly together the stress will rise to σ_P, the 'Passive Stress' before failure occurs. Thus the active stress is the stress on a wall that is moving outward and the slip within the material is in a downward and outward direction as shown in figure 3.7. Conversely, the passive stress is that which would occur on a bulldozer that was causing inward and upward failure within the material. Any horizontal stress between these two values will be stable and will cause no slip within the material. We will give further consideration to this result in §4.3.

Students first encountering the concept of the Rankine states are often dismayed by the realisation that in most cases it is not possible to evaluate the stresses that actually occur and all that can usually be achieved is the evaluation of the limits of the range of values the stress may have. This is in fact a well-known result for a more elementary situation, well within the usual A-level mechanics syllabus. Consider a block of mass M resting on a rough inclined surface at angle α to the horizontal as shown in figure 3.8. Resolving normal to the plane gives $N = Mg \cos \alpha$ and the frictional force F therefore lies within the limits

$$- \mu Mg \cos \alpha < F < \mu Mg \cos \alpha \qquad (3.4.2)$$

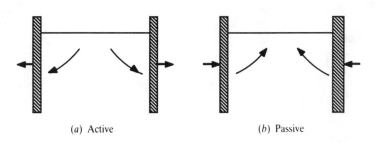

(*a*) Active   (*b*) Passive

Figure 3.7 Active and passive failure.

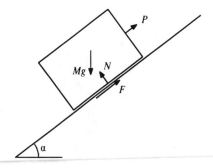

Figure 3.8 Block on an inclined surface.

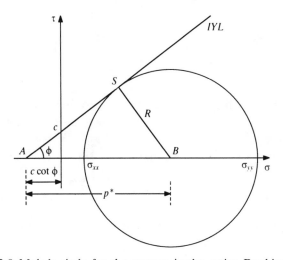

Figure 3.9 Mohr's circle for the stresses in the active Rankine state.

The force P_A required to prevent the block slipping down the plane is $Mg \sin \alpha - \mu Mg \cos \alpha$ whereas the force P_P required to pull the block up the plane is $Mg \sin \alpha + \mu Mg \cos \alpha$. Any force P in the range

$$Mg \sin \alpha - \mu Mg \cos \alpha < P < Mg \sin \alpha + \mu Mg \cos \alpha \qquad (3.4.3)$$

will be stable.

Let us consider the active state as shown in the Mohr–Coulomb diagram of figure 3.9. Defining the quantities p^* and R as in the previous section, we have

$$\sigma_{yy} = \gamma y = p^* + R - c \cot \phi \qquad (3.4.4)$$

and

$$\sigma_A = \sigma_{xx} = p^* - R - c \cot \phi \qquad (3.4.5)$$

Consideration of the right-angled triangle ASB shows that

$$R = p^* \sin \phi \qquad (3.4.6)$$

and eliminating p^* and R gives

$$\sigma_A = \frac{1 - \sin \phi}{1 + \sin \phi} \gamma y - 2c \frac{\cos \phi}{1 + \sin \phi} \qquad (3.4.7)$$

or, in the case of a cohesionless material,

$$\sigma_A = K_A \gamma y \qquad (3.4.8)$$

where K_A is defined by

$$K_A = \frac{1 - \sin \phi}{1 + \sin \phi} \qquad (3.4.9)$$

The constant K_A, which represents the ratio of horizontal and vertical stresses in a cohesionless material, is known as Rankine's *coefficient of active earth pressure*.

In the passive case, $\sigma_P > \sigma_{yy}$ and a very similar analysis gives

$$\sigma_P = \frac{1 + \sin \phi}{1 - \sin \phi} \gamma y + 2c \frac{\cos \phi}{1 - \sin \phi} \qquad (3.4.10)$$

and we can define the *coefficient of passive earth pressure*, K_P, by

$$K_P = \frac{1 + \sin \phi}{1 - \sin \phi} \qquad (3.4.11)$$

Throughout this subject, we find pairs of results representing the active and passive cases of which equations (3.4.7) and (3.4.10) are typical examples. It is fashionable to combine such pairs into a single equation by introducing the quantity κ which has the value -1 in the active case and the value $+1$ in the passive case. Equations (3.4.7) and (3.4.10) can therefore be written in the form,

$$\sigma_{xx} = \frac{1 + \kappa \sin \phi}{1 - \kappa \sin \phi} \gamma y + \frac{2\kappa c \cos \phi}{1 - \kappa \sin \phi} \qquad (3.4.12)$$

No extra information is obtained by this combination but the introduction of κ makes it possible to write computer programs of greater generality.

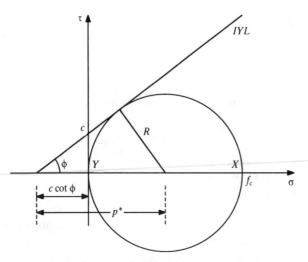

Figure 3.10 Mohr's circle for the stresses at the surface in the passive Rankine state.

Inspection of equation (3.4.7) shows that σ_A is negative for small values of y and that negative stresses occur down to a depth h_c given by,

$$h_c = \frac{2c \cos \phi}{\gamma(1 - \sin \phi)} \tag{3.4.13}$$

Most granular materials cannot support tension and cracks may therefore develop in regions in which tensile stresses are predicted. Thus in the active case cracks may form in the upper part of the material, but these will not extend to a depth greater than h_c since below this depth the material is in compression and any cracks that form will immediately close up. Thus h_c is the maximum depth for tension cracking. It should be noted that h_c is zero for a cohesionless material and it is clear from equation (3.4.10) that the horizontal stresses are always positive in the passive case.

Furthermore, in the passive case the least horizontal stress occurs at the top surface and is denoted by f_c where, from (3.4.10),

$$f_c = \frac{2c \cos \phi}{1 - \sin \phi} \tag{3.4.14}$$

The quantity f_c is known as the *unconfined yield stress* and represents the maximum compressive stress that can act along a free surface. Since on the surface, $\sigma_{yy} = 0$ and $\tau_{yx} = 0$, it is clear from figure 3.10

that f_c can be evaluated by drawing the Mohr's circle that passes through the origin and touches the internal yield locus. The circle cuts the σ axis at $\sigma = f_c$.

3.5 The angle of repose of a cohesionless material

It is a matter of common observation that the top surface of a mass of granular material need not be horizontal unlike that of a stagnant liquid. However, there exists an upper limit to the slope of the top surface and the angle between this maximum slope and the horizontal is known as the angle of repose. In this section we will relate the angle of repose to the frictional properties for the case of a cohesionless material and defer to the next section the more complicated analysis for cohesive materials.

Let us consider a planar slope inclined at angle θ to the horizontal as shown in figure 3.11. If the slope is marginally unstable some incipient slip surface will be formed such as that shown by the line AB. We will assume that this slip surface is planar and will denote by α its inclination to the horizontal. Clearly from the geometry of the figure

$$\alpha < \theta \qquad (3.5.1)$$

The justification for assuming that the slip surface is planar will be considered in more detail in §4.6.

Bearing in mind that we are confining our attention to cohesionless materials, we can say that if slip is occurring along AB the shear stress on that plane is given by

$$\tau = \mu\sigma \qquad (3.5.2)$$

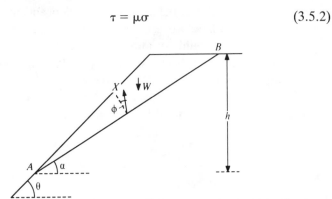

Figure 3.11 Slip plane for the failure of a slope.

Multiplying by the area of the slip plane gives

$$F = \mu N \tag{3.5.3}$$

where F is the frictional force and N the normal force. Thus the total force X on the plane is inclined to the normal by the angle ϕ where

$$\tan \phi = \mu = \frac{F}{N} \tag{3.5.4}$$

and hence X is inclined to the vertical by the angle $\alpha - \phi$.

If we now consider the material above AB we see that it is subjected to two forces, the force X and its own weight. When a non-accelerating body is subjected to only two forces, these must be co-linear. Thus X must be vertical and hence

$$\alpha = \phi \tag{3.5.5}$$

We can see from (3.5.1) that a slip plane will, therefore, be formed if

$$\theta > \phi \tag{3.5.6}$$

and, hence, that the greatest value of θ for stability is given by

$$\theta = \phi \tag{3.5.7}$$

Thus we see that the angle of repose equals the angle of internal friction ϕ. It is noteworthy that the height of the slope has not entered this analysis, from which we can deduce that the angle of repose is independent of the extent of the inclined surface.

The analysis for a conical pile of material is much more complicated, but the same result, that the angles of repose and of internal friction are the same, still applies. Similarly, the angle of the conical hollow formed in the top surface of a bunker in core flow is also equal to the angle of internal friction.

These results, however, apply only to ideal cohesionless Coulomb materials. For such a material, measuring the steepest stable slope is commonly used as a convenient method of determining the angle of friction. This method must, however, be used with the greatest caution. The easiest way of forming a pile of material is to pour the material as a narrow jet onto a horizontal surface as in figure 3.12. As material is added to the top of the pile some of it will cascade down the surface and this process can result in the segregation of coarse and fine particles, with the coarser particles travelling further down the slope.

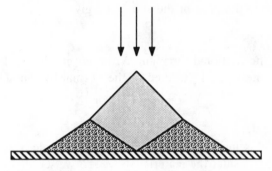

Figure 3.12 Segregation of material during pouring onto a conical pile.

Thus unless the material is composed of identical particles, the resulting pile will be non-uniform and the slope of its top surface will not necessarily represent the angle of friction of the original mixed material.

3.6 The angle of repose of a cohesive material

We can consider the limiting stability of a slope of a cohesive material by starting with the same physical situation as that considered in the previous section and illustrated in figure 3.11. Again, we will consider a slope inclined at angle θ to the horizontal and a planar slip surface AB inclined at some unknown angle α to the horizontal. We will consider unit thickness of material normal to the paper and denote the depth of point A below the top of the slope by h.

Since the material is assumed to be an ideal Coulomb material, slip will not occur along AB if

$$\tau < \mu\sigma + c \qquad (3.6.1)$$

Multiplying by the area of the plane AB, which can be seen from figure 3.11 to be h cosec α, we find that

$$F < \mu N + ch \text{ cosec } \alpha \qquad (3.6.2)$$

where F and N are the frictional and normal forces on the plane.

The volume of the material above plane AB is $\frac{1}{2}h^2(\cot \alpha - \cot \theta)$ so that the weight W of this material is given by

$$W = \frac{\gamma h^2}{2} (\cot \alpha - \cot \theta) \qquad (3.6.3)$$

Resolving horizontally, we have

$$F \cos \alpha = N \sin \alpha \qquad (3.6.4)$$

and resolving vertically gives

$$F \sin \alpha + N \cos \alpha = W \qquad (3.6.5)$$

Eliminating F, N and W from relationships (3.6.2) to (3.6.5) and rearranging, gives that slip will not occur along the plane inclined at angle α to the horizontal if

$$\frac{\gamma h}{2c} < \frac{\sin \theta \cos \phi}{\sin(\theta - \alpha) \sin(\alpha - \phi)} = f(\alpha) \qquad (3.6.6)$$

From this we can deduce that slip will not occur on *any* plane if

$$\frac{\gamma h}{2c} < f_{max}(\alpha) \qquad (3.6.7)$$

where $f_{max}(\alpha)$ is the maximum value of $f(\alpha)$. We can find this maximum value by differentiating $f(\alpha)$ with respect to α (or by inspection) and we find that $f(\alpha)$ has its maximum value when

$$\alpha = \frac{\theta + \phi}{2} \qquad (3.6.8)$$

with the corresponding maximum value being given by

$$f_{max}(\alpha) = \frac{\sin \theta \cos \phi}{\sin^2 \frac{1}{2}(\theta - \phi)} = \frac{2 \sin \theta \cos \phi}{1 - \cos(\theta - \phi)} \qquad (3.6.9)$$

Thus, the critical angle θ of a slope of a cohesive material is given by

$$\frac{\gamma h}{2c} = \frac{2 \sin \theta \cos \phi}{1 - \cos(\theta - \phi)} \qquad (3.6.10)$$

It is seen that in this case the angle of repose θ depends on the height h of the slope, and some people argue that as a result the phrase 'angle of repose' should not be used, since this is not a constant for a given material.

Certain special cases are worthy of note. First, if the material is cohesionless, the left-hand side of equation (3.6.10) becomes infinite so that $\cos(\theta - \phi) = 1$, or

$$\theta = \phi \qquad (3.6.11)$$

This is the result found in the previous section showing that the two analyses are consistent with each other. Secondly, it will be noted that vertical slopes are sometimes possible. Putting $\theta = 90°$, we find from (3.6.8) that the inclination of the slip plane is given by

$$\alpha = \tfrac{1}{2}(90 + \phi) \tag{3.6.12}$$

and from (3.6.10) that the maximum height of a vertical slope is given by

$$h = \frac{4c \cos \phi}{\gamma(1 - \sin \phi)} \tag{3.6.13}$$

This is twice the depth to which tension cracks can occur in Rankine's active state, as given by equation (3.4.13), a result which some people regard as obvious.

Equation (3.6.13) can be used to obtain a crude estimate of the cohesion of a sample of material. With ϕ being commonly about 30° we have

$$h = \frac{4\sqrt{3}\,c}{\gamma} \tag{3.6.14}$$

Thus the cohesion can be seen to be of the order of $\gamma h_m/7$ where h_m is the maximum height of a stable vertical surface.

The results of these last two sections should be qualitatively obvious to any experienced maker of sand castles. If a trench is dug in the dry, and therefore cohesionless, sand at the top of the beach, the walls will always collapse to give a trench of triangular cross-section, corresponding to a constant angle of repose. However, it is possible to dig a vertically sided trench in the damp (cohesive) sand but, if too deep a trench is dug, a wedge of material will break off the side at an angle of $(45 + \phi/2)$ to the horizontal. In good sand a vertical trench of about 20 cm depth can be dug, showing that the value of the cohesion is given approximately by

$$c \approx \frac{\gamma h_m}{7} = \frac{1500 \times 9.81 \times 0.2}{7} = 0.4 \text{ kN m}^{-2}$$

3.7 The wall failure criterion

Besides forming an internal slip plane, the material may also slip along some bounding surface such as the wall of a bunker or any other container. The criterion for the existence of incipient slip along a wall is similar to that for forming an internal slip plane and can be expressed in the form

$$\tau_w = f(\sigma_w) \tag{3.7.1}$$

where σ_w and τ_w are the normal and shear stresses on the wall.

For an ideal Coulomb material the wall yield locus is linear and can be expressed in the form,

$$\tau_w = \mu_w \sigma_w + c_w \tag{3.7.2}$$

where μ_w is the coefficient of wall friction and c_w is known as the adhesion. As before, this is an idealisation and some curvature of the wall yield locus is often found. None-the-less a 'best fit' straight line is sufficient for most purposes.

We can define an angle of wall friction ϕ_w by

$$\tan \phi_w = \mu_w \tag{3.7.3}$$

where ϕ_w is the angle between the linearised wall yield locus and the σ-axis. If the wall yield locus is curved, we will define μ_w as the slope of the locus, i.e.

$$\tan \phi_w = \mu_w = \frac{d\tau_w}{d\sigma_w} \tag{3.7.4}$$

On the other hand Arnold et al. (1980), following Jenike (1961), prefer to define the angle of wall friction as the inclination of the line drawn from the origin to the point of interest on the wall yield locus as indicated by the angle ϕ'_w on figure 3.13. This is equivalent to defining the angle of wall friction by

$$\tan \phi'_w = \mu'_w = \frac{\tau_w}{\sigma_w} \tag{3.7.5}$$

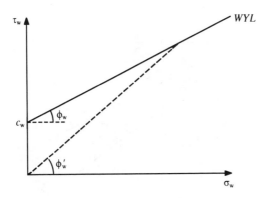

Figure 3.13 Alternative definitions of the angle of wall friction.

It is not clear what advantage is gained by this latter definition, which has the disadvantage that ϕ'_w becomes a strong function of the stress σ_w, since, unless c_w is zero, ϕ'_w must tend to 90°, and μ'_w to ∞, as $\sigma_w \to 0$. Defining ϕ_w as the inclination of the tangent to the wall yield locus results in a quantity which varies little with normal stress and is constant for a Coulomb material. Only if the material is Coulombic and adhesionless does ϕ'_w become constant. It therefore appears that ϕ_w is the more convenient definition of the angle of wall friction, and this definition will be used throughout this book.

We can draw the wall yield locus, or more strictly the two wall yield loci, since we are concerned with $|\tau_w|$, on the Mohr–Coulomb diagram as in figure 3.14. If the material is in a state of incipient internal slip, the Mohr's circle must touch the internal yield locus, which is marked *IYL* on the figure. If the *wall yield loci*, marked as *WYL*, lie within the internal yield loci, they will cut the circle at the four points A, B, C and D.

Recalling that each point on Mohr's circle represents the stresses on a plane through the point of interest, we can in principle identify the point W on the circle corresponding to the wall plane. If W lies within the arcs BC or AD, we have that

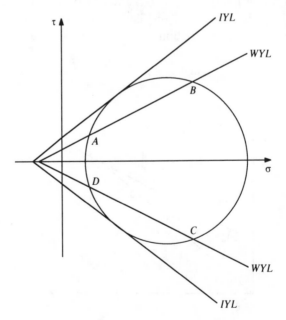

Figure 3.14 Possible positions of the wall plane.

$$|\tau_w| < \mu_w\sigma_w + c_w$$

and slip does not occur along the wall. If, however, W lies at A, B, C or D,

$$|\tau_w| = \mu_w\sigma_w + c_w$$

and the material is in a state of incipient slip along the wall. However, since the magnitude of the wall shear stress may not exceed $\mu_w\sigma_w + c_w$, we see that the point W cannot lie in the arcs AB or CD.

We see therefore that the relationship between Mohr's circle and the wall yield locus differs from that between the circle and the internal yield locus. This is because the internal yield locus imposes a limit on the shear stress on *any plane* through the material and therefore the circle must lie entirely between the internal yield loci. On the other hand, the wall yield loci impose limits on the shear stress on the *wall plane*, and it is of no concern if the shear stresses on other planes exceed this limit. Thus Mohr's circle can extend beyond the wall yield loci, though the point representing the wall plane must lie between the wall yield loci.

Since Mohr's circle cannot extend beyond the internal yield loci, any portion of a wall yield locus lying outside the internal yield loci is meaningless. Thus if the wall and internal yield loci were as indicated in figure 3.15, the section of the wall yield locus beyond the intersection point A would have no physical significance and under these circumstances the effective wall yield locus would reduce to the internal yield locus. The effective wall yield locus would then take the form shown by the dashed line in figure 3.15. This introduces the concept

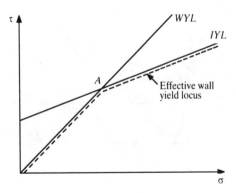

Figure 3.15 The effective wall yield locus.

of a 'fully rough wall' which we will define as a wall with yield locus identical to the internal yield locus of the bulk material, i.e. $\phi_w = \phi$ and $c_w = c$. This can be seen physically by considering a very rough wall made by coating a surface with some extremely sticky glue. A mono-layer of particles would become permanently attached to the wall and slip would occur, not along the wall, but along an *internal* slip plane one particle diameter in from the wall.

Many of the analyses in this book are concerned with cohesionless materials and under these circumstances the internal yield locus passes through the origin and hence the wall yield locus must also pass through the origin and the wall will be adhesionless. If the material is failing internally and also slipping along a wall, Mohr's circle must touch the internal yield locus and the wall plane will be at one of the points W or W' shown in figure 3.16. For the sake of clarity, only one of the wall yield loci is shown, and it should be appreciated that there exist two other possible points of intersection given by the mirror image in the σ-axis. This type of Mohr–Coulomb diagram is frequently encountered and the analysis below relates some of the angles in this figure.

By definition, the internal and wall yield loci are inclined at angles ϕ and ϕ_w to the σ-axis, as shown. If the co-ordinate of the centre A of the circle is p, we can see from the right-angled triangle OAS that the radius R of the circle is given by

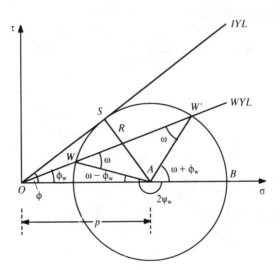

Figure 3.16 Mohr's circle for the stresses adjacent to the wall.

$$R = p \sin \phi \qquad (3.7.6)$$

Applying the sine rule to triangle OAW gives

$$\frac{\sin \phi_w}{R} = \frac{\sin \omega}{p} \qquad (3.7.7)$$

so that from (3.7.6) we have

$$\sin \omega = \frac{\sin \phi_w}{\sin \phi} \qquad (3.7.8)$$

where we require the root of this equation in the range $0 < \omega < \pi/2$.

Since the triangle AWW' is isosceles, the angle $AW'O$ is also equal to ω and elementary trigonometry gives the angle WAO as $\omega - \phi_w$ and the angle $W'AB$ as $\omega + \phi_w$.

Very often we are concerned with a vertical wall, which in view of our co-ordinate system is said to be an x-plane. We have already defined the angle ψ as the angle measured anticlockwise from the x-plane to the major principal plane. Thus, if the wall is at point W the value of ψ at the wall, which we will denote by ψ_w, is given by

$$2\psi_w = 180 + \omega - \phi_w \qquad (3.7.9)$$

and if the wall is at point W'

$$2\psi_w = 360 - \omega - \phi_w \qquad (3.7.10)$$

The corresponding values on the lower wall yield locus are

$$2\psi_w = 180 - \omega + \phi_w \qquad (3.7.11)$$

and

$$2\psi_w = \omega + \phi_w \qquad (3.7.12)$$

Jenike (1961) maintains that no real wall is as rough as that defined as a fully rough wall above. He states that 'experiments indicate that' the point W' never occurs to the left of the highest point on the circle. He does not justify this statement theoretically but he does show that there are algebraic complications in the predictions of velocity distributions if W' lies to the left of the maximum. Thus, according to Jenike and some subsequent authors, the maximum value that ϕ_w can take is that which makes the angle $W'AO$ equal to $90°$. Under these circumstances

$$\tan \phi_w = R/p \qquad (3.7.13)$$

so that with Jenike's definition of the fully rough wall the maximum value of ϕ_w is given by

$$\tan \phi_w = \sin \phi \qquad (3.7.14)$$

There is some evidence, which will be discussed in chapters 6 and 8, that Jenike's arguments are valid when flow is actually occurring along the wall. Thus the maximum shear stress that can occur during motion is somewhat less than that which can occur in a static material. Great caution must therefore be exercised when using the concept of a fully rough wall, but we can conclude with certainty that for a cohesionless material $\phi_w \leq \phi$. The concept of a fully rough wall is simply a matter of definition which in this work will be taken to be coincidence of the wall and internal yield loci. We must, however, bear in mind that this is an idealisation and that in practice it is unlikely that we will have a wall as rough as this.

4

Coulomb's method of wedges

4.1 Introduction

In this chapter we will consider a series of analyses based on the *Method of Wedges*. This method dates back to Coulomb's original work of 1773, though changing fashions in the presentation of algebra makes the following account quite different from Coulomb's original version.

We have already used the method in a simplified form in §3.6 but without any comments on its validity. An attempt is made in §4.6 to justify the use of the method and to discuss its accuracy. First, however, we will consider a few classic results in §4.2 to §4.4 and then adapt the method for some novel situations in the remaining section of this chapter.

It must be appreciated from the onset that the method is strictly applicable to two-dimensional situations only and it is therefore best used for the prediction of the force on a planar retaining wall of infinite extent. We will, however, extend the method to the case of parallel walls in §4.8. Attempts have been made to adapt the method to three-dimensional situations such as bunkers of rectangular or circular cross-section. None of these is wholly satisfactory and in this chapter we will confine ourselves to effectively two-dimensional systems. Consideration of the stress distribution in cylindrical bunkers is therefore deferred to the next chapter.

4.2 Force exerted by a cohesionless material on a vertical retaining wall

Let us first consider a vertical retaining wall supporting a cohesionless granular material with a horizontal top surface as shown in figure 4.1.

47

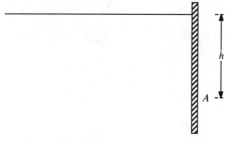

(*a*) Before the wall has broken

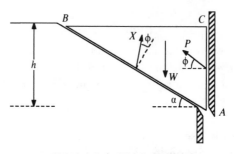

(*b*) After the wall has broken

Figure 4.1 Failure of a retaining wall.

We will pay attention to the part of the wall above the point A, which is at depth h below the top surface of the material, and consider what would happen if the wall were to break at A due to the forces exerted by the material. Under these circumstances the wall would move horizontally outwards by a small distance as shown in figure 4.1(*b*) and some motion would occur within the material. Coulomb assumed that a planar slip surface was formed through A at some unknown angle α to the horizontal and that the material above the slip plane moved bodily downwards and towards the wall. It is the assumption that the slip surface is planar that is the weak point of analyses of this type, the rest of the analysis being exact. We will consider the validity of this assumption and the likely accuracy of the predictions in detail in §4.6 below.

We have assumed that the motion of the wedge is downwards and therefore the friction forces acting both between the wedge and the rest of the material and between the wedge and the wall will be directed upwards. Since we are confining our attention to cohesionless materials the shear stress on the slip plane will be μ times the normal stress so that the resultant force X will be inclined to the normal to the plane by the angle of friction ϕ as shown in figure 4.1. Similarly the force P on the wall will, in general, be inclined to the normal by the angle of wall friction ϕ_w, but first we will confine our attention to a fully rough wall so that $\phi_w = \phi$. Thus, the wedge of sliding material is subjected to the three forces X and P and its own weight W. We will consider unit distance normal to the figure and take the symbols P and W to refer to forces per unit run of wall. The weight W of the wedge is therefore the product of the weight density γ and the area of the triangle ABC, i.e.

$$W = \frac{\gamma h^2}{2} \cot \alpha \qquad (4.2.1)$$

Resolving horizontally gives

$$P \cos \phi = X \sin (\alpha - \phi) \qquad (4.2.2)$$

and vertically gives

$$P \sin \phi + X \cos (\alpha - \phi) = W \qquad (4.2.3)$$

Eliminating X and W from these equations, yields

$$P \cos \phi = \frac{\gamma h^2}{2} \frac{\cot \alpha}{\tan \phi + \cot (\alpha - \phi)} = \frac{\gamma h^2}{2} f(\alpha) \qquad (4.2.4)$$

where,

$$f(\alpha) = \frac{\cot \alpha}{\tan \phi + \cot (\alpha - \phi)} \qquad (4.2.5)$$

Figure 4.2 shows a plot of $f(\alpha)$ and it can be seen that $f(\alpha) = 0$ at both $\alpha = \phi$ (since $\cot 0 = \infty$) and at $\alpha = 90°$ (since $\cot 90 = 0$). Between these values $f(\alpha)$ is positive and finite and, therefore, passes through a maximum which we will denote by f_{max} at angle α_m.

If we now consider an experiment in which we support the wall with some small force P_1, we can evaluate $f_1 = 2P_1 \cos \phi / \gamma h^2$ and see from figure 4.2 that a slip plane would form at either angle α_1 or α_1' and the wall would therefore move outwards. Increasing the force to P_2 would result in slip planes α_2 and α_2'; a further increase to the value P_m given by

$$P_m \cos \phi = \frac{\gamma h^2}{2} f_{max} \qquad (4.2.6)$$

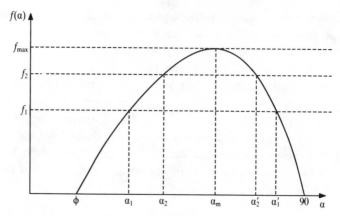

Figure 4.2 The function $f(\alpha)$.

would result in the slip plane inclined at α_m to the horizontal.

Greater values of P are not associated with any slip plane and no motion would occur. Thus the observation that the wall remains in place shows that the force exerted on it by the granular material is greater than P_m. We can therefore conclude that the *least* force exerted by a *stable* material on a retaining wall is given by

$$P_m \cos \phi = \frac{\gamma h^2}{2} f_{max} \tag{4.2.7}$$

The value of f_{max} can be obtained by setting $df/d\alpha = 0$. It is, however, more convenient to write $\tan \alpha$ as t and, recalling that $\tan \phi = \mu$, we find that equation (4.2.5) can be written in the form

$$f(t) = \frac{t - \mu}{t(1 - \mu^2) + 2t^2\mu} \tag{4.2.8}$$

The derivative df/dt is now given by

$$\frac{df}{dt} = \frac{t(1 - \mu^2) + 2t^2\mu - (t - \mu)[(1 - \mu^2) + 4t\mu]}{[t(1 - \mu^2) + 2t^2\mu]^2} \tag{4.2.9}$$

and on setting this to zero we obtain the quadratic

$$2\mu t^2 - 4\mu^2 t - \mu(1 - \mu^2) = 0 \tag{4.2.10}$$

which has solutions

$$t = \mu \pm \frac{\sqrt{(\mu^2 + 1)}}{\sqrt{2}} \tag{4.2.11}$$

or

$$\tan \alpha_m = \tan \phi \pm \frac{\sec \phi}{\sqrt{2}} \qquad (4.2.12)$$

We will show later in this section that the lower root of equation (4.2.12) has no physical meaning and hence we can say that

$$\tan \alpha_m = \tan \phi + \frac{\sec \phi}{\sqrt{2}} \qquad (4.2.13)$$

Substituting this root of equation (4.2.11) into equation (4.2.8) gives

$$f_{max} = \frac{1}{[\sqrt{(1 + \mu^2)} + \sqrt{2} \, \mu]^2} = \frac{\cos^2 \phi}{(1 + \sqrt{2} \sin \phi)^2} \qquad (4.2.14)$$

and from equation (4.2.4) the corresponding force on the wall is given by

$$P_m = \frac{\gamma h^2}{2} \frac{\cos \phi}{(1 + \sqrt{2} \sin \phi)^2} \qquad (4.2.15)$$

The normal force on the wall, $P_m \cos \phi$, is clearly the integral of the normal stress with respect to depth, i.e.

$$P_m \cos \phi = \int_0^h \sigma \, dx \qquad (4.2.16)$$

and hence we can see that

$$\sigma = \frac{dP_m}{dh} \cos \phi = \gamma h \frac{\cos^2 \phi}{(1 + \sqrt{2} \sin \phi)^2} \qquad (4.2.17)$$

and the corresponding shear stress τ is given by

$$\tau = \sigma \tan \phi = \gamma h \frac{\cos \phi \sin \phi}{(1 + \sqrt{2} \sin \phi)^2} \qquad (4.2.18)$$

Thus, we see that the normal and shear stresses on the wall increase linearly with depth as in a static liquid. However, the magnitude of these stresses differs from that in the hydrostatic case ($\sigma = \gamma h$, $\tau = 0$) and we will discuss this difference in §4.3 below.

We stated above that the lower root of equation (4.2.12) had no physical meaning. This can be shown on two grounds. First, provided

$\phi < 45°$, the value of α_m predicted from the lower root of equation (4.2.12) is negative, and can therefore be dismissed by geometric arguments, since the resulting slip plane would not intersect the top surface. Whilst this argument cannot be applied directly for the rare cases when $\phi > 45°$, it seems improbable that the lower root should suddenly assume significance when ϕ exceeds some critical value. Secondly, and perhaps more convincingly, we can calculate the corresponding value of the force X across the slip plane by substituting the lower root for α_m into equation (4.2.2). Since the lower root inevitably gives $\alpha_m < \phi$, it follows from (4.2.2) that the corresponding value of X is negative. Cohesionless materials cannot support tension and hence this prediction of X and the corresponding root for α_m are meaningless.

A very similar analysis can be performed for the partially rough wall, i.e. for the case when $\phi_w < \phi$. This has, however, been delayed to this stage because of the greater complexity of the algebra. The analysis follows exactly the same lines and is left as an exercise for the reader. It will simply be stated that equation (4.2.1) remains unchanged and that equations (4.2.2) and (4.2.3) become

$$P \cos \phi_w = X \sin (\alpha - \phi) \qquad (4.2.19)$$

and

$$P \sin \phi_w + X \cos (\alpha - \phi) = W \qquad (4.2.20)$$

respectively.

Equation (4.2.5) takes the form

$$f(\alpha) = \frac{\cot \alpha}{\tan \phi_w + \cot (\alpha - \phi)} \qquad (4.2.21)$$

and on differentiation, the critical value of t is again given by a quadratic, as in equation (4.2.10), which becomes

$$(\mu + \mu_w)t^2 - 2\mu(\mu + \mu_w)t - \mu(1 - \mu\mu_w) = 0 \qquad (4.2.22)$$

Hence, equation (4.2.13) takes the form

$$\tan \alpha_m = \tan \phi + \frac{\sec \phi}{\sqrt{(1 + k)}} \qquad (4.2.23)$$

where

$$k = \frac{\mu_w}{\mu} \qquad (4.2.24)$$

Finally, equation (4.2.17) becomes

$$\sigma = \gamma h f_{max} = \gamma h \frac{\cos^2\phi}{[1 + \sqrt{(1 + k)\sin \phi}]^2} \qquad (4.2.25)$$

It can be seen that all these results reduce to the form for the fully rough wall presented above when $\mu = \mu_w$, i.e. when $k = 1$.

4.3 Force on a retaining wall – passive analysis and discussion

For reasons that are discussed in §4.6 below, Coulomb's method is less satisfactory when dealing with passive failure. We will therefore consider passive failure only briefly and the reader is cautioned against placing too much reliance on the numerical predictions of this section. None-the-less, comparison of the passive and active cases provides an insight of considerable educative value into Coulomb's method.

Paying attention again to a vertical wall, we can consider what would happen if the inwardly directed force supporting the wall were sufficient to cause inward movement of the wall. According to Coulomb's model, a wedge of material would be thrust upwards and inwards as shown in figure 4.3, and the corresponding frictional forces would be directed downwards. Thus the forces on the wall and on the slip plane would be inclined at angles ϕ to the normal as shown. Comparison of figures 4.1 and 4.3 shows that the only difference lies in the directions of P and X which now lie at angle ϕ on the opposite side of the normal. Thus, the analysis of the passive case can be adapted from that of the active case simply be reversing the sign of ϕ.

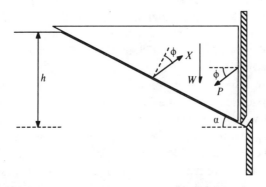

Figure 4.3 Passive failure of a retaining wall.

We find therefore that equation (4.2.5) becomes

$$f(\alpha) = \frac{\cot \alpha}{\cot (\alpha + \phi) - \tan \phi} \tag{4.3.1}$$

It can be seen from this equation that $f(\alpha)$ is positive only in the range $0 < \alpha < \phi$ and tends to infinity at both extremes of this range. This equation therefore has a minimum at α_{min}, where by analogy with equation (4.2.13),

$$\tan \alpha_{min} = \frac{\sec \phi}{\sqrt{2}} - \tan \phi \tag{4.3.2}$$

and from equation (4.2.14)

$$f_{min} = \frac{\cos^2\phi}{(1 - \sqrt{2} \sin \phi)^2} \tag{4.3.3}$$

The corresponding result for a partially rough wall is, from equation (4.2.25),

$$f_{min} = \frac{\cos^2\phi}{(1 - \sqrt{(1 + k)} \sin \phi)^2} \tag{4.3.4}$$

Recalling, from equation (4.2.4), that

$$P = \frac{\gamma h^2}{2} f(\alpha) \tag{4.3.5}$$

we can plot the force P resulting from slip along the plane α as we did in figure 4.2 but this time we will consider both the active and passive situations. The combined results are shown in figure 4.4.

Considering P to be the independent variable, we can see that if $P < P_{max}$, an active slip plane will be formed and that if $P > P_{min}$ a passive slip plane will develop. Comparison of equations (4.2.15) and (4.3.4) shows that inevitably $P_{min} > P_{max}$ so that there is a range of values of P for which no slip, either active or passive, occurs. This is the stable range and any force in the interval, $P_{max} < P < P_{min}$ can be supported without motion of either the wall or the material. The paradoxical appearance of this last expression comes naturally from mathematical convention and is deliberately retained here for its impact value. It should be remembered that P_{max} is the *greatest* force for active failure and hence the *least* force in the absence of failure.

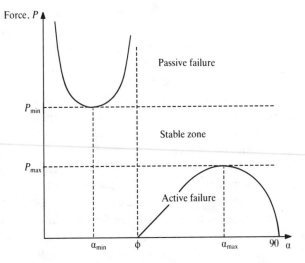

Figure 4.4 Active and passive wall forces.

Similarly P_{min} is the *lower* limit of passive failure and the *upper* limit of the stable range.

The deduction that there is a range of stable values of the force P on the wall is clearly analogous to the conclusions of Rankine's analysis which was presented in §3.4. Indeed, for the case of a smooth wall ($\phi_w = 0$ and hence $k = 0$), we have from equation (4.2.15),

$$P_{max} = \frac{\gamma h^2}{2} \frac{\cos^2\phi}{(1 + \sin\phi)^2} = \frac{\gamma h^2}{2} \frac{(1 - \sin\phi)}{(1 + \sin\phi)} \qquad (4.3.6)$$

and similarly

$$P_{min} = \frac{\gamma h^2}{2} \frac{(1 + \sin\phi)}{(1 - \sin\phi)} \qquad (4.3.7)$$

Thus, for the smooth wall

$$\frac{\gamma h^2}{2} \frac{(1 - \sin\phi)}{(1 + \sin\phi)} < P < \frac{\gamma h^2}{2} \frac{(1 + \sin\phi)}{(1 - \sin\phi)} \qquad (4.3.8)$$

and this result is entirely compatible with Rankine's result for a cohesionless material, given by equation (3.4.10).

Whereabouts in the range $P_{max} < P < P_{min}$, the force P will actually lie, depends on the elasticity of both the material and the wall and also on the method by which the material was placed behind the wall. If the wall is very flexible, so that it is displaced outwards to an appreciable extent during filling, the force P will approach the active value P_{max}. If, on the other hand, the wall is forced inwards or if the material is rammed into place, the force will increase to the passive value P_{min}. These results are illustrated in figure 4.5 which shows the dependence of P on the displacement of the wall. The passive and active limits occur whenever there is sufficient displacement, and between the two limits there is a narrow elastic regime. The extent of outward movement of the wall to give the active result and the extent of inward displacement to give the passive value, can in principle be determined from a knowledge of the elastic properties of the material. Such calculations are, however, complex and beyond the scope of the book. We will therefore content ourselves with predicting the range within which the force must lie and the reader must keep this limitation to our ambition permanently in mind.

For a fluid, the variation of the wall stress σ_f with depth is given by

$$\sigma_f = \gamma h \qquad (4.3.9)$$

whereas for a granular material supported by a smooth wall, the corresponding stress σ_s can be found by differentiating equation (4.3.7) to give

$$\sigma_s = \gamma h \frac{1 - \sin \phi}{1 + \sin \phi} \qquad (4.3.10)$$

For a rough wall, we have from equation (4.2.25) that the stress σ_r is

$$\sigma_r = \gamma h \frac{\cos^2\phi}{[1 + \sqrt{(1 + k)} \sin \phi]^2} \qquad (4.3.11)$$

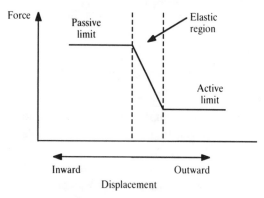

Figure 4.5 Dependence of the wall force on the displacement of the wall.

Since equation (4.3.11) reduces to equation (4.3.10) when $k = 0$, it is clear that under all circumstances

$$\sigma_f > \sigma_s > \sigma_r \tag{4.3.12}$$

The reasons for this are two-fold. First, the wall roughness provides a vertical force which partially supports the weight of the material. Thus in the case of active failure against a rough wall, σ_{yy} increases less rapidly with depth than in the smooth-walled and hydrostatic cases with the result that $\sigma_{yy} < \gamma h$. Additionally, in the active failure, $\sigma_{xx} < \sigma_{yy}$ in contrast to a liquid in which the pressure is isotropic and $\sigma_{xx} = \sigma_{yy}$. Thus, we see that in active failure the normal force on a wall is always less than that exerted by a fluid of the same density. However, the shear stresses on a rough wall generate a compressive force in the wall which is absent in the smooth-walled and hydrostatic cases and this must be borne in mind when designing the strength of a wall. Failure by buckling due to the compressive force in the wall is as common as outward failure caused by the normal force particularly when the retaining wall consists of a thin metal sheet.

4.4 Inclined walls and top surfaces

We can use the method of the previous sections to analyse the case of a partially rough wall inclined at angle η to the vertical, supporting a granular material with a planar top surface inclined at β to the horizontal, as shown in figure 4.6. The analysis follows exactly the same lines, with a quadratic again being found from which $\tan \alpha_m$ can be calculated. Though in principle the method is very similar to that

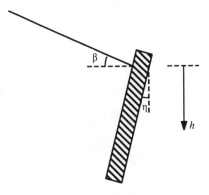

Figure 4.6 Inclined retaining wall.

of §4.2, the formal algebra becomes tedious due to the complexity of the trigonometrical relationships between the five angles involved. None-the-less, Coulomb derived the following result for the force P on the wall during active failure;

$$P = \frac{\gamma h^2}{2} \frac{\cos^2(\eta - \phi)}{\cos^2\eta \cos(\eta - \phi_w)\left\{1 + \sqrt{\left[\dfrac{\sin(\phi + \phi_w)\sin(\phi - \beta)}{\cos(\eta + \phi_w)\cos(\eta - \beta)}\right]}\right\}^2}$$

(4.4.1)

and this must be regarded as a triumph of eighteenth century algebra.

Equation (4.4.1) can be seen to reduce to equation (4.2.25) for the special case of $\beta = 0°$, $\eta = 0°$ i.e. for a horizontal fill behind a vertical wall.

It is also noteworthy that the predicted value of P from equation (4.4.1) becomes complex if $\beta > \phi$. This is yet another illustration that the slope of a top surface cannot exceed ϕ, i.e. that the angles of repose and internal friction are equal; a result which we first derived in §3.5.

Though a great many problems can be solved by the ritual application of equation (4.4.1), it is instructive to consider an alternative approach in which particular situations are analysed on an arithmetical, as opposed to an algebraic, basis.

Let us for example consider a vertical wall supporting a fill whose upper surface is inclined at $\beta = 10°$ to the horizontal. We will take $\phi = 30°$ and $\phi_w = 20°$. Considering point A, a distance h below the top surface at the wall, as in figure 4.7, we can denote by b the horizontal distance between the wall and the point of intersection C of the slip plane with the top surface. Clearly

$$b \tan \alpha - b \tan 10 = h \qquad (4.4.2)$$

and, since the area ABC is $hb/2$ we have that

$$W = \frac{\gamma h^2}{2(\tan \alpha - \tan 10)} \qquad (4.4.3)$$

Resolving horizontally and vertically as in §4.2 gives

$$P \cos 20 = X \sin(\alpha - 30) \qquad (4.4.4)$$

and

$$P \sin 20 + X \cos(\alpha - 30) = W \qquad (4.4.5)$$

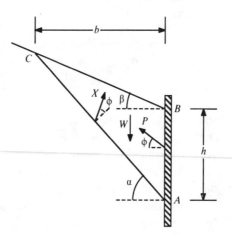

Figure 4.7 Retaining wall supporting a fill with an inclined upper surface.

Hence, eliminating X and W from these equations, we have

$$\frac{2P \cos 20}{\gamma h^2} = \frac{1}{(\tan \alpha - \tan 10)[\tan 20 + \cot(\alpha - 30)]} = f(\alpha) \quad (4.4.6)$$

There are two methods by which we can evaluate the maximum value of $f(\alpha)$. First, we could make the substitution $\tan \alpha = t$, as before, from which we find that

$$f(t) = \frac{t - 0.5776}{0.9413 \, t^2 + 0.6268 \, t - 0.1369} \quad (4.4.7)$$

which, on differentiation and setting to zero, gives the quadratic

$$t^2 - 1.11552 \, t - 0.2391 = 0 \quad (4.4.8)$$

From this we find that $t = 1.334$ and hence $\alpha_m = 53.15°$. We can then find f_{max} from equation (4.4.6) as 0.3195.

Recalling from equation (4.2.16) that $\sigma = d(P_m \cos \phi_w)/dh$, we find that

$$\sigma = 0.3195 \, \gamma h \quad (4.4.9)$$

Alternatively, we could simply evaluate $f(\alpha)$ for various values of α, a trivially easy exercise on a programmable calculator, and obtain the results presented in table 4.1.

By inspection, the maximum value of $f(\alpha)$ is about 0.32 and occurs at a value of α just over 53° as we found above.

Table 4.1

$\alpha(°)$	47	50	53	56	59
$f(\alpha)$	0.307	0.3165	0.3195	0.3171	0.3100

However, a striking feature of this table is the extreme flatness of the maximum. There is only 4% variation in $f(\alpha)$ for values of α covering the range $\pm 6°$ from the optimum value. It is therefore not necessary to evaluate $f(\alpha)$ at closely spaced intervals.

Furthermore, this observation gives credence to Reimbert's (1976) assertion that it is usually sufficiently accurate to assume that $\alpha_m = 45 + \phi/3$. In our example, this would give $\alpha_m = 55°$ and hence $f = 0.3185$, a result that differs from the exact value of f_{max} by only 0.3%. Thus, using Reimbert's rule-of-thumb, we can get an acceptably accurate result simply by substituting his recommended value of α_m into equation (4.4.6). This process avoids having to maximise $f(\alpha)$ either by numerical means or by differentiation. Unfortunately, it is not easy to quantify in a general way the expected accuracy of this method. Reimbert's assumption will be discussed in further detail in §4.6.

4.5 Numerical and graphical methods

Whenever the shape of the top surface of the fill is non-planar, the algebra associated with Coulomb's method becomes complicated. Many problems can, however, be solved by numerical or graphical methods with comparative ease.

A numerical solution can be obtained by the following procedure, which can easily be implemented on a micro-computer or a programmable calculator.

Paying attention to a particular point A on the wall a depth h below the top surface as in figure 4.8, we can assume a value for the slip plane angle α. The area above this assumed slip plane and hence the associated value of W can be found by some numerical integration technique, such as the trapezium rule, and equations (4.2.2) and (4.2.3) can then be solved to give P. This process is repeated for a series of values of α and the largest resulting value of P is the one required. The determination of the maximum value of P and the associated value α_m is facilitated by the use of the following algorithm which is

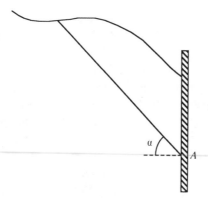

Figure 4.8 Retaining wall supporting a fill with an arbitrary upper surface.

based on fitting a parabola through three successive determinations of P.

In general this algorithm can be expressed as follows:

If we take three equally spaced values x_1, x_2 and x_3 of the independent variable x (which may be α or $\tan \alpha$) and calculate the corresponding values y_1, y_2 and y_3 of the dependent variable (which in this case is P), then the maximum value of y is given by

$$y_{max} = y_2 + \frac{(y_1 - y_3)^2}{8(2y_2 - y_1 - y_3)} \qquad (4.5.1)$$

and occurs at

$$x_m = x_2 - \frac{(y_1 - y_3)(x_2 - x_1)}{2(2y_2 - y_1 - y_3)} \qquad (4.5.2)$$

This algorithm is most accurate if the values of x are chosen so that y_2 is greater than either y_1 or y_3 and the accuracy increases as the size of the interval in x is reduced. However, we saw in §4.4 that the maximum in P is usually exceedingly flat so that comparatively large intervals in α of, say, 5° are often sufficiently accurate. The procedure is therefore to start with a comparatively small value of α and to increase α in regular steps, noting the corresponding values of P until a value of P is found which is less than the preceding one. The last three values of α and P are denoted by x_1, x_2, x_3 and y_1, y_2, y_3 and equations (4.5.1) and (4.5.2) are used to find y_{max} (i.e. P_m) and x_m (i.e. α_m). If y_{max} is considerably greater than y_2, it may be wise to repeat the calculation with a smaller interval of α.

Once P_m has been determined for a given value of h, the procedure can be repeated for other values of h and the normal stress on the wall can be estimated from $\Delta(P_m \cos \phi_w)/\Delta h$. Alternatively, an interpolating polynomial can be fitted to the results and the normal stress found by differentiation. Since only in exceptional cases does α_m vary rapidly with depth (it is after all a constant for a planar top surface) the subsequent determinations of α_m can be speeded up by starting with a value only marginally below that obtained for the previous value of h.

Alternatively, though less accurately, the problem may be solved by a graphical method. Again a series of values of α, say α_1, α_2, α_3, . . . are chosen and, by careful drawing, the area above each slip plane can be determined. This is most conveniently done by means of a planimeter. From the areas the corresponding weights W_1, W_2, W_3, . . . can be found. In general the three forces W, P and X acting on the material above the slip plane can be drawn in the form of a vector triangle as in figure 4.9(a). The vector W is vertical, P is inclined at $\eta + \phi_w$ to the horizontal and the force X is inclined at $\alpha - \phi$ to the vertical.

We can combine several such triangles into a single diagram as in figure 4.9(b). Starting with a common origin O we can draw the W vectors as the vertical lines Ow_1, Ow_2, . . . to some appropriate scale. The P vector is the line through O inclined at $\eta + \phi_w$ to the horizontal

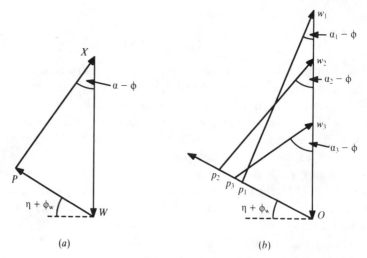

(a)   (b)

Figure 4.9 Force diagrams for a graphical solution.

and through each of the points w_1, w_2, . . . we can draw the X vector which is inclined to the vertical by $(\alpha_1 - \phi)$, $(\alpha_2 - \phi)$ . . . The points of intersections of these vectors with the P vector are denoted by p_1, p_2, . . . and the values P are given by the lengths of the vectors Op_1, Op_2, . . . Since we are concerned with the largest value of P, we are only interested in the point p lying furthest to the left which for the case shown in figure 4.9(b) is p_2. P_m is therefore given by the length Op_2. As with the numerical solution, this procedure can be repeated for a series of values of h to give the normal stress on the wall.

4.6 Accuracy of Coulomb's method of wedges

It is appropriate at this stage to assess the validity of Coulomb's method of wedges and hence to obtain an estimate of the probable accuracy of the predictions.

As mentioned in §4.1, the only assumption in the method of wedges (apart from the assumption that the material obeys Coulomb's yield criterion) is that the slip surface is planar. The rest of the analysis is rigorous. There is, however, no certainty that the slip surface is planar and in fact it is often curved as shown in figure 4.10. The exact shape of the slip surface can be calculated by the methods of chapter 7, but we can obtain a rough estimate of the amount of curvature by evaluating the slopes of the slip surface at the wall, α_w, and at the top surface of the material, α_T. If these two angles are equal we can say with fair confidence that the slip surface will be planar and if these angles are also equal to the angle α_m predicted by Coulomb's method we can deduce that the predicted stresses are accurate. If, on the other hand, the values of α_w and α_T are not the same, the slip surface will

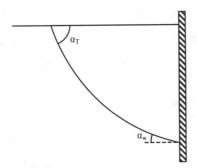

Figure 4.10 Non-planar slip surface.

clearly not be planar and it becomes important to estimate the magnitude of the error introduced by the assumption of a planar slip surface.

Figure 4.11(a) shows the Mohr's circle for the stresses adjacent to a partially rough wall. If we consider a right-hand wall the wall shear stress will be positive and if, for the present, we confine our attention to active failure, we can see that the wall stresses are given by the point W on the figure. It can be seen that the angle OCS' is $90 - \phi$ and that the angle WCO is $\omega - \phi_w$ where ω is given by equation (3.7.8) i.e.

$$\sin \omega = \frac{\sin \phi_w}{\sin \phi} \qquad (4.6.1)$$

Thus, the slip plane S' is inclined anticlockwise from the wall plane by $\frac{1}{2}(90 + \omega - \phi - \phi_w)$. From figure 4.10, this angle is $90 - \alpha_w$ and hence we have

$$\alpha_w = \frac{1}{2}(90 + \phi + \phi_w - \omega) \qquad (4.6.2)$$

The angle α_T can be evaluated from the knowledge that on the top surface, the shear stress is zero and hence the vertical and horizontal stresses are principal stresses. In active failure, the vertical stress σ_{yy} is the major principal stress and is marked by the point Y on figure

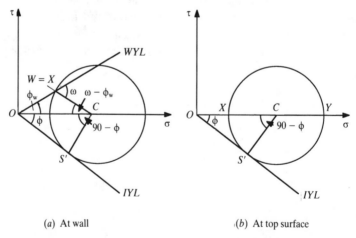

(a) At wall   (b) At top surface

Figure 4.11 Mohr's circles for determining the slip surface slope in active failure.

4.11(b), from which we can see that the slip plane S' lies at $\frac{1}{2}(90 + \phi)$ clockwise from the y-plane. Hence,

$$\alpha_T = \tfrac{1}{2}(90 + \phi) \qquad (4.6.3)$$

It should, however, be pointed out that on the top surface all the stresses are zero and that the Mohr's circle degenerates to a point. Strictly, figure 4.11(b) gives the Mohr's circle for the stresses immediately below the top surface.

Thus, we can see that in general

$$\alpha_T \neq \alpha_w \qquad (4.6.4)$$

and we can evaluate the quantity $\alpha_T - \alpha_w$ which is a measure of the curvature of the slip surface, from equations (4.6.2) and (4.6.3), giving

$$\alpha_T - \alpha_w = \tfrac{1}{2}(\omega - \phi_w) \qquad (4.6.5)$$

This is seen to be half the angle WCO of figure 4.11(a). It is clear from this figure that the angle WCO is likely to be small provided ϕ_w is considerably less than ϕ but it is of interest to evaluate these two estimates of α and to compare them with the value of α_m predicted by equation (4.2.23). The results are given in table 4.2 for the case of $\phi = 40°$ and for various values of ϕ_w.

It can be seen that for $\phi_w = 0$, all three estimates of α are identical, telling us that the method of wedges is accurate for a smooth vertical wall. As ϕ_w increases, the difference between α_T and α_w also increases but remains small until ϕ_w approaches ϕ. However, for the fully rough wall there is a considerable difference between α_T and α_w showing that the assumption of a planar slip surface is not correct under these circumstances. It should, however, be noted that under all circumstances the value of α_m lies between α_T and α_w and is always closer to α_T

Table 4.2

$\phi_w(°)$	0	10	20	30	40
$\alpha_T(°)$	65	65	65	65	65
$\alpha_w(°)$	65	62	59	54	40
$\alpha_m(°)$	65	64	63	61	60
$f(\alpha_T)$	0.217	0.201	0.186	0.171	0.156
$f(\alpha_w)$	0.217	0.201	0.184	0.158	0
$f(\alpha_m)$	0.217	0.201	0.187	0.174	0.161
$\alpha_R(°)$	58.3	58.3	58.3	58.3	58.3
$f(\alpha_R)$	0.204	0.193	0.182	0.172	0.160

than α_w. We will see in §7.7 that the curvature of the actual slip surface is mainly in the region close to the wall so that most of the slip surface is inclined to the horizontal by an angle close to α_T. Thus the value of α_m is probably a good best-fit straight line to the actual surface.

For the cases when the slip surface is not planar, we can make a rough estimate of the probable accuracy of the predictions by evaluating the function $f(\alpha)$ defined by equation (4.2.21). Recalling that the normal stress at depth h is given by

$$\sigma = \gamma h f(\alpha) \qquad (4.6.6)$$

the probable accuracy of the result can be gauged by comparing the values of $f(\alpha)$ predicted from the three estimates of α given in the upper part of table 4.2. It can be seen that for $\phi_w < 0.5\,\phi$, the three estimates are barely distinguishable, though inevitably $f(\alpha_m)$ is the largest. Even for $\phi_w = 30°$, there is only a 10% discrepancy which might be quite acceptable in many practical applications. However, the value of $f(\alpha_w)$ of zero for the fully rough wall indicates something seriously wrong with the method in this situation. The problem is, however, not as severe as seems at first sight since, as we noted in §3.7, it is unlikely that any real wall will be as rough as our idealised fully rough wall.

Thus we can conclude that, for the case of a vertical wall supporting a cohesionless fill with a horizontal top surface, Coulomb's method is sufficiently accurate for most practical purposes provided $\phi_w \lesssim 0.8\,\phi$. Similar calculations can be performed for other cases and should perhaps be done routinely to guard against a combination of circumstances that give rise to unacceptable uncertainties.

Reimbert's rule-of-thumb that $\alpha = 45 + \phi/3$ gives the results shown as α_R and $f(\alpha_R)$ in the lower part of table 4.2. The similarity between $f(\alpha_R)$ and $f(\alpha_m)$ shows that indeed Reimbert's rule is an excellent way of obtaining a quick first estimate, though it inevitably under-predicts the force on the wall to some small extent.

A similar analysis can be performed for the passive case. Now the Mohr's circles are as shown in figure 4.12 and we can deduce by similar arguments to those presented above that

$$\alpha_T - \alpha_w = \tfrac{1}{2}(\omega + \phi_w) \qquad (4.6.7)$$

Comparison with equation (4.6.5) shows that the curvature is inevitably of greater magnitude. Thus, Coulomb's method of wedges does not usually give an accurate prediction in passive failure and this analysis justifies the statement to that effect in the introduction to §4.3.

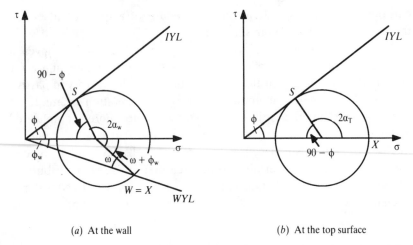

(a) At the wall   (b) At the top surface

Figure 4.12 Mohr's circles for determining the slip surface slope in passive failure.

Similarly, we can consider the applicability of the method of wedges to cohesive materials. Though the presence of cohesion does complicate the analysis to some extent, the method is quite workable for cohesive materials. However, analysis of the inclinations of the slip surface similar to the one presented above, shows that cohesive materials have greater curvature of the slip surface than cohesionless materials. In particular, for a purely cohesive material for which $c \neq 0$ and $\phi = 0$,

$$\alpha_T - \alpha_w = 45° \tag{4.6.8}$$

Thus the method should be used with caution in the presence of cohesion, and this casts some doubt on the numerical accuracy of the predictions made in §3.6 above.

There is one further uncertainty connected with Coulomb's method. Analyses of this type consider the force exerted on the wall by a slip plane of any inclination α, and it is assumed that in active failure the largest force is the one required. Strictly, we should consider all slip surfaces, planar or not, and take the largest force resulting from any of these. This is a formidable task, but for highly cohesive materials it is often assumed that the slip surface is part of a circle and a search is made for the co-ordinates of the centre of the circle that gives the largest force on the wall. This method is sometimes known as the *Swedish circle* method as it was first used for the redesign of the harbour at Gothenburg following the failure of a quay designed on the assumption of a planar slip surface.

We may conclude therefore that the method of wedges gives a reasonably accurate prediction of the wall stresses in active failure for cohesionless materials but that the method is less satisfactory in passive failure or for cohesive materials. Bearing this limitation in mind, we will consider in the following sections some more complicated geometries, partly in order to derive results that are of value in themselves but also because this method displays in a readily comprehensible form certain curious features of stress distributions whose origins are less obvious in the more rigorous analyses considered in chapter 7.

4.7 Effect of a surcharge

Circumstances arise when we must consider the effect of a stress Q applied to the top surface of the fill. Within the context of materials handling, these circumstances are less obvious than in the analogous soil mechanics case in which Q represents the stresses resulting from the weight of a building or other structure erected on the surface. We will therefore reserve until §5.7 and §5.9 a discussion of the origin of the stress Q, which is often referred to as a surcharge.

The analysis for a uniform surcharge

$$Q = Q_0 = \text{constant} \tag{4.7.1}$$

is straight-forward. Following the method of §4.2 for a vertical partially rough wall, we can see from figure 4.13 that on resolving horizontally we have

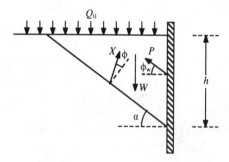

Figure 4.13 Failure of a wall due to an applied surcharge.

$$P \cos \phi_w = X \sin (\alpha - \phi) \qquad (4.7.2)$$

and resolving vertically gives

$$P \sin \phi_w + X \cos (\alpha - \phi) = W + Q_0 h \cot \alpha \qquad (4.7.3)$$

where W is given by equation (4.2.1)

$$W = \frac{\gamma h^2}{2} \cot \alpha \qquad (4.7.4)$$

On rearranging, these equations give

$$P \cos \phi_w = \left(\frac{\gamma h^2}{2} + Q_0 h \right) \frac{\cot \alpha}{\tan \phi_w + \cot (\alpha - \phi)} \qquad (4.7.5)$$

$$= \left(\frac{\gamma h^2}{2} + Q_0 h \right) f(\alpha) \qquad (4.7.6)$$

where $f(\alpha)$ is exactly the same as in equation (4.2.21).
Thus the maximum force P_m is given by

$$P_m \cos \phi_w = \left(\frac{\gamma h^2}{2} + Q_0 h \right) f_{max} \qquad (4.7.7)$$

where f_{max} is given by equation (4.2.25).
From equation (4.2.17) we have that

$$\sigma_w = \frac{dP_m}{dh} \cos \phi_w = (\gamma h + Q_0) f_{max} \qquad (4.7.8)$$

Thus we see that the wall stress increases with depth at exactly the same rate as in the case of zero surcharge but from a non-zero value of $Q_0 f_{max}$ at the top surface.

We will now consider two cases of non-uniform surcharge. The numerical accuracy of these calculations is perhaps somewhat suspect but the analysis is included since it shows in a fairly clear manner some interesting peculiarities of the stress distributions in granular materials. These will be considered in greater detail in chapter 7 but the comparative difficulty of the algebra of that chapter obscures the reasons for this unexpected behaviour. The main purpose of the two examples in this section is to prepare the reader for the analyses of §4.8 and chapter 7 and to give him a qualitative explanation of the phenomena presented there.

In our first case we will consider a vertical wall supporting a horizontal fill with a surcharge which is zero up to point B a distance

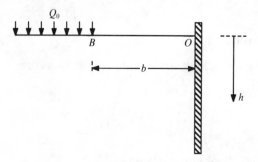

Figure 4.14 Non-uniform surcharge.

b from the wall and thereafter has the constant value of Q_0 as shown in figure 4.14. We will solve this problem for one particular case, namely for $\phi = 30°$, $\phi_w = 20°$ and $Q_0 = 0.5\gamma b$.

Measuring h downward from the top of the wall, as before, we can see intuitively that for small values of h the slip plane will cut the top surface to the right of point B. Analysis of such a slip plane is identical to that of §4.2 and we therefore have from equation (4.2.23) that

$$\tan \alpha_m = \tan \phi + \frac{\sec \phi}{\sqrt{(1 + k)}} \tag{4.7.9}$$

from which we find that

$$\alpha_m = 55.98° \tag{4.7.10}$$

Hence from equation (4.2.25),

$$f_{max} = 0.2794 \tag{4.7.11}$$

Thus for small h

$$\sigma_w = 0.2794 \, \gamma h \tag{4.7.12}$$

or

$$\frac{\sigma_w}{\gamma b} = 0.2794 \, H \tag{4.7.13}$$

where

$$H = h/b \tag{4.7.14}$$

Similarly, for very large values of h, the slip plane will cut the top surface far to the left of B and the effect of the 'missing' portion of the surcharge to the right of B will be negligible. Thus from equation (4.7.8),

$$\sigma_w = (\gamma h + 0.5\gamma b)f_{max} \qquad (4.7.15)$$

or

$$\frac{\sigma_w}{\gamma b} = 0.2794\,(H + 0.5) \qquad (4.7.16)$$

For intermediate values of h, we must resort to first principles and, resolving horizontally and vertically, we obtain, as in equation (4.7.2) and (4.7.3),

$$P \cos \phi_w = X \sin (\alpha - \phi) \qquad (4.7.17)$$

$$P \sin \phi_w + X \cos (\alpha - \phi) = W + F \qquad (4.7.18)$$

where F is the force exerted on the top surface by the surcharge, and α is the slip plane angle as shown in figure 4.13. Since the slip plane cuts the top surface a distance $h \cot \alpha$ from O we have that

$$\begin{matrix} F = Q_0(h \cot \alpha - b) & \text{if } h \cot \alpha > b \\ F = 0 & \text{if } h \cot \alpha < b \end{matrix} \qquad (4.7.19)$$

This can most conveniently be expressed in terms of the Macaulay brackets { } which have the value of the argument if this is positive and have the value zero if the argument is negative. Rearranging equations (4.7.17) to (4.7.19) gives

$$\frac{P \cos \phi_w}{\gamma b^2} = \frac{H^2 \cot \alpha + \{H \cot \alpha - 1\}}{2[\tan \phi_w + \cot(\alpha - \phi)]} = P^*(H,\alpha) \qquad (4.7.20)$$

For the case in hand, $P^*(H,\alpha)$ is plotted as a function of α in figure 4.15 for the three depths of $H = 1.1$, 1.2 and 1.3. In all three cases it is seen that the curve has two sections that intersect at $H = \tan \alpha$, with each section having its own maximum. These maxima correspond to the cases of the slip plane cutting the top surface to the left or to the right of point B. The maxima corresponding to $\tan \alpha > H$ are always at $\alpha = \alpha_m = 55.98°$ whereas the other (left-hand) maxima are at varying values of α.

As always we require the largest value of P^*, and for $H = 1.1$ this is seen to occur as usual at α_m. The maximum value is found to be 0.169. At $H = 1.2$, the two maxima have the same value of 0.201 and for $H = 1.3$, the left hand maximum, at $\alpha = 47°$ is the larger. Thus

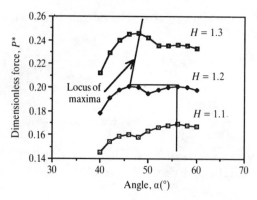

Figure 4.15 Dependence of the wall force on the slip plane inclination at various heights. Note the change in the position of the maximum.

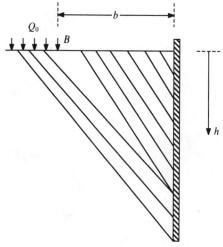

Figure 4.16 Slip plane pattern for a non-uniform surcharge.

at H of about 1.2, the slip plane angle suddenly changes from the previously constant value of $55.98°$ to about $47°$ as illustrated in figure 4.16 which shows the slip plane pattern for this example.

As we have seen, we require the right-hand maximum of figure 4.15 for $H < 1.2$ and the left-hand maximum for $H > 1.2$. The values of these maxima, P_m^*, are plotted in figure 4.17 where it is seen that there is a sudden change of slope at $H = 1.2$. The continuations of the two parts of the curve are shown dotted and correspond to the lesser, and therefore irrelevant, maximum.

The stress distribution on the wall can be found by differentiating the curve of P_m^* *vs* H numerically and is plotted in figure 4.18. It is seen that for $H < 1.2$ the results lie exactly on the line of equation (4.7.13), the relationship for zero surcharge. At $H = 1.2$ there is a discontinuity in the wall stress corresponding to the discontinuity in the slope of figure 4.17 and for large H, the stresses tend rapidly to the known asymptote of equation (4.7.16).

Figures 4.16 and 4.18 illustrate a most important feature of stress distributions in granular materials. For low H, the slip plane cuts the top surface to the right of point B and the stress on the wall is totally independent of the existence of the surcharge to the left of B. It is only at $H = 1.2$ that the effect of this surcharge is suddenly felt at the wall. This shows that the direction of the transmission of information

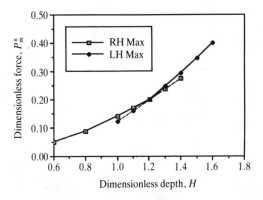

Figure 4.17 Wall force as a function of depth. Note the discontinuity of slope at $H \approx 1.2$.

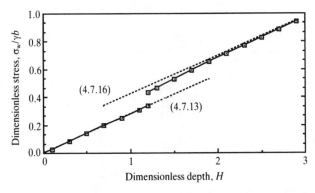

Figure 4.18 Discontinuous dependence of stress on depth.

about stresses is limited, with some parts of the material being *totally unaffected* by the presence of the surcharge. This is in marked contrast to the similar problem of predicting the temperature distribution in a conducting material, where the application of an elevated temperature on part of the boundary affects the temperature at *all* points within the material.

We will now consider the complementary problem of a uniform surcharge Q_0 extending from the top of the wall as far as the point B a distance b from the wall. Beyond B the surcharge is zero as shown in figure 4.19. As before we will consider the particular case of $\phi = 30°$, $\phi_w = 20°$ and $Q_0 = 0.5\gamma b$.

Following the method of the first part of this section, we can see from equation (4.7.10) that

$$\alpha_m = 55.98° \tag{4.7.21}$$

and from equations (4.7.7) and (4.7.11) that

$$P_m \cos \phi_w = 0.2794 \left(\frac{\gamma h^2}{2} + Q_0 h \right) \tag{4.7.22}$$

or

$$\frac{\sigma_w}{\gamma b} = 0.2794 \, (H + 0.5) \tag{4.7.23}$$

These results are, however, only valid provided the slip plane from the point of interest cuts the top surface to the right of point B. Thus these equations apply only for $H < \tan \alpha_m = 1.4817$.

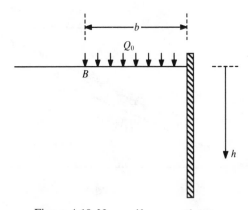

Figure 4.19 Non-uniform surcharge.

For larger values of H, we can evaluate the force F exerted by the surcharge on the top surface of the wedge as in equation (4.7.19), which now takes the form

$$\left.\begin{array}{ll} F = Q_0 h \cot \alpha & \text{if } h \cot \alpha < b \\ F = Q_0 b & \text{if } h \cot \alpha > b \end{array}\right\} \qquad (4.7.24)$$

Combining this equation with the equations of horizontal and vertical resolution, equations (4.7.17) and (4.7.18), yields

$$\frac{P \cos \phi_w}{\gamma b^2} = \frac{H^2 \cot \alpha + H \cot \alpha - \{H \cot \alpha - 1\}}{2[\tan \phi_w + \cot (\alpha - \phi)]} = P^*(H, \alpha) \tag{4.7.25}$$

which is analogous to equation (4.7.20) and where again $\{\ \}$ represent the Macaulay brackets, the value of which is zero if the argument is negative.

Taking the particular case of $H = 1.6$, we can plot $P^*(H, \alpha)$ as a function of α in figure 4.20. As in figure 4.15, we see that the curve has two sections, depending on whether the slip plane cuts the top surface to the left or right of point B. However, in this case, neither section has a maximum value and the largest value of P^* therefore occurs where the two sections meet, i.e. at the value of α such that the slip plane passes exactly through the point B. Thus the critical value of α is given by

$$H = \tan \alpha \tag{4.7.26}$$

and this result holds for a range of values of H from the value of 1.4817 to some upper limit which we have yet to determine. Within

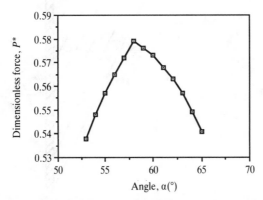

Figure 4.20 Wall force as a function of slip plane angle.

this range we can substitute equation (4.7.26) into (4.7.25) and obtain

$$\frac{P_{\mathrm{m}} \cos \phi_{\mathrm{w}}}{\gamma b^2} = \frac{H + 1}{2[\tan \phi_{\mathrm{w}} + \cot(\tan^{-1}H - \phi)]} \qquad (4.7.27)$$

from which we can calculate P_{m} directly without any need for differentiation. This result will of course be valid only if no larger value of P can be obtained at the same value of H.

Thus we see that for $H < 1.48$, the slip planes are parallel with a slope of 1.48 but that for a range of greater values of H, the slip planes pass through point B as shown in figure 4.21. Clearly for large enough values of H, the slip plane must cut the top surface to the left of B and equation (4.7.25) then takes the form,

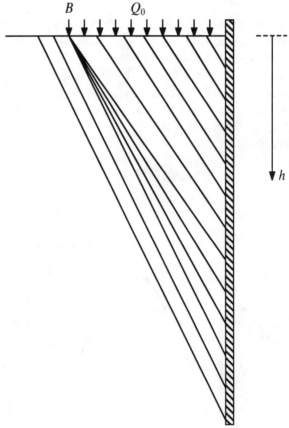

Figure 4.21 Slip plane pattern.

Table 4.3

H	1.9	2.0	2.1	2.2	2.3
$\dfrac{P_m \cos \phi_w}{\gamma b^2}$	0.7491	0.7994	0.8532	0.9103	0.9705
$\tan \alpha_m$	2.25	2.14	2.05	1.96	1.87

$$\frac{P \cos \phi_w}{\gamma b^2} = \frac{H^2 \cot \alpha + 1}{2[\tan \phi_w + \cot(\alpha - \phi)]} \tag{4.7.28}$$

As always we are concerned with the maximum value of P for a given H and this is most conveniently obtained numerically. The results are given in table 4.3 for a range of values of H together with the associated values of $\tan \alpha_m$.

Recalling that this analysis is based on the assumption that the slip plane cuts the top surface to the left of point B, we must dismiss all the results for which $H < \tan \alpha_m$ and we see therefore that the results of this table are valid only for $H \geqslant 2.08$.

We see therefore that the wall force is given by equation (4.7.22) for $H < 1.48$, by equation (4.7.27) for $1.48 < H < 2.08$ and by the results of table 4.3 for $H > 2.08$. The corresponding stresses can be obtained by differentiation, either algebraic or numerical, and are presented in figure 4.22 as are the stress distributions predicted on the assumption of a) no surcharge (equation (4.7.13)), and b) a uniform surcharge $Q_0 = 0.5\ \gamma b$ (equation (4.7.23)). It can be seen that for

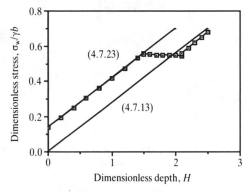

Figure 4.22 Wall stress as a function of depth.

$H < 1.48$, exact agreement with equation (4.7.23) is obtained, as expected. However, in the range $1.48 < H < 2.08$, the wall stresses actually decrease slightly with increasing depth and pass below the value given by equation (4.7.13). For values of $H > 2.08$, the wall stress again increases and approaches the result of equation (4.7.13) asymptotically from below. The corresponding pattern of slip planes is shown in figure 4.21, the most striking feature of which is the fan of slip planes emanating from the point B. We will find that this is a common feature of the exact stress analyses presented in chapter 7.

4.8 Twin retaining walls

As the final example in this chapter, we will consider twin retaining walls a distance b apart as shown in figure 4.23. We will take the case of a typical cohesionless material for which $\phi = 30°$ and $\phi_w = 20°$ and consider a uniform surcharge Q_0. As in the previous example we find that for small values of h, the slip planes cut the top surface so that $\alpha_m = 55.98°$ and the wall force is given by equation (4.7.22). On differentiating this equation we find that

$$\sigma_w = 0.2794 \, (\gamma h + Q_0) \tag{4.8.1}$$

This equation applies down to a depth H given by

$$H = \frac{h}{b} = \tan \alpha_m = 1.4817$$

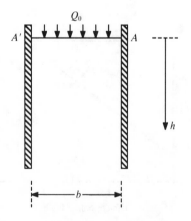

Figure 4.23 Twin retaining walls.

For a range of H greater than this value the slip planes pass through the far corner A' and the wall force is given by equation (4.7.27) which takes the form

$$\frac{P_m \cos \phi_w}{\gamma b^2} = \frac{H + 2q_0}{2[\tan \phi_w + \cot(\tan^{-1}H - \phi)]} \qquad (4.8.2)$$

where

$$q_0 = \frac{Q_0}{\gamma b} \qquad (4.8.3)$$

However, for large enough values of H the slip plane must cut the far wall and under these circumstances we must modify the force balances given by equations (4.7.17) and (4.7.18). We will consider a slip plane at angle α to the horizontal through the point B on the right-hand wall and denote by z the depth below the top surface at which this plane cuts the far wall, as shown in figure 4.24. Clearly

$$z = h - b \tan \alpha \qquad (4.8.4)$$

or

$$Z = H - \tan \alpha \qquad (4.8.5)$$

where

$$Z = \frac{z}{b} \qquad (4.8.6)$$

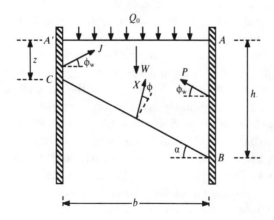

Figure 4.24 Forces during the active failure of the right-hand wall.

We will denote the force on the section of the far wall CA' by J and by symmetry we can say that J will equal the force on the right hand wall above the point $h = z$, i.e.

$$J = P(z) \tag{4.8.7}$$

Resolving horizontally and vertically gives

$$P \cos \phi_w = X \sin (\alpha - \phi) + J \cos \phi_w \tag{4.8.8}$$

$$P \sin \phi_w + J \sin \phi_w + X \cos (\alpha - \phi) = W + Q_0 b \tag{4.8.9}$$

where the weight of the material above the slip plane is given by

$$W = \frac{\gamma b}{2}(h + z) \tag{4.8.10}$$

Eliminating X from these equations gives

$$P \cos \phi_w = \frac{\dfrac{\gamma b}{2}(h + z) + Q_0 b - J \cos \phi_w [\tan \phi_w - \cot (\alpha - \phi)]}{\tan \phi_w + \cot (\alpha - \phi)} \tag{4.8.11}$$

We can now proceed to work down the right-hand wall making small increments, of say 0.1, in H and selecting various values of α. For each value of α, the corresponding value of Z is found from equation (4.8.5) and J can be found from our previous evaluations of P at smaller values of H. Thus, we can obtain P as a function of α and maximise P using the algorithm of equations (4.5.1) and (4.5.2). In fact, it is more convenient to take Z as the independent variable, using the values corresponding to the values of H at which P has already been determined.

The slip planes obtained by this method for the case of $Q_0 = 3\gamma b$ are shown as the dashed lines in figure 4.25, with their mirror images, which represent the slip planes for failure of the far wall being shown as full lines. The corresponding stress profiles are given in figure 4.26. It is seen that down to point B at a depth of $H = 1.48$, the slip planes cut the top surface and the wall force is given by equation (4.7.22). In this region the wall stress increases linearly with depth as given by equation (4.8.1). Between B and C the slip planes pass through the far corner A' forming a fan of slip planes. In this region the wall force is given by equation (4.8.2) and the corresponding stress is found to decrease with depth as in the final example of §4.7. Below C the stress

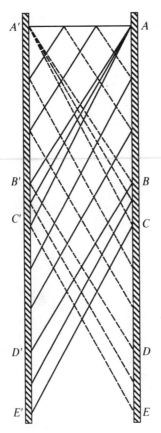

Figure 4.25 Slip planes for twin retaining walls, with surcharge $Q_0 = 3\gamma b$.

increases again but by D the slip plane intersects the far wall at B', the point at which the stress begins to fall, and a reflection of this behaviour is found in the region DE. Thus the wall stress varies in an irregular manner consisting of more or less linear portions, the first being truly linear. There are discontinuities of slope between the various portions, with the slope of alternate portions being of opposite sign as shown in figure 4.26. It can, however, be seen that at great depth the wall stress approaches a steady value, which we will show in §5.5 is equal to $\frac{1}{2}\cot \phi_w$.

For lower surcharges, these effects are less marked and for the case of zero surcharge the fan of slip planes emerging from the corner is narrow and there is no region in which the stress actually decreases with depth. The stress distribution for this case is given in figure 4.27.

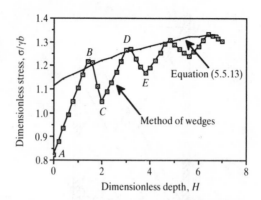

Figure 4.26 Wall stress distribution for twin retaining walls, with surcharge $Q_0 = 3\gamma b$.

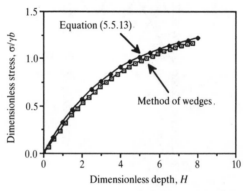

Figure 4.27 Comparison of the stress distributions predicted by the method of wedges and by the method of differential slices, with surcharge $Q_0 = 0$.

Also shown in these figures are the corresponding stress distributions predicted by the method of §5.5 and a comparison between the two predictions is made in that section.

The methods outlined in these last two sections can be used for many two-dimensional situations including inclined or irregular top surfaces and walls of arbitrary or variable slope. For example, the author knows of at least one design office that uses this method for the determination of the wall forces in storage bins in which the surface of the material is at the angle of repose as shown in figure 4.28. The reader is, however, reminded that this method can only be applied to effectively two-dimensional systems. Situations of axial symmetry must be analysed by the methods of chapters 5 and 7.

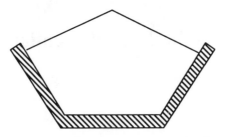

Figure 4.28 Cross-section of a grain silo.

Finally, a comment should be made about the analysis proposed by Airy in 1897. He considered the stresses in a bunker of square cross-section and used a method similar to the one described above. His analysis contains two errors. First, when considering the force balance on the material above the slip plane he neglected the force on the far wall, on the argument that, since the material was falling away from the far wall, this force would be zero. Secondly, he neglected the force on the two side walls and thus the total weight of the material was supported by the single wall under consideration. Clearly in a square bunker all four wall forces are equal, so that each wall supports a quarter of the weight of the material. Thus Airy's method over-estimates the wall force and as a result the corresponding axial stress passes through a maximum and eventually becomes negative. Despite this obviously unrealistic result, some authors still use Airy's results down to the stress maximum, giving specious arguments for assuming that the wall stress remains at the maximum value below this point. The resulting stress predictions are not in static equilibrium.

5

The method of differential slices

5.1 Introduction

The *Method of Differential Slices* is the name given by Hancock (1970) to a series of approximate analyses based on a method introduced by Janssen in 1895. The method has, however, been considerably extended since that time. Commercially these analyses are perhaps the most important in this book since they form the basis of the recommended procedures in most, if not all, of the national codes of practice for bunker design. The original version of the analysis, as presented by Janssen, is approximate and most design manuals present a set of empirically derived correction factors for use in conjunction with the predictions. More fundamental texts such as Walker (1966), however, attempt to correct the errors introduced by Janssen's approximations on a more rational basis. The purpose of this chapter is to outline Janssen's original method, to assess its accuracy and to describe the improvements and extensions that have been introduced subsequently.

In §5.2 we will outline the basic method for a cylindrical bunker and in §5.3 to §5.7 we will describe the improvements that can be made by more careful analysis. The method is extended to conical and wedge-shaped hoppers in §5.8 and Walters' analysis for the 'switch stress' is given in §5.9. Finally, in §5.10 we compare Janssen's analysis with the related analyses of Enstad and Reimbert.

5.2 Janssen's original analysis

Janssen's original analysis is best illustrated by considering the stress distribution in a cylindrical bunker containing a cohesionless granular material as illustrated in figure 5.1. Here we have taken a set of

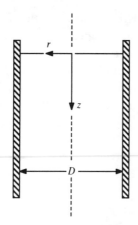

Figure 5.1 Definition of symbols.

cylindrical axes (r,z) with origin at the centre of the flat top surface of the fill.

Janssen's analysis is based on two unjustifiable assumptions:

(i) that the stresses are uniform across any horizontal section of the material, and
(ii) that the vertical and horizontal stresses are principal stresses.

It can easily be shown that neither of these is correct and §5.3 and §5.4 are devoted to the presentation of improved analyses in which these assumptions are relaxed. In this section we will accept Janssen's assumptions in order to provide a relatively simple analysis for the purpose of illustrating the principles of the method.

Let us perform a force balance on an elemental slice at depth z below the top surface as shown in figure 5.2. There is a downward force $A\sigma_{zz}$ on the top surface of the element and an infinitesimally

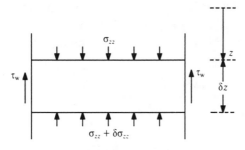

Figure 5.2 Stresses on a cylindrical element.

different upward force $A(\sigma_{zz} + \delta\sigma_{zz})$ on the base. Here A is the cross-sectional area of the bunker, given by $\pi D^2/4$. Strictly, the force on the top surface should be $\int \sigma_{zz} \, dA$ but in view of assumption (i) above, integration is unnecessary and the simple product is sufficient.

The weight of the material within the element is given by $\gamma A \delta z$ and there is an upward force $\pi D \delta z \, \tau_w$ on the side of the element due to the wall shear stress τ_w. Resolving vertically gives

$$\frac{\pi D^2}{4} \sigma_{zz} + \frac{\pi D^2}{4} \gamma \delta z = \frac{\pi D^2}{4} (\sigma_{zz} + \delta\sigma_{zz}) + \pi D \delta z \, \tau_w \qquad (5.2.1)$$

or

$$\frac{d\sigma_{zz}}{dz} + \frac{4 \tau_w}{D} = \gamma \qquad (5.2.2)$$

Since we have confined our attention to cohesionless materials we can say that

$$\tau_w = \mu_w \sigma_{rr} \qquad (5.2.3)$$

Assumption (ii) above specifies that the horizontal and vertical stresses are principal stresses and we saw in §3.3 that for a cohesionless material, the ratio of the principal stresses is a constant. Thus we can say that

$$\sigma_{rr} = K \sigma_{zz} \qquad (5.2.4)$$

where K is known as the Janssen constant. Comparison of equations (5.2.4), (3.3.16) and (3.4.9) shows that the Janssen constant is identical to Rankine's coefficient of earth pressure.

However, assumption (ii) does not specify which of the two normal stresses is the major principal stress and we must therefore consider both possibilities. We will call the case when σ_{rr} is the minor principal stress, 'the active case', and denote the corresponding value of K by K_A. When σ_{rr} is the major principal stress we have the 'passive case' and K will be denoted by K_P. From equation (3.4.9) we have

$$K_A = \frac{1 - \sin\phi}{1 + \sin\phi} \qquad (5.2.5)$$

and, from equation (3.4.11),

$$K_P = \frac{1 + \sin\phi}{1 - \sin\phi} \qquad (5.2.6)$$

The circumstances in which the active and passive solutions are appropriate are discussed in greater detail in §5.6 below. It is sufficient at this stage to note that most design manuals assume that a close approximation to the active state is achieved on filling.

In the rest of this section, we will present the analysis in terms of K, with the understanding that it takes the value of K_A or K_P as appropriate.

Substituting equations (5.2.3) and (5.2.4) into equation (5.2.2) gives

$$\frac{d\sigma_{zz}}{dz} + \frac{4\mu_w K}{D}\sigma_{zz} = \gamma \tag{5.2.7}$$

which is a simple first order differential equation for which we can write down the solution by inspection,

$$\sigma_{zz} = \frac{\gamma D}{4\mu_w K} + C\exp\left(\frac{-4\mu_w Kz}{D}\right) \tag{5.2.8}$$

where C is an arbitrary constant to be determined from the boundary conditions.

Most commonly, the top surface of the material is open to the atmosphere so that the normal stress on it may be taken to be zero. However, to preserve generality, we will assume that the top surface is subjected to a uniform surcharge Q_0, so that our boundary condition becomes

$$\sigma_{zz} = Q_0 \text{ on } z = 0 \tag{5.2.9}$$

with the expectation that in most realistic circumstances Q_0 will be zero.

Putting this boundary condition into equation (5.2.8), we obtain

$$\sigma_{zz} = \frac{\gamma D}{4\mu_w K}\left[1 - \exp\left(-\frac{4\mu_w Kz}{D}\right)\right] + Q_0\exp\left(-\frac{4\mu_w Kz}{D}\right) \tag{5.2.10}$$

From equation (5.2.4), we have that $\sigma_{rr} = K\sigma_{zz}$ and hence,

$$\sigma_{rr} = \frac{\gamma D}{4\mu_w}\left[1 - \exp\left(-\frac{4\mu_w Kz}{D}\right)\right] + Q_0 K\exp\left(-\frac{4\mu_w Kz}{D}\right) \tag{5.2.11}$$

Furthermore, since $\tau_w = \mu_w\sigma_{rr}$,

$$\tau_w = \frac{\gamma D}{4}\left[1 - \exp\left(-\frac{4\mu_w Kz}{D}\right)\right] + Q_0 K\mu_w\exp\left(-\frac{4\mu_w Kz}{D}\right) \tag{5.2.12}$$

It can be seen from equations (5.2.10) to (5.2.12), that all three stresses tend exponentially to asymptotic values at great depth and we will denote these values by σ^∞_{zz}, σ^∞_{rr} and τ^∞_w. Granted that the axial stress σ_{zz} does tend to an asymptotic value, a simple force at great depth gives

$$\frac{\pi D^2}{4}\, dz\, \gamma = \pi D\, dz\, \tau^\infty_w \qquad (5.2.13)$$

and hence,

$$\tau^\infty_w = \frac{\gamma D}{4} \qquad (5.2.14)$$

This force balance does not require the assumption that the material obeys the Coulomb yield criterion, or indeed that the material is a granular material. It simply specifies that the wall shear stress must support the weight of the material. Equation (5.2.14) is therefore undoubtedly‚ correct for all systems in which the stresses tend to asymptotic values. Thus we see that equation (5.2.12) predicts the correct value of τ_w in the limit as $z \to \infty$.

The asymptotic value of σ_{rr} is simply τ_w/μ_w and therefore the value of σ^∞_{rr} is correct for all cohesionless Coulomb materials. The asymptotic value of σ_{zz}, however, involves the Janssen constant K and its accuracy will therefore depend on the validity of the assumptions in the Janssen analysis. Similarly, the rate of approach to the asymptote depends on K and its accuracy may therefore be questioned.

Thus we may conclude that Janssen's method predicts that the stresses tend asymptotically to constant values. That for the shear stress is correct for all materials, that for the stress σ_{rr} is correct for cohesionless Coulomb materials whereas the predicted asymptotic value of σ_{zz}, and the rate of approach to the asymptotes are subject to errors of, as yet, unknown magnitude. The success of Janssen's method lies in the fact that the two quantities of greatest interest, the values of τ_w and σ_{rr} at great depth are independent of the rather questionable assumptions in the method. Unfortunately, this very convenient result has often given undue confidence in the remaining predictions of the method.

It has been seen above that the asymptotic values of τ_w and σ_{rr} are independent of the value of K and hence it follows that these are the same in the active and passive cases. The same is not, however, true for the asymptotic value of σ_{zz}. Furthermore the rate of approach to the asymptote is also a function of K and therefore differs in the active and passive cases.

Examination of equations (5.2.10) to (5.2.12) shows that the approach to the asymptote is governed by the term

$$\exp\left(\frac{-4\mu_w Kz}{D}\right)$$

so that we can define a characteristic depth z_c given by,

$$z_c = \frac{D}{4\mu_w K} \tag{5.2.15}$$

Clearly z_c is the depth over which the departure from the asymptote decreases by a factor of e. The stresses will therefore reach 90% of their final values at a depth of about $2.5z_c$.

Taking a typical material for which $\phi = 30°$ and $\phi_w = 20°$ (i.e. $\mu_w = 0.364$), we see from equation (5.2.5) that $K_A = 0.333$ and hence $z_c \approx 2D$. Thus in the active case the stresses reach their asymptotic values at a depth of about five diameters. In the passive case, $K_P = 3$, $z_c \approx 0.22D$ and the stresses rise to their final values by a depth of about one-half of a diameter. These effects are illustrated in figure 5.3 where the wall stress is plotted as a function of depth for the case of $\gamma = 8$ kN m^{-3}, $\phi = 30°$, $\phi_w = 20°$ and $D = 1.0$ m. It is seen that the shear stress rises to the same asymptote in the two cases but that the rate of approach to the asymptote is very much faster in the passive case.

Figure 5.4 shows the wall stress distribution in the same situation but with various values of the surcharge Q_0. As can be seen from

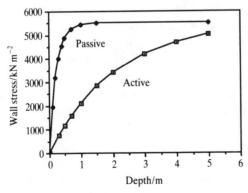

Figure 5.3 Variation of wall stress with depth in the active and passive states.

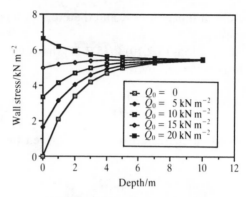

Figure 5.4 Effect of a surcharge, Q_0, on the wall stress distribution.

equation (5.2.11), the only term involving Q_0 also contains the group $\exp(-4\mu_\mathrm{w}Kz/D)$ and hence the stresses become independent of Q_0 at great depth.

5.3 Walker's improvement of Janssen's analysis

Janssen's assumption that the vertical and horizontal stresses are principal stresses is clearly erroneous, since, if σ_{rr} is a principal stress there can be no shear stress on the wall, yet it is assumed by Janssen that $\tau_\mathrm{w} = \mu_\mathrm{w}\sigma_{rr}$. Walker (1966) corrected this inconsistency to some extent by considering in greater detail the actual stress distribution in the wall region.

If we assume, as we did in the previous section, that the material is tending to slide down the wall, there will be an upward shear stress exerted by the wall on the material. Considering the material adjacent to the right-hand wall we can, therefore, take the wall shear stress to be positive and construct the Mohr–Coulomb failure diagram for the stresses at the wall as in figure 5.5. This diagram is drawn for the case of a cohesionless material in contact with a partially rough wall and, as usual, there are two points of intersection of the wall yield locus with Mohr's circle, W and W'. Point W corresponds to active failure and we can define the point W' as representing the passive case, though this is discussed in greater detail in §5.6 below. Considering first the active case, we can note that the point W represents the stresses on the wall plane which is also an r-plane and the stresses on the z-plane will be given by point Z at the other end of the diameter

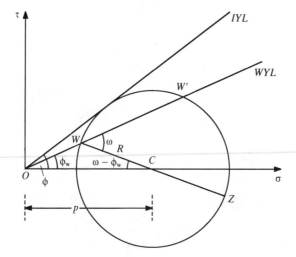

Figure 5.5 Mohr's circle for the material adjacent to the wall in active failure.

through W. From equation (3.7.8), we see that the angle CWW' is equal to ω where,

$$\sin \omega = \frac{\sin \phi_w}{\sin \phi} \tag{5.3.1}$$

From the geometry of the figure it is clear that the angle WCO is equal to $\omega - \phi_w$ and hence

$$\sigma_{rr} = p - R \cos (\omega - \phi_w) = p(1 - \sin \phi \cos (\omega - \phi_w)) \tag{5.3.2}$$

and

$$\sigma_{zz} = p + R \cos (\omega - \phi_w) = p(1 + \sin \phi \cos (\omega - \phi_w)) \tag{5.3.3}$$

As before, we will define the Janssen constant as the ratio σ_{rr}/σ_{zz} but this time we will give the constant the symbol K_w to remind us that we have evaluated it at the wall (or by Walker's analysis). Since we are considering the active case we can deduce that

$$K_{wA} = \frac{\sigma_{rr}}{\sigma_{zz}} = \frac{1 - \sin \phi \cos (\omega - \phi_w)}{1 + \sin \phi \cos (\omega - \phi_w)} \tag{5.3.4}$$

Figure 5.6 shows the Mohr–Coulomb diagram for the passive case from which it is clear that

$$K_{wP} = \frac{\sigma_{rr}}{\sigma_{zz}} = \frac{1 + \sin \phi \cos (\omega + \phi_w)}{1 - \sin \phi \cos (\omega + \phi_w)} \tag{5.3.5}$$

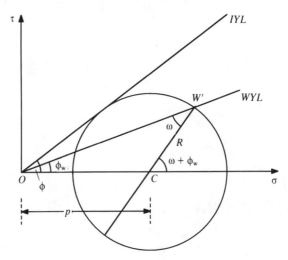

Figure 5.6 Mohr's circle for the material adjacent to the wall in passive failure.

These last two equations can be brought into common form by the use of the parameter κ defined in §3.4,

$$K_w = \frac{1 + \kappa \sin \phi \cos (\omega + \kappa \phi_w)}{1 - \kappa \sin \phi \cos (\omega + \kappa \phi_w)} \tag{5.3.6}$$

where κ takes the value -1 in the active case and $+1$ in the passive case.

The values of K_{wA} and K_{wP} can be used instead of K_A and K_P in the Janssen equations, equations (5.2.10) to (5.2.12). The experimental evidence is not sufficiently precise to show that this gives an improved prediction of the stress distribution, but this analysis is more intellectually satisfying than the original Janssen version. There seem to be no arguments in favour of the original version save the slightly simpler arithmetical calculations involved.

It is noteworthy that as the wall roughness is increased the slope of the wall yield locus steepens and the two points W and W' approach one another. For the fully rough wall W and W' become coincident and from equation (5.3.1) ω becomes 90°. Under these circumstances K_{wA} and K_{wP} are equal and are given by

$$K_w = \frac{1 - \sin \phi \cos (90 - \phi)}{1 + \sin \phi \cos (90 + \phi)} = \frac{1 - \sin^2 \phi}{1 + \sin^2 \phi} \tag{5.3.7}$$

Thus for the fully rough wall there is no distinction between the active and passive cases, though we must always bear in mind that no

Table 5.1 *Values of K_{wA} and K_A as functions of ϕ_w for $\phi = 30°$*

$\phi_w(°)$	0	10	20	25	28	30
K_{wA}	0.333	0.340	0.370	0.407	0.458	0.600
K_A	0.333	0.333	0.333	0.333	0.333	0.333

wall can actually be as rough as the 'fully-rough' wall defined in §3.7.

It is instructive to compare the values of K_A and K_{wA} for the typical case of a material with an angle of internal friction of 30°. From equation (5.2.5), K_A has the constant value of 0.333 but from (5.3.4) it can be seen that K_{wA} is also a function of the angle of wall friction. The values are given in table 5.1 from which it can be seen that for small values of ϕ_w, $K_{wA} \approx K_A$, with significant differences occurring only when ϕ_w exceeds 0.7 ϕ.

Recalling that the asymptotic wall stresses are independent of K, the above analysis confirms the numerical accuracy of Janssen's original analysis for the case of relatively smooth walls in active failure. For rougher walls the stresses approach their asymptotic values more rapidly than predicted by Janssen since in all cases $K_{wA} > K_A$. There are, however, greater discrepancies between K_P and K_{wP} particularly as $\phi_w \to \phi$.

5.4 Walker's distribution factor

In the previous section we considered Walker's correction to Janssen's analysis. Unfortunately, this correction seems to cause as many problems as it solves. Walker's analysis corrects the errors associated with Janssen's second assumption, that the vertical and horizontal stresses are principal stresses but in doing so it proves that Janssen's first assumption, that the stresses are independent of horizontal position, cannot be correct. This can be seen immediately. At $r = D/2$, $\tau_{rz} = \tau_w \neq 0$ but at $r = 0$, $\tau_{rz} = 0$ by symmetry. Thus the stresses cannot be independent of radial position except in the trivial case of $\mu_w = 0$.

If the stresses are not independent of r, the force on the top of the Janssen element must be evaluated from $\int \sigma_{zz} \, dA$. It is, however, convenient to define a mean axial stress by

$$\frac{\pi D^2}{4} \bar{\sigma}_{zz} = \int \sigma_{zz} \, dA \qquad (5.4.1)$$

and equation (5.2.2) can then be written

$$\frac{d\bar{\sigma}_{zz}}{dz} + \frac{4\tau_w}{D} = \gamma \qquad (5.4.2)$$

However, modifying equations (5.2.3) and (5.2.4) to allow for Walker's modification of §5.3 gives

$$\tau_w = \mu_w \, \sigma_{rr} = \mu_w \, K_w \, \sigma_{zz} \qquad (5.4.3)$$

where σ_{rr} and σ_{zz} must be evaluated at the wall, so that strictly we should write this equation in the form

$$\tau_w = \mu_w \, (\sigma_{rr})_w = \mu_w \, K_w \, (\sigma_{zz})_w \qquad (5.4.4)$$

Walker (1966) defines a distribution factor $\mathcal{D}$ as the ratio of the axial stress at the wall and the mean axial stress, i.e.

$$\mathcal{D} = \frac{(\sigma_{zz})_w}{\bar{\sigma}_{zz}} \qquad (5.4.5)$$

so that on substituting into equation (5.4.2) we have

$$\frac{d\bar{\sigma}_{zz}}{dz} + \frac{4\mu_w K_w \mathcal{D} \bar{\sigma}_{zz}}{D} = \gamma \qquad (5.4.6)$$

If $\mathcal{D}$ is a constant with respect to z, this equation can be integrated to give

$$\bar{\sigma}_{zz} = Q_0 \exp\left(- \frac{4\mu_w K_w \mathcal{D} z}{D}\right) + \frac{\gamma D}{4\mu_w K_w \mathcal{D}} \left[1 - \exp\left(- \frac{4\mu_{,w} K_w \mathcal{D} z}{D}\right)\right] \qquad (5.4.7)$$

and $(\sigma_{zz})_w$, $(\sigma_{rr})_w$ and τ_w can be evaluated from equations (5.4.5) and (5.4.4).

We have now improved Janssen's method so that up to the derivation of equation (5.4.6) the analysis is exact. However, there is no easy way of evaluating $\mathcal{D}$ at an arbitrary value of z and so we cannot be certain that $\mathcal{D}$ is a constant as is required for the integration of equation (5.4.6). Thus equation (5.4.7) remains suspect. It is, however, possible to evaluate $\mathcal{D}$ at great depth and Walker assumed that $\mathcal{D}$ remained constant at its great depth value.

Considering for a moment the stress state at great depth in a two-dimensional system with the y-axis directed vertically downwards, we have from appendix 2 that for a static material

$$\frac{\partial \sigma_{xx}}{\partial x} + \frac{\partial \tau_{yx}}{\partial y} = 0 \qquad (5.4.8)$$

Since at great depth all derivatives with respect to y are zero it follows that

$$\frac{\partial \sigma_{xx}}{\partial x} = 0 \qquad (5.4.9)$$

i.e. σ_{xx} is a constant with respect to horizontal position x.

By analogy with this result, Walker assumed that σ_{rr} was constant with respect to r in a cylindrical bunker at great depth. In fact this is not quite correct, as can be seen by the following argument.

It can be seen from equation (A2.6), appendix 2, that in cylindrical co-ordinates equation (5.4.9) takes the form

$$\frac{\partial \sigma_{rr}}{\partial r} + \frac{\partial \tau_{zr}}{\partial z} + \frac{\sigma_{rr} - \sigma_{zz}}{r} = 0 \qquad (5.4.10)$$

or, at great depth,

$$\frac{\partial \sigma_{rr}}{\partial r} + \frac{\sigma_{rr} - \sigma_{zz}}{r} = 0 \qquad (5.4.11)$$

Since σ_{zz} is a principal stress and σ_{rr} is not, $\partial \sigma_{rr}/\partial r \neq 0$ and σ_{rr} is therefore a function of r.

However, Walters and Nedderman (1973) have shown the error is small at least in the active state and we will therefore evaluate $\mathcal{D}$ at great depth using the usual erroneous assumption that σ_{rr} is constant.

Since all the stresses have reached their asymptotic values at great depth the force on the top and bottom surfaces of an element of radius r and thickness dz will be equal, so that a force balance gives

$$2\pi r\, \tau_{rz}\, dz = \pi r^2\, \gamma\, dz \qquad (5.4.12)$$

or

$$\tau_{rz} = \frac{\gamma r}{2} \qquad (5.4.13)$$

This result is entirely compatible with equation (5.2.14)

$$\tau_w = \frac{\gamma D}{4} \qquad (5.4.14)$$

Hence,

$$(\sigma_{rr})_w = \frac{\gamma D}{4\mu_w} \qquad (5.4.15)$$

and, from Walker's assumption that σ_{rr} is constant, we have for an arbitrary point

$$\sigma_{rr} = \frac{\gamma D}{4\mu_w} ; \ \tau_{rz} = \frac{\gamma r}{2} = \mu_w \sigma_{rr} r^* \qquad (5.4.16)$$

where $r^* = 2r/D$.

We can now evaluate σ_{zz} from the appropriate Mohr–Coulomb diagram, figure 5.7, by applying Pythagoras' theorem to the triangle *ABC* the sides of which are of lengths, $\frac{1}{2}(\sigma_{zz} - \sigma_{rr})$, τ_{rz} and $R = \frac{1}{2}(\sigma_{zz} + \sigma_{rr}) \sin \phi$, giving

$$(\sigma_{zz} - \sigma_{rr})^2 + 4 \tau_{rz}^2 = (\sigma_{zz} + \sigma_{rr})^2 \sin^2 \phi \qquad (5.4.17)$$

or, from equation (5.4.16)

$$\sigma_{zz}^2 \cos^2 \phi - 2 \sigma_{zz}\sigma_{rr}(1 + \sin^2 \phi) + \sigma_{rr}^2 (\cos^2 \phi + 4\mu_w^2 r^{*2}) = 0 \qquad (5.4.18)$$

Equation (5.4.18) is a quadratic in σ_{zz} having roots

$$\sigma_{zz} = \sigma_{rr} \frac{(1 + \sin^2 \phi) \pm 2 \sin \phi \sqrt{(1 - cr^{*2})}}{\cos^2 \phi} \qquad (5.4.19)$$

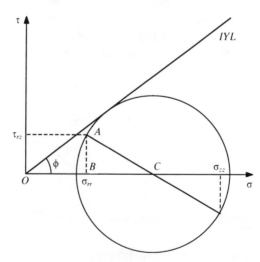

Figure 5.7 Mohr's circle for the stresses part-way across a bunker.

where $c = \tan^2 \phi_w / \tan^2 \phi$.

The two roots in equation (5.4.19) correspond to the active and passive solutions, so that this equation can more conveniently be written

$$\sigma_{zz} = \sigma_{rr} \frac{(1 + \sin^2 \phi) - 2\kappa \sin \phi \sqrt{(1 - cr^{*2})}}{\cos^2 \phi} \tag{5.4.20}$$

where κ is the parameter defined in §3.4 and takes the value -1 in active failure and $+1$ in passive failure.

The mean value of σ_{zz}, $\bar{\sigma}_{zz}$ is given by

$$\bar{\sigma}_{zz} = \frac{4}{\pi D^2} \int 2\pi r \, \sigma_{zz} \, dr$$

$$= \int \sigma_{zz} \, d(r^{*2})$$

$$= \sigma_{rr} \{1 + \sin^2\phi - \kappa(4/3c) \sin \phi \tag{5.4.21}$$

$$[1 - (1 - c)^{3/2}]\} \sec^2\phi$$

The value of $(\sigma_{zz})_w$ is given by equation (5.4.20) when $r^* = 1$. Thus

$$(\sigma_{zz})_w = \sigma_{rr}[1 + \sin^2 \phi - 2\kappa \sin \phi \sqrt{(1 - c)}] \sec^2\phi \tag{5.4.22}$$

so that from equation (5.4.5)

$$\mathcal{D} = \frac{1 + \sin^2 \phi - 2\kappa \sin \phi \sqrt{(1 - c)}}{1 + \sin^2 \phi - (4/3c)\kappa \sin \phi \, [1 - (1 - c)^{3/2}]} \tag{5.4.23}$$

For a fully rough wall, $c = 1$ and equation (5.4.23) takes the form

$$\mathcal{D}_{fr} = \frac{1 + \sin^2 \phi}{1 + \sin^2 \phi - (4/3c)\kappa \sin \phi} \tag{5.4.24}$$

It can be shown that in the active case, $\mathcal{D}$ always has a value in the range 0.6 to 1 and that provided $\phi_w < \phi - 5°$, $\mathcal{D}$ usually has a value between 0.9 and 1. Thus the effect of $\mathcal{D}$ on the predictions of Janssen's method is minimal. It makes a small change in the rate of approach to the asymptote and a small change in the asymptotic value of σ_{zz}. The asymptotic values of the wall stresses, being independent of $\mathcal{D}$ are the same as in Janssen's basic method. Under these circumstances the distribution factor is rarely included in practice, the importance of this section being to show that the inclusion of a distribution factor is usually unnecessary in active failure.

On the other hand, in the passive state, $\mathcal{D}$ can take values in the range 1 to 3 and there may therefore be good reasons for including a distribution factor in passive analyses. As so often in this subject, the standard approximations are much more appropriate for active situations. The main importance of the distribution factor arises in the stress analysis in a conical hopper, where, as we will see later, passive failure normally occurs. It must, however, always be kept in mind that the value of $\mathcal{D}$ has been evaluated only at great depth and furthermore on the erroneous assumption that σ_{rr} is independent of r.

A graphical presentation of the variation of $\mathcal{D}$ with ϕ and ϕ_w is given by Walters (1973) and this may be used as an alternative to the evaluation of $\mathcal{D}$ from equation (5.4.23) or (5.4.24).

5.5 Vertical walled bunkers of arbitrary cross-section

For reason of structural convenience, bunkers are often made with rectangular or even hexagonal cross-sections. In such situations, the wall shear stress will almost certainly vary along the perimeter of the bunker though there seems to be no established method of quantifying this variation, nor does there seem to be any body of reliable experimental information on this matter. None-the-less Janssen's method, or Walker's modification of it, is often used to predict the variation of the wall stresses with depth below the top of the fill.

In this section we will develop Janssen's analysis for the case of a vertical walled bunker of arbitrary cross-section. If we denote the cross-sectional area by A and the length of the perimeter by P, we can perform a force balance on an elementary slice and obtain, by analogy with equation (5.2.1),

$$A\sigma_{zz} + A\,\gamma\delta z = A(\sigma_{zz} + \delta\sigma_{zz}) + P\,\delta\tau z_w \qquad (5.5.1)$$

or

$$\frac{d\sigma_{zz}}{dz} + \frac{P}{A}\tau_w = \gamma \qquad (5.5.2)$$

where τ_w is the value of the wall shear stress averaged round the perimeter.

By analogy with equations (5.2.3) and (5.2.4) we can say that

$$\tau_w = \mu_w\,\sigma_w \qquad (5.5.3)$$

where σ_w is also averaged round the perimeter, and we can define a Janssen constant K by

$$\sigma_w = K \sigma_{zz} \qquad (5.5.4)$$

Eliminating τ_w and σ_w, we obtain

$$\frac{d\sigma_{zz}}{dz} + \frac{P\mu_w K}{A}\sigma_{zz} = \gamma \qquad (5.5.5)$$

and this can be seen to be similar to equation (5.2.7) but with the group A/P replacing $D/4$.

The hydraulic mean diameter D_H is a concept much used in fluid mechanics and is defined by

$$D_H = \frac{4A}{P} \qquad (5.5.6)$$

Clearly this is the relevant dimension for the case in hand, and substituting for D_H in equation (5.5.5) gives

$$\frac{d\sigma_{zz}}{dz} + \frac{4\mu_w K}{D_H}\sigma_{zz} = \gamma \qquad (5.5.7)$$

which is identical in form to equation (5.2.7). The rest of the analysis follows the same lines as in §5.2 and all the equations of that section are valid if D is replaced by D_H. All forms of K may be used and a distribution factor $\mathcal{D}$ can be incorporated, though it is doubtful whether any convincing evaluation of $\mathcal{D}$ can be made in our present state of knowledge.

For a $b \times l$ rectangular cross-section, the hydraulic mean diameter is given by

$$D_H = \frac{4A}{P} = \frac{4bl}{2(b + l)} = \frac{2bl}{(b + l)} \qquad (5.5.8)$$

For a square of side b, equation (5.5.8) reduces to

$$D_H = b \qquad (5.5.9)$$

and for long parallel walls a distance b apart

$$D_H \to 2b \text{ as } l \to \infty \qquad (5.5.10)$$

For a hexagonal bunker of side b, $P = 6b$ and $A = 3\sqrt{3}.b^2/2$, giving

$$D_H = b\sqrt{3} \qquad (5.5.11)$$

Some authors prefer the use of the hydraulic mean radius R_H defined by

$$R_\mathrm{H} = \frac{A}{P} \qquad (5.5.12)$$

This is algebraically correct but the student should be careful when making comparisons between analyses using these two conventions. It can be seen from equations (5.5.6) and (5.5.12) that the hydraulic mean diameter is *four* times the hydraulic mean radius, an unexpected result that has led to innumerable arithmetical errors. Furthermore, for a cylindrical bunker the hydraulic mean diameter is equal to the actual diameter whereas the hydraulic mean radius is *half* the actual radius. The use of the hydraulic mean diameter has therefore some mnemonic advantage and is to be preferred if only on that account.

We can obtain the stress distribution on twin retaining walls a distance b apart by substituting equation (5.5.10) into equation (5.2.11), giving

$$\sigma_\mathrm{w} = \frac{\gamma b}{2\mu_\mathrm{w}} \left[1 - \exp\left(-\frac{2\mu_\mathrm{w} K z}{b} \right) \right] + Q_0 K \exp\left(-\frac{2\mu_\mathrm{w} K z}{b} \right) \quad (5.5.13)$$

This is perhaps the most reliable prediction of this section since by symmetry σ_w will be constant with respect to horizontal distance along the wall. This result is of particular theoretical significance, since the wall stress distribution in this geometry can also be predicted by the method of wedges as described in §4.8. Comparison of the predictions of the two methods gives some insight into the reliability of these analyses. The results for a typical material with $\phi = 30°$ and $\phi_\mathrm{w} = 20°$ are shown in figure 4.27 for the case of zero surcharge and in figure 4.26 for a surcharge of $3\gamma b$. In the latter case, it can be seen that the stress increases linearly with depth for some distance below the top surface and then undergoes a series of changes in slope with each section being roughly linear and alternate sections having slopes of opposite sign. For zero surcharge the stress profile is monotonic but still consists of a series of roughly linear sections, the first being exactly linear, though this cannot be seen on the scale of figure 4.27. Equation (5.5.13), on the other hand, predicts a smooth exponential approach to the final value. Despite the fundamentally different form of the solution, the numerical differences are slight for zero surcharge and this gives confidence in both methods. On the other hand, for a comparatively large surcharge of $3\gamma b$, there are considerable differences between the two predictions. We will see in chapter 7 that the exact solution is much closer to that of the method of wedges, and this does not surprise us, since we are unable to justify the basic assumption in Janssen's method, that the stresses are uniform across any cross-section. By contrast, the assumption in the method of wedges that the slip surfaces are planar is very nearly correct.

Thus we must conclude that the agreement at zero surcharge is largely coincidental and must beware of using Janssen's method in the presence of a surcharge. Unfortunately, Janssen's method is by far the simplest way of predicting wall stress distributions and the temptation to use it is almost irresistible.

5.6 The active and passive solutions

It is appropriate at this stage to give further consideration to the meaning of the words active and passive in the context of the Janssen analysis, and this is most conveniently illustrated for the case of parallel walls which can be moved inwards and outwards. Figure 5.8 shows the Mohr–Coulomb diagram for the right-hand wall, that for the left-hand wall would be in mirror image about the σ axis. It can be seen that there are four possible positions for the wall plane in incipient failure which are labelled A, B, C and D. All four can be achieved in practice in the situations which are illustrated in figure 5.9. The heavy arrows

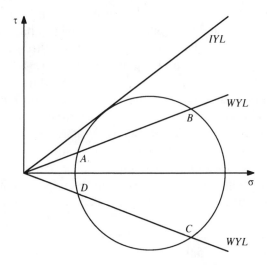

Figure 5.8 Mohr's circle illustrating four possible positions of the wall plane.

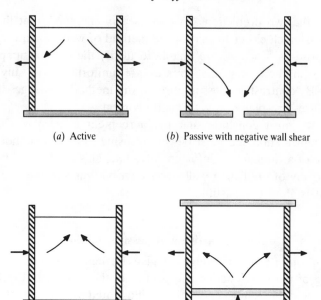

(a) Active   (b) Passive with negative wall shear

(c) True passive   (d) Active with negative wall shear

Figure 5.9 Four possible failure mechanisms.

show the motion of the walls and the light arrows show the resulting motion of the material.

If, after filling, the walls are moved apart as in figure 5.9(a), the material will slip downwards and outwards causing a positive shear stress on the right-hand wall. Since the material is failing outwards the horizontal stress must be less than the vertical stress and the wall stresses are therefore given by point A on figure 5.8. This is the traditional active situation and would occur in a bunker if the walls expanded sufficiently due to the increased stresses on filling.

Figure 5.9(b) shows material discharging from a bunker with the walls being pushed inwards sufficiently to assist the discharge. With downward motion of the material, the shear stress on the right hand wall is positive and with lateral compression the wall normal stress will have its larger value. The wall stresses are therefore given by point B of figure 5.8. This is the state we have called 'passive' in the preceding parts of this chapter.

If the walls are pushed inwards and discharge through the base is prevented as in figure 5.9(c), the material must be expelled upwards. The shear stress on the right-hand wall is now negative and the normal stress will have its greater value. The stresses are therefore given by point C of figure 5.8. This is the genuine passive solution, being the direct analogue of the passive solutions presented in §3.4 and chapter 4. The more precise authors distinguish between states B and C by calling state C the 'True Passive' state and denoting state B as 'Passive with Negative Wall Shear', though this terminology does not conform with our sign convention for shear stresses.

The rather contrived geometry of figure 5.9(d) would give the state properly called 'Active with Negative Wall Shear' but this does not seem to correspond to any physical situation of commercial importance.

Thus we see that there are *four* possible situations and the usual practice of denoting them by the *two* words 'active' and 'passive' inevitably causes confusion. States A and D are of the active type, states B and C of the passive type. A and B correspond to downwards motion along the right-hand wall and C and D to upward motion. It must, however, be recalled that as a result of our sign convention, the corresponding Mohr–Coulomb diagram for the left-hand wall is the mirror image of figure 5.8 in the σ-axis.

The true passive state (state C) has considerable importance if the walls of a bunker contract due to some external agency or if the material expands after filling. The usual Janssen analysis can be modified by reversing the sign of the wall shear stress in equation (5.2.2) giving, by analogy with equation (5.2.11),

$$\sigma_{rr} = \frac{\gamma D}{4\mu_w} \left[\exp\left(\frac{4\mu_w K z}{D}\right) - 1 \right] \qquad (5.6.1)$$

where K_P or K_{wP} must be used for K. It is seen the σ_{rr} increases exponentially with depth giving rise to very large stresses at comparatively small depths. This explains why it is not normally possible to extrude granular materials.

It must, however, be realised that equation (5.6.1) represents an unattainable limit since in most cases it is impossible to generate a sufficient wall force to achieve the true passive state. Any realistic analysis of the effects of wall contraction must therefore include elastic effects and is therefore outside the scope of this book. Thus we must place no reliance on values calculated from equation (5.6.1) but simply note that in the true passive state the wall stresses become very large indeed.

The most common situation in which the material is stressed by contraction of the walls is that of annual or diurnal changes in temperature. This is particularly important for metallic bunkers with their high coefficient of thermal expansion. On warming, the walls expand and the material slumps somewhat. On cooling the walls contract and the stresses rise some way towards those of the passive state. There is, however, some evidence that the wall stresses do not fall on subsequent expansion yet increase again on the next contraction. This is the phenomenon known as 'thermal racking' and results in a progressive increase in stress. The problem is particularly severe in chemical reactors containing a granular catalyst. Here expansion may be due to temperature changes of several hundreds of degrees, instead of the tens of degrees associated with ambient fluctuations. Pulverisation of the catalyst on shut-down is of frequent occurrence.

Finally, care must be taken in the storage of material which can expand. This is a particular problem in the storage of cereals which are normally loaded dry. Should moisture get into the silo, the grains would swell generating dangerously large stresses. This phenomenon has been used to advantage for at least the last five thousand years. Masonry blocks can be split by driving dry wooded wedges into cracks and then expanding the wedges by saturating with water. The damage tree roots can cause to the foundations of a building is another illustration of the same phenomenon.

5.7 Inclined top surfaces

Very commonly cylindrical bunkers are filled from a central loading port in the roof. Under these circumstancs the top surface of the fill will be conical with sides inclined to the horizontal at approximately the angle of repose ϕ. Similarly, material stored between parallel walls is often surmounted by a triangular wedge also inclined at the angle of repose. On the other hand, during discharge in the core flow mode, a central depression will be formed in the top surface, again at the angle of repose. The effect of these inclined top surfaces can be allowed for by either of the two approximate analyses below.

Let us first consider a cylindrical bunker of diameter D filled with a cohesionless material having a conical top surface. To preserve generality we will denote the inclination of the top surface to the horizontal by α, noting that in most cases α will be either ϕ or $-\phi$. We will take the plane of intersection of the top surface with the bunker wall as the origin of our vertical co-ordinate z, as shown in figure 5.10.

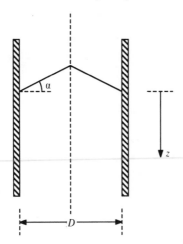

Figure 5.10 Cylindrical bunker containing material with an inclined upper surface.

In the simpler of the two analyses, it is usual to assume that the top surface has been levelled off, so that the bunker is to be treated as though the top surface was a distance h above the plane $z = 0$. The distance h will be negative in the case of a central depression. Since the volume of a cone is one-third of the base area times the height, h equals one-third of the height of the cone, i.e.

$$h = \frac{D}{6} \tan \alpha \tag{5.7.1}$$

and from equation (5.2.11) we can say

$$\sigma_{rr} = \frac{\gamma D}{4\mu_w} \left\{ 1 - \exp\left[-\frac{4\mu_w K(z + h)}{D} \right] \right\} \tag{5.7.2}$$

A distribution factor can be included in this equation if desired.

In the case of parallel walls a distance b apart, h will be equal to one-half of the height of the wedge since the area of a triangle is one-half of the base times the height. Thus

$$h = \frac{b}{4} \tan \alpha \tag{5.7.3}$$

and from equation (5.5.13) we have

$$\sigma_{rr} = \frac{\gamma b}{2\mu_w} \left\{ 1 - \exp\left[-\frac{2\mu_w K(z + h)}{b} \right] \right\} \tag{5.7.4}$$

A somewhat more rigorous analysis can be performed by noting that there can be no shear stress in the region above $z = 0$ and hence the weight of the cone is not supported by the wall and its whole weight of

$$\frac{1}{3}\gamma \frac{\pi D^2}{4} \frac{D \tan \alpha}{2}$$

must act on the surface $z = 0$. Assuming that this is equivalent to a uniform surcharge, since no other assumption is possible when using the method of differential slices, we have that

$$Q_0 = \frac{\gamma D}{6} \tan \alpha \qquad (5.7.5)$$

or for parallel walls

$$Q_0 = \frac{\gamma b}{4} \tan \alpha \qquad (5.7.6)$$

Hence from equation (5.2.11)

$$\sigma_{rr} = \frac{\gamma D}{4\mu_w}\left[1 - \exp\left(-\frac{4\mu_w K z}{D}\right)\right] \qquad (5.7.7)$$
$$+ \frac{\gamma D K}{6} \tan \alpha \exp\left(-\frac{4\mu_w K z}{D}\right)$$

and similar results can be obtained for the case of parallel walls from equations (5.5.13) and (5.7.6).

Table 5.2 compares the predictions of equations (5.7.2) and (5.7.7) for the case of a cylindrical bunker of diameter 1.0 m with $\alpha = \phi = 30°$, $\phi_w = 20°$ and $\gamma = 8.0$ kN m^{-3}. It can be seen that in this situation the differences are totally negligible compared with the likely precision to which the basic parameters are known. This is usually the case, though it is wise to confirm it in any practical situation.

5.8 Conical and wedge-shaped hoppers

Janssen's analysis has been extended to the cases of conical and wedge-shaped hoppers by Walker (1966) and Walters (1973) and the analysis presented below is closely based on their work.

Again we will take a cohesionless material of constant bulk density γ and we start by considering a conical hopper of half-angle α as

Table 5.2

Depth z (m)	0	0.1	0.2	0.3	0.5	1.0	∞
σ_{rr} (kN m^{-2}) from (5.7.2)	0.278	0.551	0.811	1.057	1.510	2.452	5.495
σ_{rr} (kN m^{-2}) from (5.7.7)	0.285	0.558	0.818	1.060	1.506	2.456	5.495

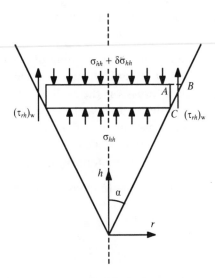

Figure 5.11 Stresses on a cylindrical element in a conical hopper.

shown in figure 5.11. In this case it is more convenient to use cylindrical co-ordinates (r,h) where h is measured vertically upwards from the apex of the hopper and r is measured radially outwards. Since, for the present, we are concerned with the stress distribution in a static material, it will be supposed that the walls continue down to the apex of the hopper. In any real storage hopper there will be an orifice at the base and the walls will terminate somewhat above the virtual apex. The effect of flow on the stress distribution will be considered in chapter 10 where it will be shown that the differences in the stress distributions in the flowing and static cases are confined to a small region immediately above the orifice.

Following Janssen's method, we will consider a force balance on an elementary slice of thickness δh at height h. To begin with we will consider a cylindrical slice as shown in figure 5.11. The fact that we seem to be omitting from consideration the material in the wedge ABC is of no consequence since in the limit as $\delta h \to 0$, the weight of this material becomes infinitesimal compared with the rest of the material in the hopper. This will be discussed further below.

The stresses acting on the element are:

(i) σ_{hh} on the lower surface,

(ii) $\sigma_{hh} + \delta\sigma_{hh}$ on the upper surface, and

(iii) $(\tau_{rh})_w$ on the side of the element.

It is important to be clear about the meaning of the term $(\tau_{rh})_w$. This is the shear stress acting on a vertical surface adjacent to the wall. It is not equal to the shear stress on the wall τ_w and the relationship between these two quantities is discussed in greater detail below.

Using the usual Janssen assumptions that the stresses are constant across any cross-section, we can relate the stresses on the surface of the element to the weight of the material within the element thus

$$\pi(h \tan \alpha)^2 \sigma_{hh} + 2\pi(h \tan \alpha)(\tau_{rh})_w \, \delta h$$
$$= \pi(h \tan \alpha)^2(\sigma_{hh} + \delta\sigma_{hh}) \qquad (5.8.1)$$
$$+ \pi(h \tan \alpha)^2 \gamma \, \delta h$$

or, in the limit as $\delta h \to 0$

$$\frac{\mathrm{d}\sigma_{hh}}{\mathrm{d}h} + \gamma = \frac{2(\tau_{rh})_w}{h \tan \alpha} \qquad (5.8.2)$$

This last equation can be compared with the basic Janssen equation, equation (5.2.2). Two differences are apparent. First, there is a difference in sign. This results solely from the fact that we are now measuring our vertical co-ordinate upwards and this difference is simply an artefact of the presentation and of no consequence. The second and more fundamental difference is that the constant diameter D of equation (5.2.2) is now replaced by the variable diameter $2h \tan \alpha$ and this difference does have important consequences for the subsequent analysis.

It is now necessary to relate the stresses σ_{hh} and $(\tau_{rh})_w$ and this is done by considering the Mohr–Coulomb failure diagram for the material adjacent to the right-hand wall as shown in figure 5.12. We will consider first the passive case (i.e. state B of §5.6) so that the wall plane is denoted by the point W' on the figure. The co-ordinates of W' are (σ_w, τ_w) and by inspection of the figure these are seen to be

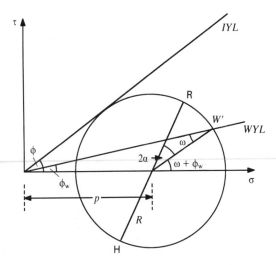

Figure 5.12 Mohr's circle for the stresses adjacent to the wall in a conical hopper.

$$\sigma_w = p + R \cos(\omega + \phi_w) = p(1 + \sin \phi \cos(\omega + \phi_w)) \quad (5.8.3)$$

$$\tau_w = R \sin(\omega + \phi_w) = p \sin \phi \sin(\omega + \phi_w) \quad (5.8.4)$$

The circumferential surface of our element is an *r*-plane which is inclined to the walls at angle α and is therefore represented on figure 5.12 by the point R at the end of the radius inclined at 2α to the radius to W'. The point H representing the *h*-plane lies at the opposite end of the diameter to R. From figure 5.12 we have

$$\sigma_{rr} = p + R \cos(\omega + \phi_w + 2\alpha) = p(1 + \sin \phi \cos(\omega + \phi_w + 2\alpha))$$
$$(5.8.5)$$

$$(\tau_{rh})_w = R \sin(\omega + \phi_w + 2\alpha) = p \sin \phi \cos(\omega + \phi_w + 2\alpha) \quad (5.8.6)$$

$$\sigma_{hh} = p - R \cos(\omega + \phi_w + 2\alpha) = p(1 - \sin \phi \cos(\omega + \phi_w + 2\alpha))$$
$$(5.8.7)$$

From equation (5.8.3) to (5.8.7), the ratio of any pair of these stresses can be evaluated. Since all these stresses are directly proportional to p, the ratios are independent of p and depend only on the angles of friction and the hopper half-angle. Thus in a given situation all these ratios are constants. In particular we can define the constant B as the ratio $(\tau_{rh})_w/\sigma_{hh}$ i.e.

$$B = \frac{\sin \phi \sin(\omega + \phi_w + 2\alpha)}{1 - \sin \phi \cos(\omega + \phi_w + 2\alpha)} \tag{5.8.8}$$

Substituting into equation (5.8.2) gives

$$\frac{d\sigma_{hh}}{dh} - \frac{2B}{h \tan \alpha} \sigma_{hh} = -\gamma \tag{5.8.9}$$

and defining m by

$$m = \frac{2B}{\tan \alpha} \tag{5.8.10}$$

gives

$$\frac{d\sigma_{hh}}{dh} - \frac{m}{h} \sigma_{hh} = -\gamma \tag{5.8.11}$$

Equation (5.8.11) can be solved by multiplying by an integrating factor which in this case turns out to be h^{-m} giving

$$h^{-m} \frac{d\sigma_{hh}}{dh} - mh^{-(m+1)} \sigma_{hh} = -\gamma h^{-m} \tag{5.8.12}$$

The right-hand side is now a perfect differential so that equation (5.8.12) can be rewritten in the form

$$\frac{d}{dh}(h^{-m}\sigma_{hh}) = -\gamma h^{-m}$$

which can be integrated to give

$$\sigma_{hh} = A h^m + \frac{\gamma h}{m - 1} \tag{5.8.13}$$

or, when $m = 1$,

$$\sigma_{hh} = Ah - \gamma h \ln h \tag{5.8.14}$$

Equation (5.8.13) can be seen to consist of two terms, a term of the form Ah^m which involves the arbitrary constant A which has to be determined from the boundary conditions and a term of the form $\gamma h/(m - 1)$ which is independent of the boundary conditions and is, in fact, the particular integral of equation (5.8.11).

The most common boundary condition is that there is no applied stress on the top surface of the fill which we will take to be at height

h_0. However, to preserve generality we will take as our boundary condition that

$$\sigma_{hh} = Q_0 \text{ on } h = h_0 \qquad (5.8.15)$$

as in §5.2, with the expectation that usually $Q_0 = 0$. Substituting into equation (5.8.13) gives A, which, on back-substitution, yields

$$\sigma_{hh} = Q_0 \left(\frac{h}{h_0}\right)^m + \frac{\gamma h}{m-1}\left[1 - \left(\frac{h}{h_0}\right)^{m-1}\right] \qquad (5.8.16)$$

A similar analysis for the active case gives

$$m_A = \frac{2 \sin \phi \sin(\omega - \phi_w - 2\alpha)}{\tan \alpha[1 + \sin \phi \cos(\omega - \phi_w - 2\alpha)]} \qquad (5.8.17)$$

so that in general

$$m = \frac{2 \sin \phi \sin (\omega + \kappa\phi_w + 2\kappa\alpha)}{\tan \alpha \{1 - \kappa \sin \phi \cos (\omega + \kappa\phi_w + 2\kappa\alpha)\}} \qquad (5.8.18)$$

where $\kappa = -1$ in the active case and $+1$ in the passive case as in §3.4.

A very similar analysis for the wedge-shaped hopper gives a result identical in form to equation (5.8.16) but with the value of m being one-half of that predicted from equation (5.8.18).

Figure 5.13 shows the dependence of σ_{hh} on h predicted from equation (5.8.16) with $Q_0 = 0$, $h_0 = 2$ m and $\gamma = 8.0$ kN m^{-3}, and various values of m as a parameter.

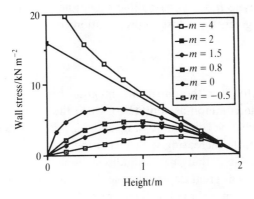

Figure 5.13 Wall stress distribution in a conical hopper, as function of height and parameter m.

Table 5.3 *Values of m for φ = 30° calculated from equations (5.8.10) and (5.8.8)*

	Parameter m		
$\alpha(°)$	10	20	30
$\phi_w(°)$			
0	3.66	2.86	2.00
10	6.41	3.11	1.73
20	5.99	2.40	1.13
30	2.64	0.64	0

For $m < 0$, it is seen that $\sigma_{hh} \to \infty$ as $h \to 0$ implying infinite stresses at the apex. Since it is a matter of common experience that conical hoppers do not inevitably fail when filled, we will investigate below the reasons why negative values of m do not occur in practice. For $m > 0$, $\sigma_{hh} \to 0$ as $h \to 0$. However, for values of m in the range $0 < m < 1$, $d\sigma_{hh}/dh$ is infinite at $h = 0$ whereas this derivative is finite at $h = 0$ for $m > 1$.

We can investigate the likely range of values of m for a material having $\phi = 30°$ for various values of ϕ_w and α and these are given in table 5.3.

It is seen that values of $m > 1$ are common, though not inevitable. Large values of m are found for narrow hoppers due mainly to the $\tan \alpha$ term in the denominator of equation (5.8.10)

With large values of m the terms in $(h/h_0)^m$ in equation (5.8.16) reduce rapidly as h decreases so that except near the top of the hopper the stress is given to a close approximation by

$$\sigma_{hh} = \frac{\gamma h}{m - 1} \tag{5.8.19}$$

which we noted was the particular integral of equation (5.8.11). This represents the common asymptote to which the stresses tend for small values of h whatever the value of h_0 and Q_0, with the rate of approach to the asymptote increasing with increasing m.

We saw in §5.2 that in a cylindrical bunker the stresses tend to constant values at great depth whereas for a conical hopper the stresses tend to an asymptote which is linear with distance from the apex. This is illustrated in figure 5.14 for $m = 4$ using three different values of

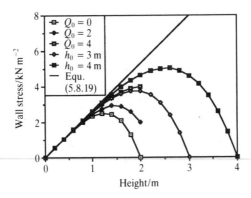

Figure 5.14 Effect of surcharge, Q_0, and height on the wall stresses in a conical hopper; parameter $m = 4$ (see text).

Q_0 at $h_0 = 2$ m and two other values of h_0 with $Q_0 = 0$. It is seen that all the curves run rapidly into the asymptote given by equation (5.8.19).

Evaluation of the parameter m_A from equation (5.8.17) shows that this commonly has negative values and we must therefore look for physical reasons why the active state does not normally occur. All granular materials are compressible to some extent. Thus the material first loaded into a hopper will be compressed as more material is placed on top of it. As the material compresses it will slide down the wall giving a positive shear stress on the right-hand wall. Moreover, as it moves downwards it passes along a converging passage so that it must be compressed laterally and extended axially. Thus we would expect $\sigma_{rr} > \sigma_{hh}$, a condition which suggests that the passive state (state B of §5.6) is likely to occur. During discharge from a mass flow hopper, lateral compression of the material must occur and the passive state seems inevitable. Thus we see that the active state is unlikely to occur on filling and cannot occur on discharge.

For wide-angle hoppers, negative values of m can occur even in the passive state. However, in such hoppers, core flow is observed with the material flowing through a self-generated flow channel. The above analysis may be appropriate in the flow channel but friction may not be fully mobilised in the surrounding stagnant zone and an elastic–plastic analysis will be needed to predict the wall stresses.

In §5.6, we introduced the concept of a distribution factor $\mathcal{D}$ and commented that it was usually ignored when considering cylindrical bunkers since in active failure the value of $\mathcal{D}$ is often close to unity.

On the other hand, we have seen that passive failure is usual in conical hoppers and under these circumstances values of $\mathscr{D}$ of up to 3 can occur. Thus we must include a distribution factor in the present analysis giving, in place of equation (5.8.9)

$$\frac{d\bar{\sigma}_{hh}}{dh} - \frac{2 B \mathscr{D}}{h \tan \alpha} \bar{\sigma}_{hh} = -\gamma \qquad (5.8.20)$$

Comparison of these equations shows that the results of the previous parts of this section still apply if m is replaced by m^* where

$$m^* = m \mathscr{D} \qquad (5.8.21)$$

It is not, however, obvious that the values of $\mathscr{D}$ evaluated at great depth in a cylindrical bunker apply in the conical case. Walters (1973), however, argues that at the stress maximum, i.e. where $d\bar{\sigma}_{zz}/dh = 0$, the distribution factor can be calculated by the same method as that used at great depth in the cylindrical case. Furthermore, he shows that the same equation applies for $\mathscr{D}$ provided the angle ϕ_w is replaced by η, defined by

$$\tan \eta = \frac{(\tau_{rh})_w}{\sigma_{rr}} = \frac{\sin \phi \sin(\omega + \phi_w + 2\alpha)}{1 + \sin \phi \cos(\omega + \phi_w + 2\alpha)} \qquad (5.8.22)$$

Thus

$$\mathscr{D} = \frac{1 + \sin^2 \phi - 2\kappa \sin \phi \sqrt{(1 - c')}}{1 + \sin^2 \phi - (4/3c')\kappa \sin \phi [1 - (1 - c')^{3/2}]} \qquad (5.8.23)$$

where

$$c' = \frac{\tan^2 \eta}{\tan^2 \phi} \qquad (5.8.23)$$

The resulting values of m^* for a cohesionless material with $\phi = 30°$ are given in Table 5.4.

Comparison with the values of m in table 5.3 shows that normally m^* is considerably larger than m and that m^* is often large enough for the stress distribution in the lower part of the hopper to be given to sufficient accuracy by the particular integral of equation (5.8.20)

$$\bar{\sigma}_{hh} = \frac{\gamma h}{m^* - 1} \qquad (5.8.24)$$

Table 5.4 *Values of m^* for $\phi = 30°$*

	Parameter m^*		
$\alpha(°)$	10	20	30
$\phi_w(°)$			
0	3.81	3.28	2.60
10	7.83	4.32	2.75
20	9.07	4.23	2.30
30	4.35	0.81	0

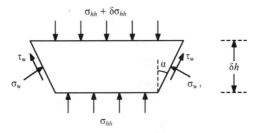

Figure 5.15 Forces on an element in the form of the frustrum of a cone.

The other stress components can be found from

$$(\sigma_{hh})_w = \mathcal{D}\bar{\sigma}_{hh} \qquad (5.8.25)$$

and the stress ratios obtained from equations (5.8.3) to (5.8.7).

Walters (1973) maintains that the cylindrical element used above is not accurate and an element in the form of the frustrum of a cone should be used as shown in figure 5.15. A force balance on such an element gives

$$\frac{d\bar{\sigma}_{hh}}{dh} - m^{**}\frac{\bar{\sigma}_{hh}}{h} = -\gamma \qquad (5.8.26)$$

where

$$m^{**} = m^* + 2(\mathcal{D} - 1) \qquad (5.8.27)$$

It is clear that when $\mathcal{D} = 1$, all three estimates of m become identical.

In fact, there are approximations implicit both in the analysis based on the cylindrical element, leading to equation (5.8.20), and in the analysis based on the conical element, which leads to equation (5.8.27). It is not clear, at least to the present author, which is the more accurate but the matter is of minor importance since, if $\mathcal{D}$ is close to unity, there is little difference between the predictions and, if $\mathcal{D}$ is not close to unity, the evaluation of $\mathcal{D}$ is suspect. It must be remembered that the method of differential slices is based on unjustifiable assumptions about the stress distribution on horizontal planes and that great accuracy must not be expected.

In the later sections of this book we will give alternative predictions of the stress distributions in a conical hopper. These are more conveniently expressed in the form of the wall stress σ_w as a function of r which we will now define as distance from the apex. Equation (5.8.16) can be put into this form by the substitutions

$$r = h \sec \alpha \qquad (5.8.28)$$

and from equations (5.8.3) and (5.8.7),

$$\sigma_w = (\sigma_{hh})_w \frac{1 + \sin \phi \cos (\omega + \phi_w)}{1 - \sin \phi \cos (\omega + \phi_w + 2\alpha)} \qquad (5.8.29)$$

5.9 Walters' switch stress analysis

Bunkers designed by the methods described in the previous sections of this chapter normally perform satisfactorily and most design manuals recommend only modest safety factors. However, occasionally such bunkers fail and do so in a manner that shows that the walls must have been subjected to very much larger stresses than expected. Such failures can occur in bunkers that have performed satisfactorily for many years and the period immediately following the initiation of discharge seems particularly hazardous.

Walters (1973) has produced an approximate analysis to show how such large stresses could occur. However, the approximations in this analysis are so gross that the method is best regarded as a qualitative explanation and no reliance should be placed on the actual numerical values. The method is illustrated by considering a particular case below and further consideration is given to the phenomenon in chapter 7.

On filling a bunker with sufficiently elastic walls, an active stress state may be formed and the stress distribution will be given by equations (5.2.10) to (5.2.12) using some active value of K such as K_A or K_{wA}. On discharge of a bunker with a central orifice, the material will flow downwards and inwards giving the passive stress state (state *B* of §5.6). The stresses are again given by equations (5.2.10) to (5.2.12) but using K_P or K_{wP}. In practice, the stress state will probably be intermediate between the active and passive state both on filling and on discharge, with the actual stress distribution being determined by random factors such as the details of the filling process and the elastic properties of the material. Our assumption of a completely active state on filling and a totally passive state on discharge is an extreme situation which may occur only rarely.

It is usually found that on filling, the material compacts slightly and dilates on discharge, though the actual density change $\Delta\rho$ is usually small. On initiation of discharge, a boundary will be formed between the flowing material and that which has not yet started to move. From continuity we can show that this boundary will move upwards at a velocity $V\rho/\Delta\rho$ where V is the velocity immediately below the boundary. Since velocities in discharging bunkers are usually small, this boundary will take some time to propagate through the bunker.

Walters considers the rather artificial situation shortly after the initiation of discharge in which there is a horizontal plane boundary dividing the dilated flowing material in the lower part of the bunker from the compacted stationary material in the upper part. We will show in §7.13 that the boundary cannot be a horizontal plane but no other assumption is possible if we wish to use an analysis of the Janssen style. The incorrect assumption about the shape of the boundary is one of the factors that make the numerical predictions of this analysis unreliable.

Let us take a tall cylindrical bunker containing a material with angle of internal friction of 30°. For simplicity, we will use Janssen's original method so that from equations (5.2.5) and (5.2.6), $K_A = \frac{1}{3}$ and $K_P = 3$. Let the boundary between the active and passive zones be at depth H below the top surface, as shown in figure 5.16, and let us assume that H is large compared with the diameter D. There is no surcharge on the top surface and therefore the stresses in the upper part of the bunker are given by equations (5.2.10) and (5.2.11),

$$\sigma_{zz} = \frac{\gamma D}{4\mu_w K_A}\left[1 - \exp\left(-\frac{4\mu_w K_A z}{D}\right)\right] \tag{5.9.1}$$

$$\sigma_{rr} = \frac{\gamma D}{4\mu_w}\left[1 - \exp\left(-\frac{4\mu_w K_A z}{D}\right)\right] \tag{5.9.2}$$

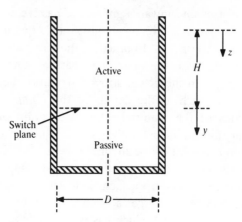

Figure 5.16 The switch plane in a cylindrical bunker.

Since K_A is small the stresses will rise slowly to their asymptotes but with H large they will be close to their asymptotic values of

$$\sigma_{zz}^\infty = \frac{\gamma D}{4\mu_w K_A} \tag{5.9.3}$$

$$\sigma_{rr}^\infty = \frac{\gamma D}{4\mu_w} \tag{5.9.4}$$

at the base of the active zone. These stresses are sketched in figure 5.17. It should be noted that in our case $K_A = \frac{1}{3}$ so that $\sigma_{zz}^\infty = 3\,\sigma_{rr}^\infty$.

We can now treat the flowing zone as a cylindrical bunker subject to a surcharge imposed by the stagnant material above, i.e.

$$Q_0 = \sigma_{zz}^\infty = \frac{\gamma D}{4\mu_w K_A} \tag{5.9.5}$$

Taking y to represent distance below the boundary, as in figure 5.16, and assuming that the material is in the passive state, we have from (5.2.10)

$$\sigma_{zz} = \frac{\gamma D}{4\mu_w K_P} + \frac{\gamma D}{4\mu_w}\left(\frac{1}{K_A} - \frac{1}{K_P}\right)\left[1 - \exp\left(-\frac{4\mu_w K_P y}{D}\right)\right] \tag{5.9.6}$$

Since K_P is large, σ_{zz} tends rapidly to its new asymptote of $\gamma D/4\mu_w K_P$ as shown in figure 5.17.

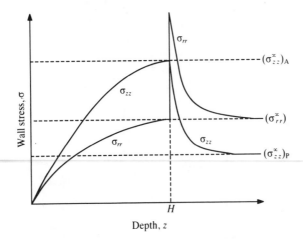

Figure 5.17 Wall stresses predicted by Walters' switch stress analysis.

As always $\sigma_{rr} = K\sigma_{zz}$, and hence

$$\sigma_{rr} = \frac{\gamma D}{4\mu_w} + \frac{\gamma D}{4\mu_w}\left(\frac{K_P}{K_A} - 1\right)\left[1 - \exp\left(-\frac{4\mu_w K_P y}{D}\right)\right] \quad (5.9.7)$$

and σ_{rr} tends rapidly to the same asymptote as before (neglecting any changes in the value of γ).

It should be noted that immediately below the boundary, i.e. at $y = 0$ the axial stress is given by

$$\sigma_{rr} = \frac{\gamma D}{4\mu_w} \frac{K_P}{K_A} \quad (5.9.8)$$

a value which is K_P/K_A times the value immediately above the boundary. With $K_P = 3$ and $K_A = \frac{1}{3}$, we have a nine-fold change in the normal stress on the wall at this section, but the elevated stress dies away rapidly with distance below the boundary as shown in figure 5.17.

Though we have emphasised that the assumptions in this analysis make the magnitudes of the predictions unreliable, the model is qualitatively reasonable and illustrates that a narrow stress peak of considerable magnitude can travel up the wall on the initiation of discharge. One advantage that is sometimes claimed for core flow bunkers is that these stress peaks occur in the flowing core and that the surrounding stagnant material shields the walls to some extent from these effects.

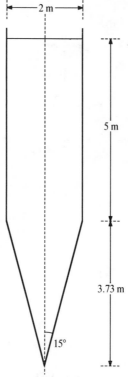

Figure 5.18 Conical plus cylindrical bin.

The methods of this section can be used to predict the complete wall stress distribution in a bin consisting of a cylindrical bunker surmounting a conical hopper as shown in figure 5.18. We will treat this by means of a numerical example and consider a bunker of diameter 2 m and height 5 m attached to a hopper of half angle 15°. The material properties will be taken to be $\phi = 30°$, $\phi_w = 20°$ and $\gamma = 8.0$ kN m^{-3}.

We will assume active failure in the bunker. From equation (3.7.8) we have $\omega = 43.16°$ and from equation (5.3.4), $K_{wA} = 0.370$. The distribution factor, $\mathcal{D}$, is given by equation (5.4.23) as 0.946. Thus we can evaluate the asymptotic value of the wall stress from equation (5.2.11) as $\gamma D/4\mu_w = 10.99$ kN m^{-2}. The exponential coefficient in equation (5.4.7), $4\mu_w K_{wA}\mathcal{D}/D = 0.2548$ m^{-1} and hence

$$\sigma_w = 10.99 \left[1 - \exp(-0.2548\, z)\right] \qquad (5.9.9)$$

At a depth of 5 m, $\sigma_w = 7.92$ kN m^{-2} and the corresponding value of σ_{zz} is $\sigma_w/K_{wA}\mathscr{D} = 22.6$ kN m^{-2}.

Within the hopper we must assume passive failure and can evaluate the angle η from equation (5.8.22) as 27.16°, giving from equation (5.8.23), $\mathscr{D} = 1.625$. From equation (5.8.8) we have $B = 0.4858$ and from equation (5.8.10), $m = 3.626$. Thus from equation (5.8.21) we have $m^* = 5.893$. It may be noted that equation (5.8.27) gives $m^{**} = 7.143$ but for the moment we will accept the former value. From equation (5.8.16) noting that $Q_0 = 22.6$ kN m^{-2} and that $h_0 = D/2 \tan \alpha = 3.732$ m we have

$$\sigma_{hh} = 22.6 \left(\frac{h}{3.732}\right)^{5.893} + \frac{8.0\,h}{4.893}\left[1 - \left(\frac{h}{3.732}\right)^{4.893}\right] \quad (5.9.10)$$

The ratio $\sigma_w/(\sigma_{hh})_w$ is given from equation (5.8.29) as 1.162 so that $\sigma_w/\sigma_{hh} = 1.162\ \mathscr{D} = 1.889$. Hence, from equation (5.9.10),

$$\sigma_w = 3.09\,h + 31.2 \left(\frac{h}{3.732}\right)^{4.893} \quad (5.9.11)$$

Equations (5.9.9) and (5.9.11) are plotted in figure 5.19. The large stress peak at the junction between the bunker and the hopper can be seen. Had we taken the value of m^{**} calculated above, this would not have affected the value of the peak stress but would have made the decay towards the final linear asymptote rather more rapid and also made a minor change in the value of the asymptotic stress distribution.

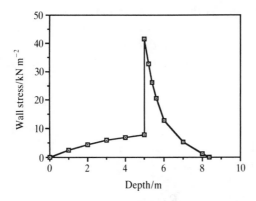

Figure 5.19 The stress peak at the transition between the cylindrical and conical parts of a bin.

It is, however, clear that for structural design the value of the peak stress is the most important factor and very often such bins are strengthened in the vicinity of the bunker/hopper transition. We have, however, emphasised above that the numerical values resulting from the above analysis must not be taken too precisely as the method is based on the unwarranted assumption that the switch plane is horizontal. Further consideration will be given to this matter in §7.13.

5.10 Analyses of Enstad and Reimbert

Enstad (1975) gives an analysis for the stress distribution in a conical hopper which is similar in concept to Walker's analysis described in the preceding sections. In the simplest version of Walker's analysis it is assumed that the vertical stress is constant across any horizontal cross-section. Enstad, on the other hand, assumes that the minor principal stress is constant across a spherical surface spanning the hopper. This surface is chosen so that the minor principal stress acts normal to it both at the wall and on the centre-line. It is not possible to say a priori, which of these two assumptions is the more accurate. The relative advantages can only be assessed by comparison with experimental results or with exact stress analyses. We will make this latter comparison in §7.5 which shows that in many circumstances, Enstad's assumption is the more reliable, justifying the greater complexity of his analysis.

We have already seen that the passive analysis is the more appropriate for a discharging conical hopper and this version will be presented here. The analyses for the active case and for the wedge-shaped hopper are left as an exercise for the student, though the final result for the wedge-shaped hopper will be quoted below.

In passive failure, the major principal stress is inclined at an angle β to the normal to the wall, where, from figure 3.16,

$$\beta = \tfrac{1}{2}(\omega + \phi_w) \qquad (5.10.1)$$

Thus Enstad's spherical surface has the geometry shown in figure 5.20. The radius R of this surface can be found by applying the sine rule to triangle OAB giving

$$R = r \frac{\sin \alpha}{\sin (\alpha + \beta)} \qquad (5.10.2)$$

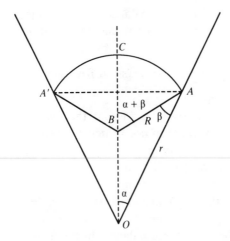

Figure 5.20 The Enstad element.

The volume under the surface is the sum of the volumes of the cone OAA' and the spherical cap ACA'. The former has volume

$$\tfrac{1}{3}\pi \, (r \sin \alpha)^2 \, r \cos \alpha$$

and the volume of the latter is given by

$$\int_0^{\alpha+\beta} \pi \, (R \sin \eta)^2 \, R \sin \eta \, d\eta = \frac{\pi R^3}{3} (\cos^3(\alpha + \beta) - 3 \cos (\alpha + \beta) + 2)$$

Thus the total volume V of the material under the spherical surface is given by

$$V = \frac{\pi r^3}{3} \sin^3\alpha \left[\cot \alpha + \frac{\cos^3(\alpha + \beta) - 3 \cos (\alpha + \beta) + 2}{\sin^3(\alpha + \beta)} \right] \qquad (5.10.3)$$

which we will write as

$$V = \frac{\pi r^3}{3} \sin^2\alpha \, f(\alpha,\beta) \qquad (5.10.4)$$

where

$$f(\alpha,\beta) = \cos \alpha$$
$$+ \frac{[\cos^3(\alpha + \beta) - 3 \cos (\alpha + \beta) + 2]\sin \alpha}{\sin^3(\alpha + \beta)} \qquad (5.10.5)$$

Enstad performs a force balance on an element bounded by two such surfaces. The volume of the material within the element δV is given by

$$\delta V = \pi\, r^2 \sin^2\alpha\, f(\alpha,\beta)\, \delta r \qquad (5.10.6)$$

and its weight is $\gamma\delta V$.

The stress on the lower surface of the element is assumed to be constant and equal to the minor principal stress σ_3. Since this is taken to be constant, the resulting vertical force is equal to the product of the stress and the projected area of the surface $\pi(r \sin \alpha)^2$. The force on the upper surface differs from this by an amount $\pi \sin^2\alpha\, \delta(r^2\sigma_3)$. On the edge of the element, which is of area $2\pi r \sin \alpha\, \delta r$, there are stresses σ_w and τ_w inclined at α to the horizontal and vertical respectively. Hence, a force balance on the element gives

$$\pi r^2\gamma \sin^2\alpha\, f(\alpha,\beta)\, \delta r + \pi \sin^2\alpha\, \delta(r^2\sigma_3)$$
$$= 2\pi r \sin \alpha\, \delta r(\sigma_w \sin \alpha + \tau_w \cos \alpha) \quad (5.10.7)$$

For a cohesionless material

$$\tau_w = \sigma_w \tan \phi_w \qquad (5.10.8)$$

and from equations (3.3.15) and (5.8.3)

$$\sigma_3 = p(1 - \sin \phi) \qquad (5.10.9)$$

and

$$\sigma_w = p(1 + \sin \phi \cos 2\beta) \qquad (5.10.10)$$

Thus on substituting these relationships into equation (5.10.7), we have

$$\pi r^2 \gamma \sin^2 \alpha f(\alpha,\beta) + \pi \sin^2 \alpha\, (1 - \sin \phi)\,\frac{d(pr^2)}{dr}$$
$$= 2\pi r \sin^2 \alpha\, p\, (1 + \tan \phi_w \cot \alpha)$$
$$\times (1 + \sin \phi \cos 2\beta) \qquad (5.10.11)$$

or

$$\frac{dp}{dr} + \gamma\,\frac{f(\alpha,\beta)}{1 - \sin \phi} = \frac{2p}{r}\left(\frac{(1 + \tan \phi_w \cot \alpha)(1 + \sin \phi \cos 2\beta)}{1 - \sin \phi} - 1\right)$$
$$(5.10.12)$$

This is of the same form as equation (5.8.11), and if we define X and Y by

$$X = 2\left[\frac{(1 + \tan \phi_w \cot \alpha)(1 + \sin \phi \cos 2\beta)}{1 - \sin \phi} - 1\right] \quad (5.10.13)$$

and

$$Y = \frac{f(\alpha, \beta)}{1 - \sin \phi} \quad (5.10.14)$$

we have

$$\frac{dp}{dr} - X\frac{p}{r} = -Y\gamma \quad (5.10.15)$$

which by analogy with the method of §5.8 has solution

$$p = \frac{\gamma Y r}{X - 1} + \left(\frac{Q_0}{1 - \sin \phi} - \frac{\gamma Y r_1}{X - 1}\right)\left(\frac{r}{r_1}\right)^X \quad (5.10.16)$$

and σ_w can be found from equation (5.10.10). In evaluating the arbitrary constant in equation (5.10.16), we have assumed that the top surface of the material is part of a sphere of radius r_1 and that a uniform surcharge Q_0 acts normal to this surface. Clearly, the necessary assumption that the top surface is spherical is less satisfactory than Walker's assumption that the surface is horizontal.

Rearranging the above equations gives

$$X = \frac{2 \sin \phi}{1 - \sin \phi}\left[1 + \frac{\sin (2\beta + \alpha)}{\sin \alpha}\right] \quad (5.10.17)$$

and

$$Y = \frac{\sin \beta \sin^2(\alpha + \beta) + 2\left[1 - \cos (\alpha + \beta)\right] \sin \alpha}{(1 - \sin \phi) \sin^3(\alpha + \beta)} \quad (5.10.18)$$

A similar analysis for the wedge-shaped hopper gives

$$X = \frac{\sin \phi}{1 - \sin \phi}\left[1 + \frac{\sin (2\beta + \alpha)}{\sin \alpha}\right] \quad (5.10.19)$$

and

$$Y = \frac{\sin \beta \sin (\alpha + \beta) + (\alpha - \beta) \sin \alpha}{(1 - \sin \phi) \sin^2(\alpha + \beta)} \quad (5.10.20)$$

It is appropriate at this stage to quote a result given by Reimbert and Reimbert (1976) which in our notation becomes

$$\sigma_w = \frac{\gamma D}{4\mu_w}\left[1 - \left(\frac{4\mu_w K_A z}{D} + 1\right)^{-2}\right]$$

(5.10.21)

There seems to be no theoretical derivation of this result which follows from the comment that the experimental measurements of the frictional force on the wall 'can be represented with proper accuracy by a branch of a hyperbola'. Comparison with equation (5.2.11) shows that Reimbert's result predicts the correct asymptotic value of σ_w but that the rate of approach to the asymptote is somewhat more rapid than that predicted by the simple Janssen analysis.

6

Determination of physical properties

6.1 Introduction

In the previous chapters we have assumed that we know the physical properties of the material such as the bulk density and the parameters of the yield functions. In practice these have to be determined experimentally and in this chapter we will consider the methods available for this. However, little attention will be paid to the details of the experimental techniques since these have been adequately discussed elsewhere, for example in the *Powder Testing Guide* by Svarovski (1987) or the Report of the European Federation of Chemical Engineers on the *Mechanics of Particulate Solids*, EFCE Working Party (1989). Furthermore, since many of the properties are measured with commercially available equipment, the manufacturer's handbook is usually the best source of information on the experimental method. Instead, we will concentrate on the interpretation of the experimental results.

We will consider the determination of the material density and the related topic of consolidation in §6.2 and pay attention to the measurements of particle size and particle size distributions in §6.3. In §6.4 we will consider the percolation of fluid through granular media and present relationships for the permeability. Finally in §6.5 to §6.7 we will discuss the measurement of the failure properties of the material and discuss these in terms of the *Critical State Theory* as expounded by Schofield and Wroth (1968). The interpretation of yield data is somewhat controversial and it is hoped that the present account is a dispassionate discussion of the various alternatives. Much of the material in the later sections is common to the field of solids handling and soil mechanics and additional information can be obtained from the standard texts of the latter subject such as Atkinson and Bransby (1978), Capper and Cassie (1963) and Terzaghi (1943).

6.2 Density and compressibility

The solid density ρ_s i.e. the actual density of the particle, can readily be measured by the standard techniques of liquid displacement. Problems can arise if the particle is porous or contains internal voids, as often occurs in agricultural products such as seeds. Under these circumstances the volume of the particle is a somewhat subjective quantity and there is corresponding uncertainty about the associated density. Fortunately, these factors are rarely of importance in the determination of the stress and velocity distributions in granular materials.

The bulk density ρ_b and its reciprocal, which is known as the specific volume v_s, can in principle be measured by filling a vessel of known volume with the material and weighing the contents. However, if this is attempted, it is soon realised that a range of bulk densities can occur depending on the manner in which the vessel is filled. The bulk density is found to increase with the application of elevated stresses and also when the sample is vibrated. As mentioned in chapter 1, the bulk density is clearly related to the void fraction ε, defined as the volume fraction of the material occupied by the interstitial medium, by

$$\rho_b = (1-\varepsilon)\rho_s + \varepsilon\rho_g \qquad (6.2.1)$$

where ρ_g is the density of the interstitial medium, which is usually air. In this case $\rho_g \ll \rho_s$ and we can say with sufficient accuracy that

$$\rho_b = (1-\varepsilon)\rho_s \qquad (6.2.2)$$

Within the soil mechanics literature it is more usual to define the specific volume V as the ratio of the volume of a sample to the volume of the solid within it and to work in terms of the voids ratio e defined as the ratio of the volume of the interstitial phase to that of the solid phase. Thus

$$e = \frac{\varepsilon}{1-\varepsilon} \qquad (6.2.3)$$

and

$$V = v_s\rho_s = 1+e \qquad (6.2.4)$$

We have noted above that the bulk density is a function of the stress state and we will first consider the case when all three principal stresses are equal. This is known as an isotropic stress state and the uniform stress is known as the isotropic pressure p_i, i.e.

$$\sigma_1 = \sigma_2 = \sigma_3 = p_i \tag{6.2.5}$$

Figure 6.1 shows the variation in the bulk density of an initially loose sample as the isotropic pressure is first increased along the section AB, then decreased along section BC, and finally increased again, along section $CDEF$. We see that on increasing the pressure the material compresses. This is due to rearrangement of the particles which in most cases are effectively incompressible and therefore reflects the change in the void fraction ε. However, on reducing the pressure, very little expansion is observed, showing that these voidage changes are effectively irreversible. On increasing the pressure again, the density remains effectively constant until the original line is reached and the material continues to compress as if no stress reduction had taken place. Finally, at very high pressures, crushing of the materials can occur and irreversible changes in the nature of the material take place.

The section of the curve $ABDE$ is often called the primary consolidation curve and it is found experimentally that along this line the specific volume is roughly proportional to the logarithm of the pressure. This relation is often written in the form

$$v_s = v_0 - \lambda \ln p_i \tag{6.2.6}$$

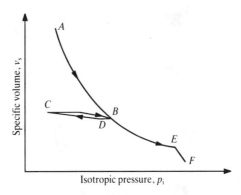

Figure 6.1 Dependence of the specific volume on isotropic pressure.

though purists argue that this form is inadmissible since one cannot take the logarithm of a dimensional quantity. Perhaps it would be better to write this equation in the form

$$v_s = v_0 - \lambda \ln (p_i/p_0) \qquad (6.2.7)$$

where p_0 is some reference pressure which is almost universally taken to be 1 kN m^{-2} and v_0 is the specific volume at that pressure. Under these circumstances equation (6.2.6) can be used provided p_i is measured in kN m^{-2}.

For the decompression branch BC we can write

$$v_s = v_0' - \lambda' \ln p_i \qquad (6.2.8)$$

and it is usual to ignore any hysteresis between the lines BC and CD. Clearly v_0' depends on the pressure at the start of the decompression process and $\lambda' \ll \lambda$. Very commonly it is assumed that no decompression takes place and therefore that $\lambda' = 0$.

These results can be idealised as in figure 6.2 in which v_s is plotted against $\ln p_i$. The primary consolidation line ABE and the decompression branch BC are now straight with the latter being almost horizontal.

Equation (6.2.6) implies that the specific volume tends to infinity, and hence the density to zero, as the pressure is reduced to zero. This is contrary to observation since we know that the void fraction in a fluidised bed at minimum fluidisation, where the inter-particle pressure is zero, is less than unity. This void fraction is usually denoted by ε_{mf} and the corresponding specific volume v_{mf} is then the largest stable specific volume for a static bed. We can therefore add the highest

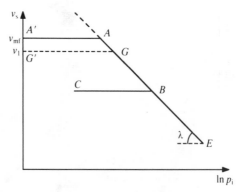

Figure 6.2 Idealised dependence of the specific volume on isotropic pressure.

permissible decompression line $A'A$ to figure 6.2. In practice, the inevitable consolidation that takes place during the process of filling a vessel results in an initial specific volume v_1 at low p_i which is less than v_{mf}. Under these circumstances, the material will be effectively incompressible until the pressure rises to the appropriate value on the primary consolidation line. For such a material compression will follow the line $G'GE$.

In practice it is very difficult to subject a material to an isotropic stress state and bulk densities are often measured in an *Oedometer*. This consists of a cylindrical cell with a moveable piston. A carefully weighed sample is placed in the cell and a stress σ applied *via* the piston. By measuring the position of the piston the dependence of v_s on σ can be determined.

The stress state in the oedometer cannot be precisely determined but if we assume that the walls are relatively smooth and that the material is in a state of active failure the two horizontal components of the stress may approximate to $K_A\sigma$ where K_A is Rankine's coefficient of earth pressure. Under these conditions the *effective* isotropic pressure, defined by

$$p_i = \tfrac{1}{3}(\sigma_1+\sigma_2+\sigma_3) \tag{6.2.9}$$

will be given approximately by

$$p_i = \left(\frac{1+2K_A}{3}\right)\sigma \tag{6.2.10}$$

Since $K_A < 1$, p_i is directly proportional to but somewhat less than σ. Thus if we plot the dependence of v_s on $\ln \sigma$ in figure 6.2 we will get a line parallel to that for isotropic compression, but displaced somewhat to the right. Thus the slope λ can readily be evaluated from the results of an oedometer test, though v_0 cannot be determined directly.

Equation (6.2.6) can also be expressed in the form,

$$\frac{1}{1-\varepsilon} = \frac{1}{1-\varepsilon_0} - \lambda\rho_s \ln p_i \tag{6.2.11}$$

and, since ε never differs greatly from 0.5, this equation can be expanded by the binomial theorem to give

$$\varepsilon = \varepsilon_0 - \frac{\lambda\rho_s}{4}\ln p = \varepsilon_0 - \Lambda \ln p_i \tag{6.2.12}$$

where $\Lambda = \lambda\rho_s/4$. It should be borne in mind that though equation (6.2.12) is only approximately the same as equation (6.2.6), this latter equation is empirical and there is no reason to believe that it is any more accurate than equation (6.2.12). Indeed, it could be that one form is better for some materials and the other form for other materials.

Jenike (1961) correlates his results in the form

$$v_s = v_0 p_i^{-\beta} \tag{6.2.13}$$

Taking logarithms of this relationship and noting that $v_s/v_0 \approx 1$ so that $\ln(v_s/v_0) \approx 1 - v_s/v_0$ we have

$$v_s = v_0 - v_0\beta \ln p_i \tag{6.2.14}$$

We see that this is similar in form to equation (6.2.6) and therefore that $\beta = \lambda/v_0 = 4\Lambda(1-\varepsilon_0)$. Jenike also notes that equation (6.2.13), like equation (6.2.6), predicts that v_s tends to infinity as p_i tends to zero and proposes the modification

$$v_s = v_0(p_i+1)^{-\beta} \tag{6.2.15}$$

where p_i must now be measured in lbf/ft^2 and v_0 is the specific volume at zero pressure. This formulation is even more offensive to the purists than equation (6.2.6), since here the quantity 1 is specific to a particular set of units.

When subjected to a general stress state the specific volume and hence the void fraction will be a function of all three principal stresses σ_1, σ_2 and σ_3, i.e.

$$\varepsilon = f(\sigma_1, \sigma_2, \sigma_3) \tag{6.2.16}$$

In principle the voidage must depend on the three stress invariants I_1, I_2 and I_3 which are three independent functions symmetric in the principal stresses. There is some degree of freedom in how these quantities are defined but the following forms are convenient for our purpose:

$$I_1 = \tfrac{1}{3}(\sigma_1+\sigma_2+\sigma_3) \tag{6.2.17}$$

$$I_2 = \tfrac{1}{3}[(\sigma_1-\sigma_2)^2+(\sigma_2-\sigma_3)^2+(\sigma_3-\sigma_1)^2] \tag{6.2.18}$$

and

$$I_3 = \sigma_1\sigma_2\sigma_3 \tag{6.2.19}$$

It should be noted that the first invariant is identical to the effective isotropic pressure defined by equation (6.2.9) and that the second invariant is some measure of the root mean square (RMS) average of the three largest shear stresses.

It is unlikely that experimental evidence will be available in the foreseeable future to permit a thorough investigation of the relationship between the voidage and the three stress invariants. Usually, only the major and minor principal stresses, σ_1 and σ_3, are known and it is assumed that the voidage depends on p and q where

$$p = \tfrac{1}{2}(\sigma_1 + \sigma_3) \tag{6.2.20}$$

and

$$q = \tfrac{1}{2}(\sigma_1 - \sigma_3) \tag{6.2.21}$$

These quantities are seen to be closely related to the first and second invariants and are the best that can be achieved in the absence of information about the third principal stress. Furthermore, it can be seen that p is the co-ordinate of the centre of the Mohr's circle and that q is its radius R. Sometimes these quantities are given the symbols σ and τ and are known as the Sokolovskii variables. This notation will not be used here to avoid confusion with our use of σ and τ as the normal and shear stresses on the slip plane.

It may be noted from equation (3.3.5) that for a cohesionless Coulomb material at incipient failure,

$$q = p \sin \phi \tag{6.2.22}$$

and this can be compared with the more familiar form

$$\tau = \sigma \tan \phi \tag{6.2.23}$$

In the field of soil mechanics the voidage is commonly measured as a function of p and q and results of the form shown in figure 6.3 are obtained. It should be noted that when $q = 0$ the results are the two-dimensional equivalent of the isotropic consolidation results of figure 6.1 so that we expect the intercepts to be of the form

$$\varepsilon = \varepsilon_0 - \Lambda \ln p \tag{6.2.24}$$

We will see in §6.5 that when measuring the failure properties of a material, we subject it to a variety of combinations of normal and shear stresses (σ, τ) on the failure plane and that there are no facilities

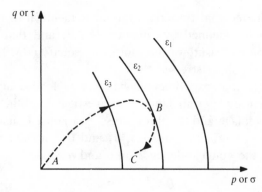

Figure 6.3 Dependence of the voidage, ε, on the stress parameters (σ, τ) or (p, q).

for measuring any other stress component. Thus it is convenient if we can regard the voidage as a function of σ and τ. Since for a cohesionless Coulomb material at incipient failure, $\sigma = p - q \sin \phi = p \cos^2 \phi$ and $\tau = q \cos \phi$, the dependence of ε on σ and τ is qualitatively similar to its dependence on p and q and hence we have labelled the axes of the sketch of figure 6.3 as σ or p and τ or q.

The stress states to which the material is subjected during the course of an investigation using a shear cell can be sketched on axes of σ and τ as in figure 6.3. If, for example, the locus of these states was given by the dashed line ABC we would find that compression would occur over the section AB resulting in a voidage ε_2 at B, but that no dilation would occur over the section BC. The voidage would therefore remain at ε_2 until the end of the process. In other words the final voidage is the smallest equilibrium voidage along the stress state locus.

It must not, however, be thought that voidages can never increase, and that all material therefore becomes progressively more dense. Dilation can occur in the presence of shear strain, as will be discussed in §6.5.

6.3 Particle size

Measuring the diameter of a single spherical particle gives rise to little difficulty. If the particle is large, its diameter can be measured with a micrometer and if it is small, a travelling microscope can be used. This latter process has been automated using video cameras and computerised image analysis techniques. Many such devices are commercially available and often contain software which can also analyse particle shape and size distributions. Particle size can also be determined by sieving and various fluid mechanical techniques based on the measurements of terminal velocities, sedimentation rates or permeability. Another favoured technique is that of the Coulter counter in which a dilute suspension is passed between electrodes and the particle size is inferred from the change in electrical resistance.

Granular materials are rarely composed of perfectly spherical particles and for all other cases some arbitrary choice of the measure of particle size has to be selected. Perhaps the best measure of size is the equivalent spherical diameter D_s. This is defined as the diameter of the sphere having the same volume as the particle,

$$D_s = \left(\frac{6V}{\pi}\right)^{1/3} \tag{6.3.1}$$

where V is the volume of the particle. A complementary parameter λ_s known as the shape factor, can be defined by

$$\lambda_s = \frac{A}{\pi D_S^2} \tag{6.3.2}$$

where A is the surface area of the particle. The shape factor is the ratio of the surface area to that of the equivalent sphere and the reciprocal of λ_s is sometimes called the sphericity.

In most of the devices using image analysis the particle is placed on a plane and viewed normal to that plane. The particle will clearly rest in an orientation which corresponds to a local potential energy minimum, and often at the absolute potential minimum, i.e. with its centre of gravity at the lowest possible position. Thus such techniques cannot distinguish between a disc and a sphere and there is no facility for measuring the volume of a particle. Instead, the projected area A_p and the associated perimeter P_p are measured and an area equivalent diameter D_A and shape factor λ_A, defined by

$$D_A = \sqrt{\left(\frac{4A_p}{\pi}\right)} \tag{6.3.3}$$

and

$$\lambda_A = \frac{P_p}{\pi D_A} \tag{6.3.4}$$

are evaluated instead of D_S and λ_S.

Other measures of the particle size are in common use including the *feret diameter*. This is the extent of the particle measured between the two extreme tangents parallel to an arbitrary direction as illustrated in figure 6.4. The variation of feret diameter with angular position can give a great deal of insight into the shape of the particle.

Whatever measure of particle size is chosen, it is most unlikely that all the particles will have the same size. For certain agricultural products such as seeds there may be only a small variation of size and some industrial processes aim to produce uniform particles. Normally, however, there is a wide range of particle sizes particularly when the granular material has been produced by crushing a coarser material.

A particle size distribution by number $p_n(D)$, where D is any measure of the particles size, can be defined such that $p_n(D)\,\mathrm{d}D$ is the number fraction of the particles having sizes between D and $D + \mathrm{d}D$. We can also define a cumulative particle size distribution $P_n(D)$ which is the fraction of the particles having size up to D. Clearly,

$$P_n(D) = \int_0^D p_n(D)\,\mathrm{d}D \tag{6.3.5}$$

Alternatively, we can define distributions by mass or weight, $p_m(D)$ $\mathrm{d}D$ being the fraction by mass of the sample having sizes between D and $D + \mathrm{d}D$, and $P_m(D)$ being the fraction by mass of the sample having sizes up to D. As before,

$$P_m(D) = \int_0^D p_m(D)\,\mathrm{d}D \tag{6.3.6}$$

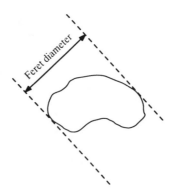

Figure 6.4 Definition of the feret diameter.

The cumulative distribution by mass P_m can be measured directly as the mass fraction of the material that falls through a sieve of appropriate aperture.

It is very rare indeed that the particle size distributions can be predicted from first principles. Usually the size distribution must be determined experimentally either by sieving or some automated method such as the Coulter counter or the image analysis techniques mentioned above. The first of these methods gives the distribution by mass and the two later methods the distribution by number, and we will derive the relationship between these two quantities below.

Assuming that all the particles are of the same shape, the mass of a particle will be proportional to the cube of the size

$$m = \lambda D^3 \qquad (6.3.7)$$

where λ depends on the density, the shape and the chosen measure of particle size, being $\pi \rho_s / 6$ for a sphere. Thus if the sample contains N particles, the number in the range D to $D + dD$ will be given by

$$dN = N p_n (D) \, dD \qquad (6.3.8)$$

and their mass will be

$$dm = \lambda D^3 N p_n (D) \, dD \qquad (6.3.9)$$

The distribution by mass is defined as the ratio of dm to the total mass, i.e.

$$p_m(D) \, dD = \frac{dm}{m} \qquad (6.3.10)$$

$$= \lambda D^3 N p_n (D) \, dD / \int_0^\infty \lambda D^3 N p_n (D) \, dD$$

i.e.

$$p_m(D) = D^3 p_n(D) / \int_0^\infty D^3 p_n (D) \, dD \qquad (6.3.11)$$

It is perhaps clearer to think of this equation in the form

$$p_m(D) = K_n D^3 p_n(D) \qquad (6.3.12)$$

where K_n is a normalisation factor required to ensure that

$$\int_0^\infty p_m(D) \, dD = 1 \qquad (6.3.13)$$

By the same token

$$p_n(D) = K'_n D^{-3} p_m(D) \qquad (6.3.14)$$

where K'_n is evaluated by the requirement that

$$\int_0^\infty p_n(D)\, dD = 1 \qquad (6.3.15)$$

As mentioned above, there is rarely any reason to expect that the particle size distribution will take any particular form, though the products of a crystallisation process may conform to the Avrami distribution,

$$\left. \begin{aligned} p_n(D) &= \frac{1}{D_{max}} & D < D_{max} \\ p_n(D) &= 0 & D > D_{max} \end{aligned} \right\} \qquad (6.3.16)$$

or, from equation (6.3.11),

$$\left. \begin{aligned} p_m(D) &= \frac{4D^3}{D_{max}^4} & D < D_{max} \\ p_m(D) &= 0 & D > D_{max} \end{aligned} \right\} \qquad (6.3.17)$$

Other popular size distributions include the Rosin–Rammler distribution and the log-normal distribution. It should be noted that the ordinary normal distribution is rarely satisfactory and can never be exact, since it implies that there are some particles of negative diameter.

The Rosin–Rammler distribution can be written as

$$P_m(z) = 1 - \exp(-z^k) \qquad (6.3.18)$$

or, on differentiation,

$$p_m(z) = k\, z^{k-1} \exp(-z^k) \qquad (6.3.19)$$

where

$$z = \frac{D}{D_R} \qquad (6.3.20)$$

The parameters k and D_R are chosen to give a best fit to the experimentally measured distribution and it can be noted from equation (6.3.18) that D_R is the diameter below which the fraction $(1 - e^{-1}) = 0.632$ of the sample lies.

The values of k and D_R can be determined by plotting η against $\ln D$ where η is defined by

$$\eta = \ln[-\ln(1-P_m)] \qquad (6.3.21)$$

the slope of the resulting line is $1/k$ and D_R is the value of D at which $\eta = 0$.

However, the analysis of the size distribution is greatly simplified if it happens to conform to the log-normal distribution which can be expressed as

$$p_n(x) = \frac{\exp(-x^2/2\sigma^2)}{\sigma\sqrt{(2\pi)}} \qquad (6.3.22)$$

where x is defined by

$$x = \ln D - \ln D_g \qquad (6.3.23)$$

Here σ is the standard deviation of x and D_g is the geometric mean diameter. From this it follows that

$$P_n(x) = \frac{1}{2}\left[1 + \mathrm{erf}\left(\frac{x}{\sqrt{2}\sigma}\right)\right] \qquad (6.3.24)$$

where erf represents the error function.

The prime advantage of the log-normal distribution is that it can be shown, see for example Kay and Nedderman (1985), that if the particles are log-normally distributed with respect to number, they are also log-normally distributed with respect to mass and moreover with the same standard distribution. Thus, by analogy, with equation (6.3.22),

$$p_m(x_m) = \frac{\exp(-x_m^2/2\sigma^2)}{\sigma\sqrt{(2\pi)}} \qquad (6.3.25)$$

and

$$P_m(x_m) = \frac{1}{2}\left(1 + \mathrm{erf}\left(\frac{x_m}{\sqrt{2}\sigma}\right)\right) \qquad (6.3.26)$$

where

$$x_m = \ln D - \ln D_{gm} \qquad (6.3.27)$$

and D_{gm} is the geometric mean by mass.

It is possible to check whether an experimental distribution conforms to the log-normal distribution by plotting ln D against y where y is defined so that

$$P = \tfrac{1}{2}\left(1 + \mathrm{erf}\left(\frac{y}{\sqrt{2}}\right)\right) = \frac{1}{\sqrt{(2\pi)}} \int_{-\infty}^{y} \exp\left(-\frac{t^2}{2}\right) dt \qquad (6.3.28)$$

and special graph paper is commercially available for this purpose. This graph paper has a logarithmic ordinate and an abscissa which is linear in y but marked with the corresponding values of P. If the distribution is log-normal the experimental points will lie on a straight line. Alternatively, use can be made of tabulation of the errors function such as those of Abramowitz and Stegun (1965) or standard statistical tables such as Lindley and Miller (1953) which give P as a function of y.

When the cumulative distribution by number, P_n, is used as the abscissa, the line passes through the points,

$$P_n = 0.5 \qquad \text{i.e. } y = 0, \quad x = 0 \qquad \text{i.e. } D = D_g$$
$$P_n = 0.1587 \quad \text{i.e. } y = -1, x = -\sigma \qquad \text{i.e. } D = D_g e^{-\sigma}$$
$$\text{and} \qquad P_n = 0.8413 \quad \text{i.e. } y = 1, \quad x = +\sigma \qquad \text{i.e. } D = D_g e^{\sigma}$$

from which the values of D_g and σ can readily be evaluated. These data are equivalent to the expression

$$\ln D = \ln D_g + \sigma y \qquad (6.3.29)$$

If the cumulative distribution by mass, P_m, is used as the abscissa, the geometric mean by mass D_{gm} is found at $P_m = 0.5$ ($y = 0$) instead of D_g and the points $P_m = 0.8413$ and 0.1587 ($y = \pm 1$) give $D = D_{gm} e^{\pm \sigma}$.

The main utility of the log-normal distribution lies in the convenience with which the mean values of powers of the size D can be evaluated. It can be shown (Kay and Nedderman, 1985) that the mean value of D^n is given by

$$D^n = D_g^n \exp(\sigma^2 n^2 / 2) \qquad (6.3.30)$$

From this we can evaluate the arithmetic mean D_a by putting $n = 1$, giving

$$D_a = D_g \exp(\sigma^2 / 2) \qquad (6.3.31)$$

the surface mean, D_s, by putting $n = 2$, giving

$$D_S^2 = D_g^2 \exp(2\sigma^2) \qquad (6.3.32)$$

and the volumetric mean, D_V, by putting $n = 3$, giving

$$D_V^3 = D_g^3 \exp(9\sigma^2/2) \qquad (6.3.33)$$

A useful measure of the mean diameter when evaluating the permeability is the volume–surface mean diameter D_{VS} defined by

$$D_{VS} = D_V^3/D_S^2 = D_g \exp(5\sigma^2/2) \qquad (6.3.34)$$

It can also be shown that the geometric means by number and mass are related by

$$D_{gm} = D_g \exp(3\sigma^2) \qquad (6.3.35)$$

Example 6.1 We can illustrate the use of the relationships derived above by considering the sieve analysis presented in table 6.1

The cumulative distribution by mass is the mass fraction below a specified value of D and therefore can be obtained by summation of the values of table 6.1 starting from the bottom. The results are given as rows one and two of table 6.2 and the later rows give the values

Table 6.1

Size range (μm)			Wt. (Mass) % in range
	+	500	2
− 500	+	400	4
− 400	+	300	11
− 300	+	200	40
− 200	+	150	21
− 150	+	100	16
− 100			6

Table 6.2

D (μm)	100	150	200	300	400	500
P_m	0.06	0.22	0.43	0.83	0.94	0.98
ln D	4.605	5.011	5.298	5.704	5.992	6.215
y	−1.55	−0.77	−0.18	0.95	1.56	2.06
η	−2.78	−1.39	−0.58	0.57	1.03	1.36

of ln D, y (evaluated from the tables of Lindley and Miller) and η (evaluated from equation (6.3.21))

Figure 6.5 shows a plot of ln D vs y and it can be seen that the relationship is closely linear. A best-fit straight line gives

$$\ln D = 5.32 + 0.433\,y$$

with a correlation coefficient of 0.997. This last value shows that the log-normal distribution is a good, but not perfect, fit to the data. By analogy with equation (6.3.29) we see that ln D_{gm} = 5.32 (D_{gm} = 204 μm) and σ = 0.433.

Hence from equation (6.3.35) we have

$$\ln D_g = \ln D_{gm} - 3\sigma^2 = 5.32 - 3 \times 0.433^2 = 4.76$$

from which we find that D_g = 116 μm. The cumulative distribution by number can be drawn on the figure since it is a line of slope 0.433 passing through the point (y = 0, ln D = 4.76).

Thus we can find the cumulative fractions with respect to both number and mass for any value of D either by reading directly off the graph or by using equation (6.3.29). Taking the value D = 250 μm, for example, gives

$$y_m = (\ln 250 - 5.32)/0.433 = 0.464 \quad \text{i.e. } P_m = 0.68$$

and

$$y_n = (\ln 250 - 4.76)/0.433 = 1.759 \quad \text{i.e. } P_n = 0.96$$

Thus while only 4% of the particles have a diameter greater than 250 μm, their mass amounts to 32% of the total.

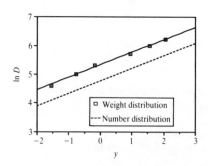

Figure 6.5 The log-normal distribution.

The other means can be evaluated from equations (6.3.30) to (6.3.34). For example, the arithmetic mean is given from equation (6.3.31) by

$$\ln D_a = \ln 116 + 0.433^2/2 \quad \text{i.e. } D_a = 127 \ \mu\text{m}$$

and the volumetric mean is found from equation (6.3.33)

$$3 \ln D_v = 3 \ln 116 + 9 \times 0.433^2/2 \quad \text{i.e. } D_v = 154 \ \mu\text{m}.$$

Alternatively the size distribution can be analysed in terms of the Rosin–Rammler distribution by plotting $\ln D$ *vs* η as in figure 6.6. This gives a best-fit straight line of equation

$$\ln D = 5.58 + 0.380 \ \eta$$

with a correlation coefficient of 0.990, from which we see that $D_R = 265 \ \mu\text{m}$ and $k = 1/0.380 = 2.63$. Comparison of figures 6.5 and 6.6 suggests that these data are better fitted by the log-normal distribution than the Rosin–Rammler distribution and this is confirmed by the lower value of the correlation coefficient in the latter case. However, though the Rosin–Rammler distribution is a reasonably good fit to the data, it is less convenient since we cannot proceed to evaluate the other means except by the tedious numerical integration of equation (6.3.15).

6.4 Permeability

The rate at which the interstitial fluid migrates through a granular material under the influence of a pressure gradient is determined by the permeability. This topic was first studied by Darcy (1856) who obtained the empirical relationship,

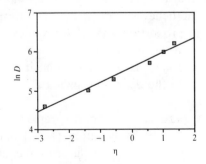

Figure 6.6 Plot to evaluate the parameters in the Rosin–Rammler distribution.

$$U = k \frac{\mathrm{d}P}{\mathrm{d}x} \qquad (6.4.1)$$

where U is the superficial velocity, i.e. the volumetric flow area divided by the total area, k is the permeability and $\mathrm{d}P/\mathrm{d}x$ is the pressure gradient in the interstitial fluid. The superficial velocity U should not be confused with the interstitial velocity U_i defined as the volumetric flow rate divided by the cross-sectional area of the voids. Clearly,

$$U = U_i \varepsilon \qquad (6.4.2)$$

where ε is the void fraction.

Equation (6.4.1) was put on a firmer theoretical basis for spherical particles of diameter d by Kozeny (1927) and Carman (1937) who derived the result

$$\frac{\mathrm{d}P}{\mathrm{d}x} = \frac{180\mu_f U (1-\varepsilon)^2}{d^2 \varepsilon^3} \qquad (6.4.3)$$

where μ_f is the viscosity of the interstitial fluid. The fact that the pressure gradient is directly proportional to $\mu_f U/d^2$ follows directly from dimensional analysis and the factor $(1-\varepsilon)^2/\varepsilon^3$ is derived by arguments that are perhaps not totally convincing but can be found in most standard texts on fluid mechanics, e.g. Kay and Nedderman (1985). The numerical factor 180 is empirical. Despite the somewhat unconvincing derivation, this equation gives excellent results at low flow rates and thereby gives a theoretical prediction of the permeability k,

$$k = \frac{d^2 \varepsilon^3}{180\mu_f (1-\varepsilon)^2} \qquad (6.4.4)$$

It should be noted that the permeability increases as the square of the particle diameter, and behaves as $\varepsilon^3/(1-\varepsilon)^2$. This term is very sensitive to ε, being roughly proportional to ε^5 for values of ε in the usual range $0.3 < \varepsilon < 0.7$. It should be noted that other definitions of the permeability are sometimes used, in particular it is not uncommon to omit the viscosity from the definition, giving a modified permeability which is equal to $k\mu_f$.

For non-spherical particles the term d is replaced by d/λ_S where λ_S is the shape factor defined by equation (6.3.2) and for a granular material containing a range of particle sizes the appropriate mean diameter is the volume/surface mean D_{VS} defined by equation (6.3.34).

At higher flow rates inertial effects and hence the density ρ_f of the interstitial fluid become important and the Ergun (1952) equation is generally regarded to be the best correlation available. This can be written as

$$\frac{dP}{dx} = \frac{150\mu_f U(1-\varepsilon)^2}{d^2\varepsilon^3} + \frac{1.75\rho_f U^2(1-\varepsilon)}{d\varepsilon^3} \qquad (6.4.5)$$

or

$$\frac{dP}{dx} = E_1 U + E_2 U^2 \qquad (6.4.6)$$

where

$$E_1 = \frac{150\mu_f(1-\varepsilon)^2}{d^2\varepsilon^3} \qquad (6.4.7)$$

and

$$E_2 = \frac{1.75\rho_f(1-\varepsilon)}{d\varepsilon^3} \qquad (6.4.8)$$

Again a shape factor can be introduced for non-spherical particles and the volume/surface mean is appropriate for particles of mixed sizes. At low velocities the second term in this equation becomes negligible and the equation reduces to the form of equation (6.4.3). The difference in the numerical factors 180 and 150 is not easily explained away and reflects the precision of these expressions. Neither the Carman–Kozeny equation (6.4.3) nor the Ergun equation (6.4.5) contains any factor to account for the differing packing arrangements that can occur within the material. It is well-known that the permeability of a regular packing differs from that of a random packing of the same voidage. Both of these equations are appropriate for the sort of random packing usually obtained when a material is poured into a vessel and should be used with caution in other circumstances.

Equation (6.4.5) can be rearranged into the form

$$\frac{dP}{dx} = \frac{\mu_f U(1-\varepsilon)^2}{d^2\varepsilon^3}\left(150 + 1.75\frac{\rho_f U d}{\mu_f(1-\varepsilon)}\right) \qquad (6.4.9)$$

and it is therefore convenient to define a Reynolds number *Re* by

$$Re = \frac{\rho_f U d}{\mu_f(1-\varepsilon)} \qquad (6.4.10)$$

so that equation (6.4.9) can be written either as

$$\frac{dP}{dx} = \frac{\mu_f U (1-\varepsilon)^2}{d^3 \varepsilon^3} (150 + 1.75\, Re) \qquad (6.4.11)$$

or as

$$\frac{dP}{dx} = \frac{\mu_f^2 (1-\varepsilon)^3}{\rho_f d^3 \varepsilon^3} (150\, Re + 1.57\, Re^2) \qquad (6.4.12)$$

Because of the difficulties in the determination of the shape factor, it is often necessary to measure the permeability directly. This can be done by filling a tube with the material and measuring the pressure gradient for a range of flow rates. Because of end effects it is wise to measure the pressure gradient between tappings some distance in from the ends of the bed. The results are best analysed by plotting $(1/U)(dP/dx)$ against U and this should give a straight line from which the slope E_2 and intercept E_1 can be determined.

It is advantageous to measure the voidage of the bed and this can be determined from measurements of the mass and total volume. It is then possible to adapt the values of E_1 and E_2 for different voidages using equations (6.4.7) and (6.4.8).

Sometimes measured values of the permeability and void fraction are used as a means of determining the average particle diameter. For this it is necessary to make some assumption about the shape factor λ_S. It is convenient if the measurements can be made for several flow rates in the range over which the two terms in the Ergun equation are of comparable magnitude. As can be seen from equations (6.4.7) and (6.4.8), both these terms can be used to give an estimate of the diameter, and comparison of the two predictions gives a useful impression of the accuracy of the results.

6.5 Measurement of the internal failure properties

Within the field of soil mechanics the failure properties are commonly measured in a shear box, a tri-axial tester or a true tri-axial tester. The shear box consists of a rectangular cell with hinged sides and free lid as shown in figure 6.7. A load N is placed on the lid and the force F required to induce shear is measured, as is the vertical displacement of the lid, which gives a measure of the dilation of the material. In the tri-axial tester a cylindrical sample is surrounded by a rubber membrane and held between two anvils. The fluid surrounding the

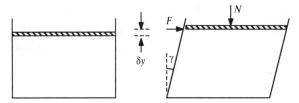

Figure 6.7 Simple shear box.

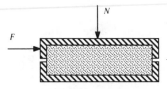

Figure 6.8 Jenike shear cell.

sample is subjected to a pressure P and the axial load N, applied through the anvils, is either increased or decreased until the material deforms. The true tri-axial tester consists of an ingenious arrangement of sliding plates enclosing a rectilinear space whose three dimensions can be varied independently. The stresses on the plates are recorded as functions of the displacements.

While these methods have considerable advantages at the high stress levels appropriate in soil mechanics, they are not ideally suited to the lower stress levels of interest in solids handling, and particularly for the low stress levels obtaining near the outlet of a hopper. Commonly, one of three designs of shear cell is used. These are known as the Jenike, the Peschl and the annular (or Walker) shear cells. Though differing considerably in geometric detail, all three cells operate in basically the same manner. The Jenike cell is effectively a split cylindrical container which is filled with the material as sketched in figure 6.8. A vertical load N is applied and the force F required to induce shear in the horizontal direction is measured. In the Walker cell, the material is held in an annular trough, the lid is loaded with a force N and the torque required to induce shear is measured. In all cases the velocities are small, an annular shear cell being typically rotated at 1.5 revolutions per hour. The prime advantage of the annular shear cell over the Jenike cell is that unlimited shear strain can be induced.

In all three cells it is assumed that the shear plane is horizontal and that the stresses are uniform across this plane. Thus in the Jenike cell the normal and shear stresses on the shear plane can be obtained from

$$\sigma = \frac{N}{A} \tag{6.5.1}$$

and

$$\tau = \frac{F}{A} \tag{6.5.2}$$

where A is the cross-sectional area of the cell. In the annular cell, equation (6.5.1) still applies but it is more convenient to work in terms of the internal and external radii r_1 and r_2 of the trough. The normal force is related to the normal stress by

$$N = \int_{r_1}^{r_2} 2\pi r \, \sigma \, dr = \pi(r_2^2 - r_1^2) \, \sigma \tag{6.5.3}$$

and the torque is related to the shear stress by

$$\Gamma = \int_{r_1}^{r_2} 2\pi r \, \tau \, r \, dr = \frac{2\pi}{3} (r_2^3 - r_1^3) \, \tau \tag{6.5.4}$$

Let us first consider a test performed in a shear box such as that shown in figure 6.7. The material is loaded into the box and sheared by the application of a force F. The resulting shear strain γ and the elevation of the lid δy are recorded. From this latter quantity the volumetric strain ε_V, defined as $-\delta V/V$, where V is the volume of the material, can be deduced. Figure 6.9, which is adapted from Atkinson and Bransby (1978), shows idealised results.

The results shown by the solid lines are obtained if the material is loaded into the cell in a dense state. Initially there is a period OA in which elastic effects dominate and a discussion of these effects is beyond the scope of this book. There is then a period ABC in which the volumetric strain increases more-or-less linearly with the shear strain. The slope of this part of the curve is denoted by $\tan v$ where v is called the angle of dilation. Beyond C, which corresponds to a shear strain of about 0.02 radians, the rate of dilation decreases and eventually becomes zero. The observation that a compacted material expands on initiation of shear is sometimes called *Reynolds Principle of Dilatancy*.

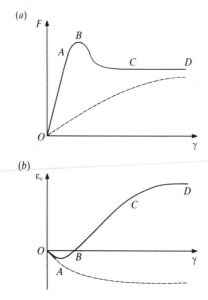

Figure 6.9 Variation of (*a*) shear force and (*b*) volumetric strain with shear strain.

The force F, required to cause shear, rises rapidly with shear strain, passes through a peak at B and then falls as the material weakens due to its decreasing density. Eventually beyond D the material shears at constant force and constant density. It should, however, be realised that some dilation has occurred before the peak force is reached so that F_{max} is a measure of the strength of the material at a density somewhat less than its initial value.

On the other hand, when the material is loaded into the shear box in a loose state, it compresses on shear and the force F rises to its asymptotic value without passing through a maximum, as shown by the dotted lines of figure 6.9.

As mentioned above, the shear box is not convenient at the low stress levels of interest in solids handling and one of the shear cells is normally used instead. The operation of these cells is most conveniently described in terms of the annular cell; the operation of the other designs is similar except that in the case of the Jenike cell the motion is linear instead of rotational.

The results of a typical annular shear cell test are shown in figure 6.10. At the beginning of a test a sample of the material is filled

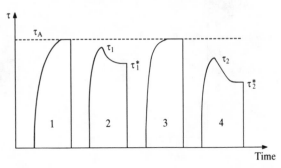

Figure 6.10 Variation of shear stress during an annular shear cell test.

loosely into the cell, the lid is replaced and loaded with a force N_A. The corresponding normal stress σ_A is found from equation (6.5.3) making allowance for the weight of the lid. The trough is then rotated and the torque is observed to rise to a value Γ_A from which we can calculate the shear stress τ_A using equation (6.5.4). This process is marked as 1 on figure 6.10 and is seen to be similar to the behaviour of a loose sample in the shear box. There are normally no facilities for measuring the change in volume in a conventional annular shear cell but we can assume that compaction has taken place and that the material now has been consolidated to a density which we will denote by ρ_A. This is the density in equilibrium with the stress combination (σ_A, τ_A) as discussed in §6.2. In principle the density ρ_A could be measured and the corresponding void fraction ε_A evaluated, but in practice this is rarely done.

The motion of the trough is now stopped and the load N_A is replaced by a *lesser* load N_1. The material does not dilate for the reasons explained in §6.2 and the voidage remains at ε_A. The trough is then rotated and the shear stress rises to a value τ_1 and then falls to τ_1^*, shown as process 2 in figure 6.10. This behaviour is seen to be similar to the behaviour of a compacted material in the shear box. It is usual to ignore any dilation that has taken place before the peak shear stress is reached and the combination (σ_1, τ_1) is taken to represent a point on the incipient yield locus for the material when it has void fraction ε_A. It must be remembered that this is the voidage the material had at the start of the shear process and not the voidage when shearing under the peak stress. The final steady stress combination (σ_1, τ_1^*) corresponds to a point on the failure locus for the equilibrium voidage which we will denote by ε_1^*. It is, however, not possible to measure this voidage, since dilation only takes place in a narrow region of unknown width in the immediate vicinity of the failure plane. Indeed we ought not to talk about a failure plane, but rather a narrow failure zone which is commonly about ten particle diameters in width.

Since the material is now in a non-uniform state, it must be re-consolidated before re-use and this can be done by re-applying the normal load N_A and rotating the trough until a steady state is achieved; process 3 of figure 6.10. A new load N_2 can then be applied and the peak τ_2 and final τ_2^* shear stresses determined. The former gives a second point on the yield locus for the material at its equilibrium voidage. By this means the complete yield locus for ϵ_A can be obtained and then, by consolidating to different stresses σ_B etc., the yield loci at ϵ_B etc. can be obtained as in figure 6.11. The points (σ, τ^*), however, lie on a single line since these were obtained after prolonged shear when the density had reached the value in equilibrium with the current stress state. This density is therefore independent of the consolidating stress. Thus on figure 6.11 we find that we have a family of incipient yield loci, each corresponding to a particular consolidating stress, but a single yield locus for sustained yield. The incipient yield loci terminate at the corresponding consolidation stress i.e. on the sustained yield locus, since if the material is subjected to a normal stress greater than the consolidation stress, the material will compress and the void fraction decrease. Thus the sustained yield locus is sometimes called the termination locus.

Very commonly, the termination locus is found to be effectively straight and its inclination to the σ axis is denoted by the angle Δ. It

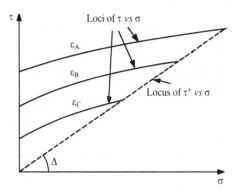

Figure 6.11 The loci of peak and asymptotic shear stresses.

is also often found that the incipient yield loci are geometrically similar, so that when expressed in terms of the reduced stresses defined by

$$\sigma_R = \sigma/\sigma_c \qquad (6.5.5)$$

and

$$\tau_R = \tau/\sigma_c \qquad (6.5.6)$$

they give a single curve as shown by the full line in figure 6.12. It must, however, be emphasised that there is no fundamental reason why the termination locus should be straight nor the yield loci geometrically similar. These observations represent a fortunate simplifying factor in the subsequent analysis. Materials for which the results conform to these two effects are sometimes called 'simple' materials.

It is clearly convenient to fit algebraic expressions to the incipient yield loci. If these are straight, expressions of the Coulomb type,

$$\tau = \mu\sigma + c \qquad (6.5.7)$$

can be used and μ and c can be determined as functions of the consolidating stress σ_c. For a 'simple' Coulomb material, μ is independent of σ_c, and c is directly proportional to σ_c.

Sometimes the yield loci show significant curvature and more complicated expressions must be used. The relationship

$$\left(\frac{\tau}{C}\right)^n = \frac{\sigma}{T} + 1 \qquad (6.5.8)$$

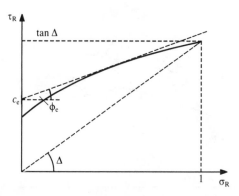

Figure 6.12 Definition of the equivalent angle of internal friction, ϕ_e, and the equivalent cohesion, C_e.

gives good results and C, T and n are determined as functions of the consolidation stress. Here C is the intercept on the τ-axis and is analogous to the cohesion, T is the intercept on the (negative) σ-axis and is referred to as the ultimate tensile strength and n is an index which controls the curvature of the line. An equation of this type was first used in the Warren Spring Laboratory and equation (6.5.8) is sometimes called the *Warren Spring Failure Criterion*. When $n = 1$, this equation reduces to the Coulomb form with $C = c$ and $T = c\cot \phi$ and for a 'simple' Warren Spring material, n is constant and C and T are directly proportional to σ_c.

In obtaining best-fit parameters from experimental results it is found that the values of T and n are strongly interrelated, mainly because the curvature of the line is most marked at low values of σ. Thus to obtain reliable values of the parameters it is convenient if an independent determination of T can be made, and apparatus is available to measure this quantity directly. From geometrical arguments, it can be shown (Nedderman, 1978) that n cannot be greater than 2 and any greater value resulting from a best-fit determination must therefore be used with caution. There seem to be no materials for which n is less than 1 and hence we can say that n is confined to the range, $1 \leq n \leq 2$.

Whether the incipient yield loci are straight or curved, they can be used in conjunction with Mohr's circle to give the limiting stress distributions in a static material using the methods described in the previous chapters. The analysis for a curved yield locus is inevitably more complicated than for a straight yield locus and therefore straight lines of the form

$$\tau = c_e + \sigma \tan \phi_e \qquad (6.5.9)$$

are often fitted over the range of stresses of interest. The slope μ_e ($= \tan \phi_e$) and intercept c_e of the 'best-fit' straight line can be used as the 'equivalent' coefficient of friction and the 'equivalent' cohesion. This normally causes little inaccuracy since the curvature of the loci is normally confined to very low stresses and at higher stresses a line of the Coulomb type often lies within the scatter of the experimental data.

Experimental difficulties are sometimes experienced when testing relatively coarse free-flowing materials as the results do not always conform to the pattern shown in figures 6.11 and 6.12. In particular, if the material is consolidated under a stress σ_c and then sheared at the same normal stress a peak shear stress is sometimes observed, and similarly if sheared under a normal stress marginally less than σ_c, peak shear stresses greater than τ_c can be found. This may be due to the voidage under the consolidating conditions being rather higher than can be maintained under static conditions. Thus on stopping the motion the material consolidates spontaneously to a greater density. The incipient yield locus, therefore, corresponds to a voidage less than that corresponding to the consolidation conditions. It is not clear that spontaneous consolidation is the only possible explanation for these effects and anomalously low values of the consolidation shear stress are also found in materials that show no detectable volume change on the cessation of shear. Under these circumstances the European Working Party on the *Mechanics of Particulate Solids* recommends that the stress state on consolidation is evaluated from the Mohr's circle which passes through the consolidation point and touches the remaining portion of the yield locus. The justification for this procedure seems questionable as will be discussed in the following section.

Other difficulties are encountered due to the parameters being somewhat time-dependent, so that the strength of the material depends on the duration of the consolidation process and the length of time elapsing between consolidation and shear. This effect may in part be due to spontaneous consolidation, but a more important factor seems to be chemical or physical changes due, for example, to the formation of cementing bridges between the particles, resulting from evaporation of salt-laden films of moisture. Thus Jenike recommends measuring both the instantaneous yield properties and those obtained as a result of 'time-consolidation'. The yield loci obtained in 'time-consolidation' tests tend to show greater cohesion and more curvature of both the incipient yield loci and the termination locus. It is these results that are of importance when starting the discharge from a hopper used for long-term storage.

A parameter of importance in hopper design is the unconfined yield stress f_c, defined in §3.4. The value of f_c can be obtained by drawing the Mohr's circle that passes through the origin and which touches the incipient yield locus under consideration, as shown in figure 3.10. Each yield locus has its own value of f_c and we can therefore obtain f_c as a function of the consolidating stress σ_c. These results form the basis of Jenike's flow function, discussed in the following sections.

The results of figures 6.3 and 6.11 can be combined as in figure 6.13. This figure can be interpreted as sections through the $(\sigma, \tau, \varepsilon)$ space within which the material is stable. In the soil mechanics literature

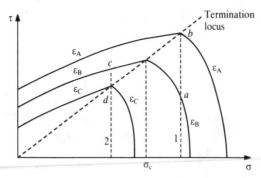

Figure 6.13 Incipient yield loci and consolidation surfaces.

it is usual to denote the surface of this space represented by the yield loci as the *Hvorslev Surface* and the consolidation loci are said to define the *Roscoe Surface*. These two surfaces intersect at a line which is called either the *Critical State Locus* or the *Critical Voids Ratio Line*. The projection of this line onto the (σ, τ) plane is the termination locus as defined above. As we saw in figure 6.9, a material approaches the critical state when subjected to sufficient shear and the branch of soil mechanics known as the *Critical State Theory* studies the behaviour at, and the approach to the critical state. Further details can be found in the pioneering work of Schofield and Wroth (1968).

Let us consider what happens when we subject a material to a stress greater than its consolidation stress σ_c and then we apply an increasing shear stress, as shown for the case of ε_B by line 1 of figure 6.13. The material will remain unchanged until the shear stress rises to a value given by point a on the consolidation line for ε_B and, thereafter, the material will compress and strengthen until it reaches the termination locus at point b. Thereafter, it will continue to shear at constant void fraction. Such a material is said to be under-consolidated and its behaviour on shear is given by the dashed lines of figure 6.9. If, however, the material is subjected to a stress less than the consolidating stress as shown by line 2, the shear stress can be increased up to point c before any strain takes place. At c the material will shear and dilate and as a result it will weaken. It will not be possible to maintain the shear stress corresponding to point c and the material will approach the termination locus from above along the line cd. Such materials are said to be over-consolidated and their stress history is similar to that shown by the full lines of figure 6.9.

When an under-consolidated material is sheared, it compresses and strengthens and consolidation is more or less uniform throughout the material. On the other hand, an over-consolidated material dilates and weakens on shear, so that further shear takes place preferentially in zones which have previously been sheared. As a consequence catastrophic failure can occur with shear being confined to a few failure zones. This is the type of behaviour we considered in §3.2 and figure 3.1 and, furthermore, it provides the justification for analyses based on the method of wedges.

6.6 Interpretation of shear cell results

Though the measurement of the failure properties of a material is experimentally difficult due to the low values of the stresses involved, the procedure described in the previous section is well-established and similar results are obtained from all types of shear cell. Furthermore, the meaning of the incipient yield loci is not in dispute. Opinions are, however, divided about the meaning of the sustained yield (or termination) locus.

Traditionally, it is assumed that the Mohr's circle describing the stress state during a consolidation or steady flow process touches the appropriate incipient yield locus *at its end point*, as shown in figure 6.14 for the particular case of a Coulomb material. This interpretation gives rise to two important concepts, the *flow function* and the *effective yield locus* which form the basis of conventional silo design. These two concepts are defined below and an alternative interpretation is given in the later stages of this section.

The principal stresses during consolidation can be found by considering figure 6.14 from which we see that

$$\tau_c = \sigma_c \tan \Delta \tag{6.6.1}$$

$$R = \tau_c \sec \phi \tag{6.6.2}$$

$$p = \sigma_c + \tau_c \tan \phi \tag{6.6.3}$$

and

$$\sigma_{c1}, \sigma_{c3} = p \pm R \tag{6.6.4}$$

These equations can be rearranged to give

$$\sigma_{c1} = \sigma_c [1 + \tan \Delta (\tan \phi + \sec \phi)] \tag{6.6.5}$$

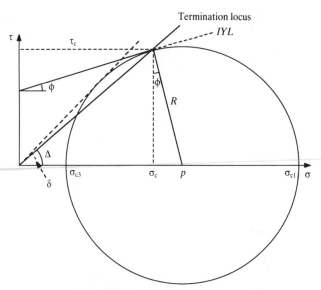

Figure 6.14 Mohr's stress circle during steady flow according to the Jenike's model; see also figure 6.16.

and

$$\sigma_{c3} = \sigma_c[1 + \tan \Delta(\tan \phi - \sec \phi)] \qquad (6.6.6)$$

Similar expressions can be used when the yield locus is curved if ϕ is replaced by ϕ_e where ϕ_e is the equivalent angle of friction, defined by equation (6.5.9), *evaluated at the end point* of the incipient yield locus.

We can therefore express the relationship we have determined experimentally between the consolidating stress σ_c and the unconfined yield stress f_c in the form of a plot of f_c *vs* σ_{c1} as in figure 6.15. This relationship is known as the flow function and forms the basis of Jenike's analysis of arching given in §10.8. Since the magnitude of the cohesion, and hence f_c tends to increase with time, it is usual to determine the flow function after various durations of time-consolidation in addition to that developed instantaneously.

For a 'simple' material all the incipient yield loci are geometrically similar and, as we saw in the previous section, the results for such materials lie on a single line when using the reduced stresses σ_R and τ_R. Thus the unconfined yield stress and the major principal consolidating stress are both directly proportional to the consolidating stress and the flow function becomes a straight line which, if

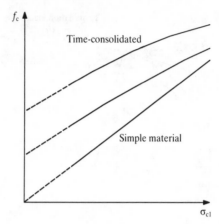

Figure 6.15 The dependence of the unconfined yield stress on the principal consolidating stress.

extrapolated, would pass through the origin. No such constraint is placed on the flow functions for time-consolidation and these normally show some curvature and, if extrapolated, have a positive intercept on the f_c axis as shown in figure 6.15. The flow function is commonly abbreviated to the symbol *FF*, but this symbol should be used with caution as sometimes it is used to denote the function relationship, $f_c = FF(\sigma_c)$ and sometimes the ratio, $FF = f_c/\sigma_c$.

The concept of the *effective yield locus* can be introduced by considering the tangent from the origin to one of the Mohr's circles in steady flow. Denoting the angle between this tangent and the σ axis by δ, we see from figure 6.14 that

$$\sin \delta = \frac{R}{p} \tag{6.6.7}$$

which from equations (6.6.1) to (6.6.3) becomes

$$\sin \delta = \frac{\sigma_c \tan \Delta \sec \phi}{\sigma_c(1 + \tan \Delta \tan \phi)} = \frac{\sin \Delta}{\cos (\Delta - \phi)} \tag{6.6.8}$$

where ϕ is replaced by ϕ_e if the yield locus is curved. It can be seen that δ is independent of σ_c and therefore all the Mohr's circles for a 'simple' material in steady flow have a common tangent which Jenike calls the effective yield locus. Since this passes through the origin, such materials behave as if they were cohesionless with an effective angle of internal friction δ. Thus for *flowing* materials the angle δ replaces the angle ϕ in all the analyses for the stress distributions in static materials presented in this book. The fact that many cohesive materials are effectively cohesionless when flowing provides a very welcome simplification to the analysis of discharging silos.

The interpretation given above does not have universal support. The *Critical State Theory* of Schofield and Wroth predicts that the yield loci must become horizontal as they approach the termination locus. This reconciles the principle of normality, discussed in §8.3 below, with the observation of steady incompressible flow. Schofield and Wroth's idealised cohesive material, 'Cam Clay', has a yield locus in p, q co-ordinates (defined by equations (6.2.20) and (6.2.21)) which is part of an ellipse with the consolidation point lying at the end of the minor axis. Whilst this seems to correlate the results for many clay-like materials adequately, it does not seem to be appropriate for the relatively free-flowing materials of interest in solids handling. Prakesh and Rao (1988) recommend the use of the yield function,

$$q = \frac{\alpha p}{\alpha - 1} \sin \phi \left[1 - \frac{1}{\alpha} \left(\frac{p}{p_c} \right)^\alpha \right] \qquad (6.6.9)$$

where p_c is the value of p at the consolidation point. This has the desired result that $dq/dp = 0$ at $p = p_c$. Experimentally, α turns out to be large (>10) in which case the numerical differences between equation (6.6.9) and the Coulomb yield criterion, equation (6.2.22), are small.

The argument that the yield locus must become horizontal at its termination point is supported by the principle of co-axiality discussed in chapter 8. Briefly, this states that during flow the principal axes of the stress and strain rate must be coincident. In steady flow in a shear cell, the direct strain rates $\dot{\varepsilon}_{xx}$ and $\dot{\varepsilon}_{yy}$ are zero, where x and y are Cartesian axes in the direction of motion and in the vertical direction respectively. The former result follows from the fact that the material in the shear zone lies between rigid blocks and the latter from the observation that dilation has ceased. The principal strain rate directions must therefore be at $\pm 45°$ to the vertical and from the principle of co-axiality the principal stress directions must also be at $\pm 45°$ to the vertical. Hence, the measured stress combination (σ_c, τ_c) in steady flow must lie at the highest point of Mohr's circle as shown in figure 6.16. The fact that this Mohr's circle is not tangential to any particular incipient yield locus is of no consequence, since incipient yield loci are not relevant for steady flow. Thus it would seem preferable to consider

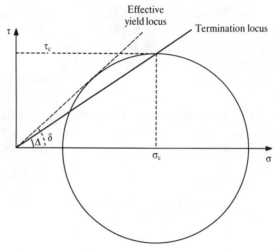

Figure 6.16 Alternative interpretation of the Mohr's circle for steady flow; see also figure 6.14.

the Mohr's circle with its centre at σ_c to that recommended by the European Working Party.

From Figure 6.16 we see that

$$\frac{\tau_c}{\sigma_c} = \tan \Delta = \sin \delta \qquad (6.6.10)$$

and the effective angle of friction can therefore be determined from the termination locus alone. On this model the principal stresses are given by

$$\sigma_{c1}, \sigma_{c3} = \sigma_c \pm \tau_c = \sigma_c(1 \pm \tan \Delta) \qquad (6.6.11)$$

and consequential changes occur in the flow function.

It is seen that the interpretation of the shear cell results is somewhat uncertain and further work is needed to clarify the situation. However, when the results of shear tests are used to predict the velocity distributions using the principle of co-axiality, it seems only reasonable to use this principle when analysing these results. On the other hand, the empirical constants in Jenike's design procedure have been determined on the basis of the more traditional analysis and this should therefore be used in conjunction with that design procedure.

6.7 Wall frictional properties

The angle of wall friction can be estimated by placing a sample of the material onto the horizontal surface in question and tilting until slip occurs. It may be necessary to confine the material in a box of negligible weight as shown in figure 6.17 and care must be taken to distinguish between the static and dynamic angles of friction. Such techniques, though convenient, are not very reproducible and it is usually preferable to adapt one of the shear cells described in §6.5 for the purpose of measuring wall frictional properties.

Any of the three basic shear cells can be adapted for wall friction measurement by replacing either the lid or the base with a sample of the wall material. The test procedure is the same as for internal failure properties. Again peak and plateau stresses are observed but usually the difference between the two is very much less than in the case of internal friction measurement. As a result, only the plateau value is usually recorded.

In internal failure, it is not clear whether the measured stresses during steady flow lie on the incipient yield locus since the failure plane need not coincide with the horizontal plane in the shear cell. However, in the case of wall friction the measured stresses are clearly those acting on the wall plane. Thus if we were to make a fully rough wall by sticking a mono-layer of the material onto the surface, the measured values of σ_w and τ_w recorded in steady flow would be the same as the values σ_c and τ_c. Hence Δ represents the value of ϕ_w for a fully rough wall so that, for a fully rough wall,

$$\tan \phi_w = \tan \Delta = \sin \delta \qquad (6.7.1)$$

and for all other walls we have that,

$$\tan \phi_w < \sin \delta \qquad (6.7.2)$$

Thus on this model, the tan of the angle of wall friction during flow can never be greater than the sine of the effective angle of internal friction. This is a more restrictive requirement than that deduced

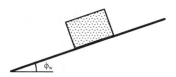

Figure 6.17 Simple method of determining the angle of wall friction.

qualitatively in §3.7, namely that the angle of wall friction must be less than or equal to the angle of incipient internal friction. For a cohesionless material, ϕ and δ are the same and these arguments provide some justification for Jenike's observation that 'experiments show that' the tan of the angle of wall friction does not exceed the sine of the angle of internal friction, i.e.

$$\tan \phi_w \leq \sin \phi \qquad (6.7.3)$$

We will see in chapter 8 that the effective angle of wall friction of a stagnant zone boundary within a silo satisfies the criterion that $\tan \phi_w = \sin \delta$, a result which supports the arguments presented above.

7

Exact stress analyses

7.1 Introduction

In this chapter we consider some more rigorous methods of predicting the stress distributions in granular materials. Many of these analyses involve the numerical solution of differential equations and the results are therefore subject to round-off errors. These can, however, be minimised by careful computation. Apart from this factor, most of the analyses are exact for a Coulomb material in incipient failure. One must, however, continuously bear in mind that the Coulomb model is a mathematical idealisation of the behaviour of real materials and that some materials may not conform, even approximately, to this concept. Furthermore, as emphasised in earlier chapters, analysis of incipient failure can only give the limits within which the stresses must lie. No attempt will be made to consider elastic–plastic analyses.

Many of the analyses presented in this chapter are relevant also to the field of soil mechanics. These analyses have been considered in detail by Sokolovskii, whose book (1965) summarises his work and provides the foundation for the topics of this chapter. Jenike and Johanson (1962) adapted these techniques to the problems of solids handling and their work has been extended by several workers, notably Hancock (1970), Horne (1977), Wilms (1984) and Benink (1989). The present chapter is based heavily on these works to which the interested reader is referred for greater detail and tabulations of results.

In §7.2, we set up the basic equations for the stress distribution and in §7.3 and §7.4 we derive particular solutions relevant at great depth in cylindrical and conical bunkers. The latter solution is known as the *Radial Stress Field* and forms the basis of Jenike's design procedure, discussed in §10.8. These solutions are compared with the approximate solutions derived in earlier chapters and §7.5 uses this comparison to assess the accuracy of the more traditional analyses.

More general solutions are most conveniently obtained by the *Method of Characteristics* and this method is presented for plane strain situations in §7.6 and used in the succeeding sections to analyse the single retaining wall (with and without surcharge), inclined walls, inclined top surfaces and twin retaining walls. The method is then extended to systems of axial symmetry in §7.11. Finally, in §7.12 and §7.13, the concept of the stress discontinuity is introduced and used to analyse the 'switch stresses' that occur at the bunker/hopper transition and at the boundary between active and passive regions.

7.2 Basic equations

When a material is in a state of either incipient or actual yield, the stresses must, by definition, satisfy some yield criterion. Of the various criteria that have been proposed, Coulomb's criterion, described in §3.2 is by far the best known and widely used and this provides the basis for the analyses presented in this chapter. An alternative, known as the Conical yield function, will be discussed in chapter 9.

As shown in §3.3 and figure 3.4, the Coulomb yield criterion can be conveniently expressed in the form

$$\sigma_{xx} = p^*(1+\sin\phi\cos 2\psi) - c\cot\phi \tag{7.2.1}$$

$$\sigma_{yy} = p^*(1-\sin\phi\cos 2\psi) - c\cot\phi \tag{7.2.2}$$

$$\tau_{yx} = -\tau_{xy} = p^*\sin\phi\sin 2\psi \tag{7.2.3}$$

Here x and y can represent any set of orthogonal axes, p^* is the distance from the centre of Mohr's circle to the point at which the Coulomb line touches the σ axis and ψ is the angle measured anticlockwise from the x-axis to the major principal stress direction. For a cohesionless material p^* equals p, the mean of the major and minor principal stresses.

The equations for static equilibrium can be obtained by resolving the forces in the co-ordinate directions and this was done in §2.4 for a set of Cartesian axes with the y-axis directed vertically downwards. The resulting equations, equations (2.4.4) and (2.4.5), are

$$\frac{\partial\sigma_{xx}}{\partial x} + \frac{\partial\tau_{yx}}{\partial y} = 0 \tag{7.2.4}$$

$$\frac{\partial \sigma_{yy}}{\partial y} + \frac{\partial \tau_{yx}}{\partial x} = \gamma \tag{7.2.5}$$

The forms for other co-ordinate systems are given in appendix 2. Substituting from equations (7.2.1) to (7.2.3) gives

$$(1 + \sin \phi \cos 2\psi) \frac{\partial p^*}{\partial x} - 2p^* \sin \phi \sin 2\psi \frac{\partial \psi}{\partial x} + \sin \phi \sin 2\psi \frac{\partial p^*}{\partial y}$$

$$+ 2p^* \sin \phi \cos 2\psi \frac{\partial \psi}{\partial y} = 0 \tag{7.2.6}$$

and

$$\sin \phi \sin 2\psi \frac{\partial p^*}{\partial x} + 2p^* \sin \phi \cos 2\psi \frac{\partial \psi}{\partial x} + (1 - \sin \phi \cos 2\psi) \frac{\partial p^*}{\partial y}$$

$$+ 2p^* \sin \phi \sin 2\psi \frac{\partial \psi}{\partial y} = \gamma \tag{7.2.7}$$

It should be noted that the cohesion c does not appear explicitly in these equations which are therefore the same for both cohesive and cohesionless materials. However, the boundary conditions normally depend on the magnitude of the cohesion and therefore the solutions to these equations are functions of the cohesion. As explained in §6.6, many materials are effectively cohesionless in steady flow and under these circumstances the effective angle of friction δ is used instead of the static angle of friction ϕ, and p instead of p^*.

Equations (7.2.6) and (7.2.7) can be expressed in any co-ordinate system. The forms in the co-ordinate systems most frequently encountered in the study of granular materials are given at the end of this section, without derivation, since this follows exactly the lines given above for the Cartesian system. There is, however, one point of difficulty in axi-symmetric systems which requires explanation.

In axi-symmetric systems the circumferential stress σ_{xx} appears in the equations of static equilibrium. By symmetry, this must be a principal stress but its magnitude cannot be obtained directly from the Mohr–Coulomb analysis since this is effectively a two-dimensional concept. The Mohr–Coulomb analysis merely states that

$$\sigma_1 \geq \sigma_{xx} \geq \sigma_3 \tag{7.2.8}$$

where σ_1 and σ_3 are the major and minor principal stresses of the Mohr's circle that touches the Coulomb yield locus.

However, Haar and von Karman have argued that σ_{xx} must equal either σ_1 or σ_3. This result, which is known as the *Haar–von Karman hypothesis* was originally derived by considerations of minimising the strain energy but can be seen physically by considering the slow discharge of a material from a conical hopper as shown in figure 7.1. Let us consider an element of material with edges parallel to the co-ordinate axes (r,θ,χ). In cross-section, view *A*, this element is shown as approximately square but as discharge proceeds it becomes compressed laterally and extended axially as shown by the dashed lines. Since shear is taking place in the r,θ plane the stresses σ_{rr} and $\sigma_{\theta\theta}$ must lie on a Mohr's circle that touches the Coulomb line and the major and minor principal stresses for this Mohr's circle are σ_1 and σ_3. The side elevation, shown as view *B*, is similar, showing that σ_{xx}

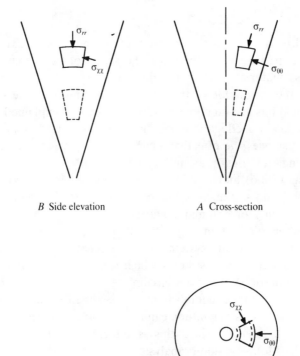

B Side elevation   *A* Cross-section

C Plan

Figure 7.1 Distortion of an element as it passes through a conical hopper.

must also lie on a circle touching the yield criterion. However, we see from the plan view C that in the θ,χ plane, the element merely contracts without distortion and the Mohr's circle for this plane need not touch the Coulomb line. Thus we conclude that whenever shear occurs in two mutually perpendicular planes, two of the three Mohr's circles must touch the Coulomb line and must therefore be identical. The third Mohr's circle degenerates into a point. Thus the circumferential stress, being a principal stress, must equal one of the other principal stresses, i.e.

$$\sigma_{xx} = \sigma_1 \text{ or } \sigma_3 \qquad (7.2.9)$$

or

$$\sigma_{xx} = p^*(1 + \kappa \sin \phi) - c \cot \phi \qquad (7.2.10)$$

where κ takes the values ± 1 depending on whether σ_{xx} is the major or minor principal stress. It can be seen from figure 7.1 that in converging flow, $\sigma_{xx} > \sigma_{rr}$ and therefore in this case $\kappa = +1$. A more rigorous mathematical proof that $\kappa = +1$ in converging flow is given in §7.3. However, to preserve generality, we will retain κ in the equations listed below. As implied by its name, the Haar–von Karman hypothesis is not subject to rigorous proof and alternative methods of evaluating σ_{xx} are given in chapter 9.

In the equations below we use the symbol ψ to denote the angle measured anticlockwise from the x-direction in Cartesian co-ordinates (or the r-direction in cylindrical co-ordinates) to the major principal stress direction. When using polar or spherical co-ordinates we will denote the angle between the r-direction and the major principal stress direction by ψ^*. Comparison of figure A2(a) with A2(b), or A2(c) with A2(d) shows that

$$\psi^* = \psi + 90 - \theta \qquad (7.2.11)$$

Basic stress equations

Polar co-ordinates r, θ with the $\theta = 0$ line directed vertically upwards
The equations of static equilibrium are

$$\frac{\partial \sigma_{rr}}{\partial r} + \frac{\sigma_{rr} - \sigma_{\theta\theta}}{r} + \frac{1}{r}\frac{\partial \tau_{\theta r}}{\partial \theta} + \gamma \cos \theta = 0 \qquad (7.2.12)$$

$$\frac{1}{r}\frac{\partial \sigma_{\theta\theta}}{\partial \theta} + \frac{\partial \tau_{\theta r}}{\partial r} + \frac{2\tau_{\theta r}}{r} - \gamma \sin \theta = 0 \qquad (7.2.13)$$

The stress components are given by

$$\sigma_{rr} = p^*(1 + \sin \phi \cos 2\psi^*) - c \cot \phi \qquad (7.2.14)$$

$$\sigma_{\theta\theta} = p^*(1 - \sin \phi \cos 2\psi^*) - c \cot \phi \qquad (7.2.15)$$

$$\tau_{\theta r} = -\tau_{r\theta} = p^* \sin \phi \sin 2\psi^* \qquad (7.2.16)$$

Hence

$$(1 + \sin \phi \cos 2\psi^*) \frac{\partial p^*}{\partial r} - 2p^* \sin \phi \sin 2\psi^* \frac{\partial \psi^*}{\partial r} + \frac{\sin \phi \sin 2\psi^*}{r} \frac{\partial p^*}{\partial \theta}$$

$$+ \frac{2p^* \sin \phi \cos 2\psi^*}{r} \frac{\partial \psi^*}{\partial \theta} + \frac{2p^* \sin \phi \cos 2\psi^*}{r} + \gamma \cos \theta = 0 \qquad (7.2.17)$$

$$\sin \phi \sin 2\psi^* \frac{\partial p^*}{\partial r} + 2p^* \sin \phi \cos 2\psi^* \frac{\partial \psi^*}{\partial r} + \frac{(1 - \sin \phi \cos 2\psi^*)}{r} \frac{\partial p^*}{\partial \theta}$$

$$+ \frac{2p^*}{r} \sin \phi \sin 2\psi^* \frac{\partial \psi^*}{\partial \theta} + \frac{2p^* \sin \phi \sin 2\psi^*}{r} - \gamma \sin \theta = 0 \qquad (7.2.18)$$

Cylindrical co-ordinates r, z, χ with the axis of symmetry, z, directed vertically downwards

The equations of static equilibrium are

$$\frac{\partial \sigma_{rr}}{\partial r} + \frac{\sigma_{rr} - \sigma_{\chi\chi}}{r} + \frac{\partial \tau_{zr}}{\partial z} = 0 \qquad (7.2.19)$$

$$\frac{\partial \sigma_{zz}}{\partial z} + \frac{\partial \tau_{zr}}{\partial r} + \frac{\tau_{zr}}{r} = \gamma \qquad (7.2.20)$$

The stress components are given by

$$\sigma_{rr} = p^*(1 + \sin \phi \cos 2\psi) - c \cot \phi \qquad (7.2.21)$$

$$\sigma_{zz} = p^*(1 - \sin \phi \cos 2\psi) - c \cot \phi \qquad (7.2.22)$$

$$\tau_{zr} = -\tau_{rz} = p^* \sin \phi \sin 2\psi \qquad (7.2.23)$$

$$\sigma_{\chi\chi} = p^*(1 + \kappa \sin \phi) - c \cot \phi \qquad (7.2.24)$$

Hence

$$(1 + \sin \phi \cos 2\psi) \frac{\partial p^*}{\partial r} - 2p^* \sin \phi \sin 2\psi \frac{\partial \psi}{\partial r} + \sin \phi \sin 2\psi \frac{\partial p^*}{dz}$$

$$+ 2p^* \sin \phi \cos 2\psi \frac{\partial \psi}{\partial z} + \frac{p^*}{r} \sin \phi \, (\cos 2\psi - \kappa) = 0 \qquad (7.2.25)$$

$$\sin \phi \sin 2\psi \frac{\partial p^*}{\partial r} + 2p^* \sin \phi \cos 2\psi \frac{\partial \psi}{\partial r} + (1 - \sin \phi \cos 2\psi) \frac{\partial p^*}{\partial z}$$

$$+ 2p^* \sin \phi \sin 2\psi \frac{\partial \psi}{\partial z} + \frac{p^*}{r} \sin \phi \sin 2\psi = \gamma \qquad (7.2.26)$$

Spherical co-ordinates, r, θ, χ with the $\theta = 0$ line vertically upwards
The equations of static equilibrium are

$$\frac{\partial \sigma_{rr}}{\partial r} + \frac{2\sigma_{rr} - \sigma_{\theta\theta} - \sigma_{xx}}{r} + \frac{1}{r} \frac{\partial \tau_{\theta r}}{\partial \theta} + \frac{\tau_{\theta r}}{r} \cot \theta + \gamma \cos \theta = 0 \quad (7.2.27)$$

$$\frac{\partial \tau_{\theta r}}{\partial r} + \frac{1}{r} \frac{\partial \sigma_{\theta\theta}}{\partial \theta} + \frac{3\tau_{\theta r}}{r} + \frac{(\sigma_{\theta\theta} - \sigma_{xx})}{r} \cot \theta - \gamma \sin \theta = 0 \qquad (7.2.28)$$

The stress components are given by equations (7.2.14 to 16) and (7.2.24) and on substitution into equations (7.2.27) and (7.2.28) we have

$$(1 + \sin \phi \cos 2\psi^*) \frac{\partial p^*}{\partial r} - 2p^* \sin \phi \sin 2\psi^* \frac{\partial \psi^*}{\partial r} + \frac{1}{r} \sin \phi \sin 2\psi^* \frac{\partial p^*}{\partial \theta}$$

$$+ \frac{2p^*}{r} \sin \phi \cos 2\psi^* \frac{\partial \psi^*}{\partial \theta} + \frac{p^* \sin \phi}{r} (3 \cos 2\psi^* - \kappa + \sin 2\psi^* \cot \theta)$$

$$+ \gamma \cos \theta = 0 \qquad (7.2.29)$$

$$\sin \phi \sin 2\psi^* \frac{\partial p^*}{\partial r} + 2p^* \sin \phi \cos 2\psi^* \frac{\partial \psi^*}{\partial r} + \frac{1}{r} (1 - \sin \phi \cos 2\psi^*) \frac{\partial p^*}{\partial \theta}$$

$$+ \frac{2p^*}{r} \sin \phi \sin 2\psi^* \frac{\partial \psi^*}{\partial \theta} + \frac{p^*}{r} \sin \phi \, (3 \sin 2\psi^* - \cos 2\psi^* \cot \theta$$

$$- \kappa \cot \theta) - \gamma \sin \theta = 0 \qquad (7.2.30)$$

7.3 Particular solutions

There are very few analytic solutions to the stress equations presented in the previous section, and almost always we must resort to numerical methods. For this, the *Method of Characteristics* outlined in §7.6 below is often the most convenient. However, there do exist a few situations in which solutions can be obtained by more straightforward means. We can obtain inspiration about the form of these solutions by considering the approximate analyses of Janssen and Walker presented in §5.2 and §5.8. In these analyses the full solution tends to an asymptote given by the particular integral of the governing differential equation. Since the exact stress equations given in §7.2 are non-linear, their solutions cannot be divided in particular integral and complementary functions but, none-the-less, we can look for solutions to these equations which are analogous to the particular integrals of the equations resulting from the approximate analyses.

In the Janssen analysis the stresses tend to constant values at great depth and in the Walker analysis they tend to an asymptote which is linear with distance from the apex of the cone or wedge. Thus in the later parts of this section we will look for solutions in which the stresses are independent of depth, in the expectation that these will be the asymptotes to which the stresses tend at great depth in parallel-sided and cylindrical bunkers. In §7.4 we will consider the asymptotes appropriate to conical and wedge-shaped geometries. Because of the great importance of these latter two solutions they have acquired the name of *The Radial Stress Field* even though they are not principally concerned with radial stresses.

Parallel planes

Janssen's analysis for material confined between parallel planes, presented in §5.5, shows that at great depth the stresses become independent of the vertical co-ordinate. Thus we can obtain the exact solution for this asymptote by putting all the derivatives with respect to y equal to zero in equations (7.2.6) and (7.2.7). In fact it turns out to be more convenient to start with equations (7.2.4) and (7.2.5) which become

$$\frac{d\sigma_{xx}}{dx} = 0 \qquad (7.3.1)$$

$$\frac{d\tau_{yx}}{dx} = \gamma \qquad (7.3.2)$$

Equation (7.3.1) has solution σ_{xx} is constant and equation (7.3.2) integrates to

$$\tau_{yx} = \gamma x + A \qquad (7.3.3)$$

where A is an arbitrary constant. If we take our origin of co-ordinates on the mid-plane, $A = 0$ by symmetry, and if the spacing of the planes is $2a$, we find that the wall shear stress is given by

$$(\tau_{yx})_w = \gamma a \qquad (7.3.4)$$

From the wall yield criterion, equation (3.7.2), and the result that σ_{xx} is constant, we have

$$\sigma_{xx} = (\sigma_{xx})_w = \frac{(\tau_{yx})_w - c_w}{\mu_w} = \frac{\gamma a - c_w}{\mu_w} \qquad (7.3.5)$$

Thus, from equations (7.2.3) and (7.2.1), we have

$$\gamma x = \tau_{yx} = p^* \sin \phi \sin 2\psi \qquad (7.3.6)$$

and

$$(\gamma a - c_w) \cot \phi_w = \sigma_{xx} = p^*(1 + \sin \phi \cos 2\psi) - c \cot \phi \quad (7.3.7)$$

which can be readily solved for p^* and ψ as functions of x. We can therefore evaluate σ_{yy} from equation (7.2.2), giving, after some manipulation,

$$[(\gamma a - c_w) \cot \phi_w - \sigma_{yy}]^2 + (2\gamma x)^2 =$$
$$[(\gamma a - c_w) \cot \phi_w + 2c \cot \phi + \sigma_{yy}]^2 \sin^2 \phi \qquad (7.3.8)$$

This is seen to be a quadratic in σ_{yy}, the two solutions of which correspond to the active and passive cases.

Despite the existence of the analytic solution presented above, it is worthwhile considering a numerical solution to equations (7.2.6) and (7.2.7) as this gives a simple introduction to the methods we must use in the more complicated analyses of the following sections. Readers who are interested in understanding the solution of the radial stress field equations are recommended to master this section before proceeding to §7.4.

Putting derivatives with respect to y equal to zero in equations (7.2.6) and (7.2.7) gives

$$(1 + \sin \phi \cos 2\psi) \frac{dp^*}{dx} - 2p^* \sin \phi \sin 2\psi \frac{d\psi}{dx} = 0 \qquad (7.3.9)$$

and

$$\sin \phi \sin 2\psi \frac{dp^*}{dx} + 2p^* \sin \phi \cos 2\psi \frac{d\psi}{dx} = \gamma \qquad (7.3.10)$$

where the partial derivatives with respect to x have become ordinary derivatives since we no longer have any dependence on y.

Multiplying equation (7.3.9) by $\cos 2\psi$ and adding equation (7.3.10) multiplied by $\sin 2\psi$ gives

$$\frac{dp^*}{dx} = \frac{\gamma \sin 2\psi}{\sin \phi + \cos 2\psi} \qquad (7.3.11)$$

and similarly we find that

$$\frac{d\psi}{dx} = \frac{\gamma(1 + \sin \phi \cos 2\psi)}{2p^* \sin \phi (\sin \phi + \cos 2\psi)} \qquad (7.3.12)$$

These equations are slightly more convenient if we define dimensionless quantities as follows

$$x = aX \qquad (7.3.13)$$

$$p^* = \gamma aP \qquad (7.3.14)$$

giving

$$\frac{dP}{dX} = \frac{\sin 2\psi}{\sin \phi + \cos 2\psi} \qquad (7.3.15)$$

and

$$\frac{d\psi}{dX} = \frac{(1 + \sin \phi \cos 2\psi)}{2P \sin \phi (\sin \phi + \cos 2\psi)} \qquad (7.3.16)$$

Given suitable boundary conditions these equations can be solved by many standard numerical integration techniques. Of these the simplest, Euler integration, is illustrated in example (7.3.1) below. This method is, however, of limited accuracy and Runge–Kutta integration is recommended if greater accuracy is required. Details of these techniques can be found in many standard texts on numerical methods. Alternatively, the reader may prefer to use one of the many computerised integration packages commercially available.

Equations (7.3.15) and (7.3.16) require two boundary conditions and normally these are the known values of ψ at the centre-line and at the wall. On the centre-line ψ equals 90° in the active case and 0° in the passive case and at the wall we can determine ψ from the wall angle of friction. Bearing in mind our co-ordinate system, the wall at $x = a$, i.e. $X = 1$ is the left-hand wall on which the shear stress will be negative. The appropriate Mohr's circle for the active failure of a cohesionless material adjacent to the wall is given in figure 7.2, from which it can be seen that

$$\text{on } X = 1, \psi = \psi_w = 90 - \tfrac{1}{2}(\omega - \phi_w) \qquad (7.3.17)$$

where ψ is measured in degrees and ω is given by equation (3.7.8).

These boundary conditions are inconvenient since they are not at the same value of X and we have to proceed by trial and error. We can guess a value for P on $X = 0$, integrate the equations up to $X = 1$ and compare the resulting value of ψ with that given by equation (7.3.17). If the value is not correct we must restart with a new value of P. Such a procedure is known as a shooting technique. Selecting a value for P fixes the initial slope of the trajectory, which we integrate to see whether we have hit our target.

Example 7.3.1 As an example we will start with a related, though less realistic, problem in which we know the value of P at $X = 0$ and wish to find the angle of wall friction to which this corresponds. Let us consider active failure in a cohesionless material with $\phi = 30° = \pi/6$ radians and assume that at $X = 0$, $\psi = 90° = \pi/2$ radians and $P = 5.0$. Since equations (7.3.15) and (7.3.16) are not amenable to analytic

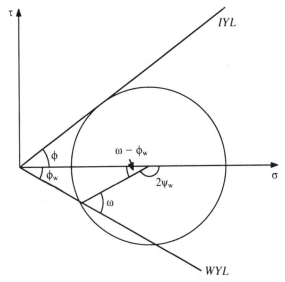

Figure 7.2 Mohr's circle for the evaluation of the stress parameter ψ at the wall.

solution, we will attempt a numerical solution using simple Euler integration. Since the values of P and ψ are known at $X = 0$ we can evaluate dP/dX and $d\psi/dX$ at $X = 0$ and, taking a step length $\triangle X$ we can deduce the values at $X = X_1 = \triangle X$ as follows

$$P_1 = P_0 + \frac{dP}{dX} \triangle X \qquad (7.3.18)$$

$$\psi_1 = \psi_0 + \frac{d\psi}{dX} \triangle X \qquad (7.3.19)$$

It should be noted that the values of ψ in the derivative $d\psi/dX$ are inevitably in radians and therefore radians must be used in equation (7.3.19).

We can now evaluate the derivatives at X_1 and find P_2 and ψ_2 by the same method. We will illustrate the method using a step length of 0.1. Thus at $X = 0$,

$$\frac{dP}{dX} = \frac{\sin 2\psi}{\sin \phi + \cos 2\psi} = \frac{\sin \pi}{\sin \pi/6 + \cos \pi} = 0$$

$$\frac{d\psi}{dX} = \frac{(1 + \sin \phi \cos 2\psi)}{2P \sin \phi (\sin \phi + \cos 2\psi)}$$

$$= \frac{(1 + \sin \pi/6 \cos \pi)}{2 \times 5.0 \times \sin \pi/6 \times (\sin \pi/6 + \cos \pi)} = -0.200$$

Thus from equations (7.3.18) and (7.3.19)

$$P_1 = P_0 + \frac{dP}{dX} \triangle X = 5.0 + 0 \times 0.1 = 5.0$$

$$\psi_1 = \psi_0 + \frac{d\psi}{dX} \triangle X = \frac{\pi}{2} - 0.200 \times 0.1 = 1.5508 \text{ radians}$$

and, of course,

$$X_1 = \triangle X = 0.1$$

Repeating the procedure with the new values of P and ψ at X_1 gives

$$\frac{dP}{dX} = -0.0801 \text{ and } \frac{d\psi}{dX} = -0.2005$$

Thus,

$$P_2 = P_1 + \frac{dP}{dX} \triangle X = 5.0 - 0.0801 \times 0.1 = 4.9920$$

$$\psi_2 = \psi_1 + \frac{\mathrm{d}\psi}{\mathrm{d}X}\,\triangle X = 1.5508 - 0.2005 \times 0.1 = 1.5308 \text{ radians}$$

$$X_2 = 2\,\triangle X = 0.2$$

This process can be repeated up to $X_{10} = 1$ giving

$$P_{10} = 4.596\,98, \psi_{10} = 1.349\,21 \text{ radians}$$

Euler integration is not particularly accurate unless very small step lengths are taken, and as an illustration of this we can compare the results presented above with the values obtained by 100 steps of length 0.01:

$$P = 4.541\,77, \text{ and } \psi = 1.343\,21 \text{ radians on } X = 1$$

and with 500 steps of 0.002:

$$P = 4.536\,51, \text{ and } \psi = 1.342.60 \text{ radians on } X = 1$$

These figures give some impression of the accuracy of the method. Clearly if only a rough estimate is required, a handful of large steps is sufficient but for more accurate work many small steps are needed and this makes the use of a computer inevitable. As mentioned above, the Runge–Kutta method gives more accurate results with less effort, and, using only ten steps of fourth order Runge–Kutta integration we find that $P = 4.535\,18$ and $\psi = 1.342\,44$ on $X = 1$ with both of these being accurate to at least the fifth significant figure.

We can find the angle of wall friction from the calculated value of ψ_w since

$$\tan \phi_w = -\frac{\tau_w}{\sigma_w} = \frac{\sin \phi \sin 2\psi_w}{1 + \sin \phi \cos 2\psi_w} \tag{7.3.20}$$

and hence using $\psi_w = 1.342\,44$ radians we find that $\phi_w = 21.801°$.

Example 7.3.2 As an alternative to the unrealistic specification of P on $X = 0$ used in the previous example, let us evaluate the stresses for the case when the wall angle of friction is 20°.

We have already shown that $P_0 = 5.0$ gives $\phi_w = 21.801°$, and we can repeat the calculation with a different value of P_0. Taking $P_0 = 5.5$ gives $\psi_w = 1.368\,92$ radians and hence $\phi_w = 19.9983°$. Linear interpolation between these values of ϕ_w suggests that we try $P_0 = 5.4953$ which yields $\phi_w = 19.9988°$. Proceeding in this manner, we find eventually that when $P_0 = 5.4949$, $\phi_w = 20.0002°$ which is well beyond the accuracy to which ϕ_w could be measured. The resulting profiles of P and ψ are given in rows two and three of table 7.1. From these we can evaluate the stress components from equations (7.2.1) to (7.2.3) and compare them with the analytic solution of equations (7.3.4) and (7.3.5). It is seen that the agreement between the numerical and analytic results is excellent.

This last example has been somewhat artificial since we know that the asymptotic shear stress $\tau_w = \gamma a$ and that $\phi_w = 20°$. Thus we could have started our calculation at $X = 1$ with the boundary conditions

$$\psi_w = 90 - \tfrac{1}{2}(\omega - \phi_w) = 78.4199°$$

and

$$P = \frac{1}{\sin \phi \sin 2\psi_w} = 5.085\ 14$$

This would have given the stress distribution in a single stage without the necessity of using a shooting technique. The results of this calculation, again using ten fourth order Runge–Kutta steps, differ from those in table 7.1 only in the sixth significant figure and agree with the known value of $\psi = 90°$ on the centre-line to at least seven significant figures. We can conclude therefore that Runge–Kutta integration gives results to greater precision than the original data justify.

Table 7.1

X	0.0	0.2	0.4	0.6	0.8	1.0
		Numerical solution				
P	5.4949	5.4803	5.4357	5.3588	5.2449	5.0851
ψ	1.5708	1.5343	1.4969	1.4579	1.4158	1.3687
$\tau_{yx}/\gamma a$	0.0000	0.2000	0.4000	0.6000	0.8000	1.0000
$\sigma_{xx}/\gamma a$	2.7475	2.7475	2.7475	2.7475	2.7475	2.7475
$\sigma_{yy}/\gamma a$	8.2424	8.2131	8.1240	7.9702	7.7423	7.4227
		Analytic solution				
$\tau_{yx}/\gamma a$	0.0000	0.2000	0.4000	0.6000	0.8000	1.0000
$\sigma_{xx}/\gamma a$	2.7475	2.7475	2.7475	2.7475	2.7475	2.7475
$\sigma_{yy}/\gamma a$	8.2424	8.2130	8.1238	7.9698	7.7414	7.4222

Cylindrical bunkers

The solution for the asymptotic stresses in a cylindrical bunker can be obtained by a similar analysis. Putting derivatives with respect to z equal to zero in equation (7.2.20) gives

$$\frac{d\tau_{zr}}{dr} + \frac{\tau_{zr}}{r} = \gamma \qquad (7.3.21)$$

which has solution, subject to $\tau_{zr} = 0$ on the axis of symmetry, of

$$\tau_{zr} = \frac{\gamma r}{2} \qquad (7.3.22)$$

However, a similar procedure with equation (7.2.19) gives

$$\frac{d\sigma_{rr}}{dr} + \frac{\sigma_{rr} - \sigma_{xx}}{r} = 0 \qquad (7.3.23)$$

which cannot be integrated directly since the relationship between σ_{rr} and σ_{xx} is as yet unknown. We must therefore resort to a numerical method.

From equations (7.2.25) and (7.2.26), putting derivative with respect to z equal to zero, we have

$$(1 + \sin\phi\cos 2\psi)\frac{dp^*}{dr} - 2p^*\sin\phi\sin 2\psi\frac{d\psi}{dr}$$

$$+ \frac{p^*}{r}\sin\phi\,(\cos 2\psi - \kappa) = 0 \qquad (7.3.24)$$

$$\sin\phi\sin 2\psi\frac{dp^*}{dr} + 2p^*\sin\phi\cos 2\psi\frac{d\psi}{dr} + \frac{p^*}{r}\sin\phi\sin 2\psi = \gamma \qquad (7.3.25)$$

Rearranging as in the Cartesian case above, we have

$$\frac{dP}{dR} = \frac{\sin 2\psi - \dfrac{P}{R}\sin\phi\,(1 - \kappa\cos 2\psi)}{\sin\phi + \cos 2\psi} \qquad (7.3.26)$$

and

$$\frac{d\psi}{dR} = \frac{(1 + \sin\phi\cos 2\psi) - \dfrac{P}{R}\sin\phi\,(1 + \kappa\sin\phi)\sin 2\psi}{2P\sin\phi\,(\sin\phi + \cos 2\psi)} \qquad (7.3.27)$$

where $R = r/a$, $P = p^*/\gamma a$ and a is the radius of the cylinder.

These two equations can be solved by the techniques described in the earlier parts of this section. As before, the probable boundary conditions are specified values of ψ on the centre-line and at the wall.

One problem, however, arises on starting the integration at the centre-line, $R = 0$, since the denominators in the groups $(1-\kappa \cos 2\psi)/R$ and $\sin 2\psi/R$ are zero. To prevent infinite gradients, the numerators must also be zero. By symmetry ψ on the centre-line is either $90°$ (in the active case) or $0°$ (passive). In both cases no problem occurs with the group $\sin 2\psi/R$ but it is apparent that the group $(1-\kappa \cos 2\psi)/R$ requires that $\kappa = -1$ when $\psi = 90°$ i.e. in the active case, and $\kappa = +1$ when $\psi = 0°$, confirming the result we derived qualitatively in §7.2.

Since both the numerator and the denominator are zero we must evaluate these groups using l'Hopital's rule. This gives

$$\frac{1 - \kappa \cos 2\psi}{R} = 2\,\kappa \sin 2\psi \frac{d\psi}{dR} = 0 \qquad (7.3.28)$$

and

$$\frac{\sin 2\psi}{R} = 2 \cos 2\psi \frac{d\psi}{dR} = 2\,\kappa \frac{d\psi}{dR} \qquad (7.3.29)$$

Thus on the centre-line, equation (7.3.26) becomes

$$\frac{dP}{dR} = \frac{\sin 2\psi}{\sin \phi + \cos 2\psi} = 0 \qquad (7.3.30)$$

and equation (7.3.27) becomes

$$\frac{d\psi}{dR} = \frac{1 + \sin \phi \cos 2\psi - 2P \kappa \sin \phi(1 + \kappa \sin \phi)\dfrac{d\psi}{dR}}{2P \sin \phi\,(\sin \phi + \cos 2\psi)} \qquad (7.3.31)$$

On rearrangement, and putting $\psi = 0°$ or $90°$, equation (7.3.31) becomes

$$\frac{d\psi}{dR} = \frac{\kappa}{4P \sin \phi} \qquad (7.3.32)$$

Equations (7.3.30) and (7.3.32) must be used instead of equations (7.3.26) and (7.3.27) when R is equal or very close to zero.

It is noteworthy that the corresponding equations in Cartesian co-ordinates, equations (7.3.15) and (7.3.16) reduce to

$$\frac{dP}{dX} = 0 \quad \text{and} \quad \frac{d\psi}{dX} = \frac{\kappa}{2P \sin \phi}$$

on $X = 0$. It is a common feature of many topics that on the centre-line, the gradients in axial symmetry are one-half those in the equivalent two-dimensional situation.

The asymptotic stress distribution in a cylindrical bunker has been obtained by solving equations (7.3.26) and (7.3.27) for the same boundary conditions as in example 7.3.2 and the results are given in table 7.2. Examination of the results of this table show that $\tau_{zr} = \gamma r/2$ as predicted by equation (7.3.22) and that σ_{rr} is not a constant unlike σ_{xx} in the two-dimensional case. This result casts doubt on the accuracy of Walker's distribution factor, presented in §5.4, since that analysis was based on the assumption that σ_{rr} was constant. Walker's analysis is reconsidered in §7.5.

Fully rough walls

One problem has, however, been glossed over in the earlier part of this section. Equations (7.3.15) and (7.3.16) contain the factor $(\sin \phi + \cos 2\psi)$ in the denominator. Thus if at any stage in the calculation $\cos 2\psi$ should happen to equal $-\sin \phi$ (i.e. if $2\psi = 90+\phi$), the denominator will be zero and the derivatives will become infinite. Both the Euler and Runge–Kutta integration schemes become unworkable under these circumstances. In both the Cartesian and cylindrical cases, the condition that $2\psi = 90+\phi$ can occur only at a fully rough wall, i.e. when $\phi = \phi_w$, and therefore is of rare occurrence. However, in these circumstances we can readily obtain a solution by changing our independent variable from X to ψ. Taking the reciprocal of equation (7.3.16) gives

$$\frac{dX}{d\psi} = \frac{2P \sin \phi \, (\sin \phi + \cos 2\psi)}{1 + \sin \phi \cos 2\psi} \tag{7.3.33}$$

and dividing equation (7.3.15) by (7.3.16) yields

$$\frac{dP}{d\psi} = \frac{2P \sin \phi \sin 2\psi}{1 + \sin \phi \cos 2\psi} \tag{7.3.34}$$

Table 7.2

R	0.0	0.2	0.4	0.6	0.8	1.0
$\tau_{zr}/\gamma a$	0.0000	0.1000	0.2000	0.3000	0.4000	0.5000
$\sigma_{rr}/\gamma a$	1.4210	1.4192	1.4139	1.4048	1.3916	1.3737
$\sigma_{zz}/\gamma a$	4.2630	4.2435	4.1842	4.0815	3.9285	3.7114

Now, we can integrate these equations taking steps in ψ subject to the boundary conditions that on $\psi = 90°$, $X = 0$ and P is some assumed value P_0. Figure 7.3 shows the results for a material for which $\phi = 30°$ and various values of P_0. It can be seen that all the curves pass through maxima at the critical value of $\psi = \frac{1}{2}(90+\phi) = 60°$.

When $P_0 = 4.5$ we find that $X = 1$ when $\psi = 75.0°$ which, from (7.3.20) gives $\phi_w = 23.8°$. When $P_0 = 3.464$, the curve touches the line $X = 1$ at the critical value of $60°$ showing that this is the value of P_0 for the fully rough wall. For values of $P_0 < 3.464$ the line never reaches $X = 1$ as illustrated by the case of $P_0 = 2.5$, showing that such values of P_0 are unrealisable. Any attempt to use the original integration scheme starting with values of $P_0 < 3.464$ would become unworkable before X reached 1. Similar effects occur in cylindrical symmetry and can be resolved by the same technique.

Cohesive materials

Finally we must consider the procedure for cohesive materials. Since the cohesion c does not appear in the differential equations the only factor affected is the nature of the boundary conditions. That on the axis of symmetry, $\psi = 90°$ (or $0°$ in the passive case), remains unchanged but on the wall we have

$$- \tau_w = \sigma_w \tan \phi_w + c_w \qquad (7.3.35)$$

i.e.

$$p_w^* \sin \phi \sin 2\psi_w = p_w^* (1 + \sin \phi \cos 2\psi_w) \tan \phi_w + c_w \quad (7.3.36)$$

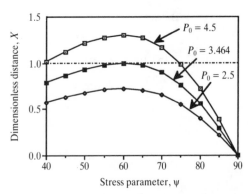

Figure 7.3 Variation of the stress parameter ψ with dimensionless distance X. Note that for values of $P_0 < 3.464$ the solution does not extend up to $X = 1$.

and we must select values of p^* on the centre-line so that the predicted values of ψ_w *and* p_w^* satisfy this equation.

7.4 The radial stress field

The conical hopper

We saw in §5.8 that Walker's analysis for the stress distribution in a cohesionless material contained in a conical hopper, suggests that the stresses tend to an asymptote which is linear with distance from the apex. We can therefore look for an exact solution to this asymptote by starting with equations (7.2.29) and (7.2.30) and making the assumption that all the stress components are proportional to r and are some, as yet unknown, function of θ. At the same time it is convenient to make the equations dimensionless by defining q by

$$p = \gamma r q \qquad (7.4.1)$$

and noting that q and ψ^* are functions of θ only. In this equation we have used p instead of p^* since this analysis is valid only for cohesionless materials.

Substituting into equations (7.2.29) and (7.2.30) and putting both $\dfrac{\partial q}{\partial r}$ and $\dfrac{\partial \psi^*}{\partial r}$ equal to zero, we have

$$(1 + \sin \phi \cos 2\psi^*)\, q + \sin \phi \sin 2\psi^* \frac{dq}{d\theta} + 2q \sin \phi \cos 2\psi^* \frac{d\psi^*}{d\theta}$$

$$+ q \sin \phi\, (3 \cos 2\psi^* - \kappa + \sin 2\psi^* \cot \theta) + \cos \theta = 0 \qquad (7.4.2)$$

$$q \sin \phi \sin 2\psi^* + (1 - \sin \phi \cos 2\psi^*) \frac{dp}{d\theta} + 2q \sin \phi \sin 2\psi^* \frac{d\psi^*}{d\theta}$$

$$+ q \sin \phi\, (3 \sin 2\psi^* - \cos 2\psi^* \cot \theta - \kappa \cot \theta) - \sin \theta = 0 \qquad (7.4.3)$$

which can more conveniently be written as

$$A \frac{dq}{d\theta} + B \frac{d\psi^*}{d\theta} + C = 0 \qquad (7.4.4)$$

and

$$D \frac{dq}{d\theta} + E \frac{d\psi^*}{d\theta} + F = 0 \qquad (7.4.5)$$

where

$$A = \sin \phi \sin 2\psi^* \qquad (7.4.6)$$

$$B = 2q \sin \phi \cos 2\psi^* \qquad (7.4.7)$$

$$C = q[1 + \sin \phi(4 \cos 2\psi^* - \kappa + \sin 2\psi^* \cot \theta)] + \cos \theta \quad (7.4.8)$$

$$D = 1 - \sin \phi \cos 2\psi^* \qquad (7.4.9)$$

$$E = 2q \sin \phi \sin 2\psi^* \qquad (7.4.10)$$

and

$$F = q \sin \phi \, (4 \sin 2\psi^* - \cos 2\psi^* \cot \theta - \kappa \cot \theta) - \sin \theta \quad (7.4.11)$$

Rearranging equations (7.4.4) and (7.4.5) gives

$$\frac{dq}{d\theta} = \frac{CE - BF}{BD - AE} \qquad (7.4.12)$$

and

$$\frac{d\psi^*}{d\theta} = \frac{AF - CD}{BD - AE} \qquad (7.4.13)$$

Equations (7.4.12) and (7.4.13) could be expressed in more direct form by substituting from equations (7.4.6) to (7.4.11) but there is no advantage in doing so as it is already apparent that we will require to use a computer to solve these equations and the set of equations (7.4.6) to (7.4.13) is perfectly convenient for computer implementation.

It is, however, worth evaluating the denominator in equations (7.4.12) and (7.4.13), thus,

$$BD - AE = 2q \sin \phi \, (\cos 2\psi^* - \sin \phi) \qquad (7.4.14)$$

Like the denominator of equation (7.3.12) this can become zero and we must expect the same difficulties when this occurs. The denominator in equations (7.4.12) and (7.4.13) is zero when $\cos 2\psi^* = \sin \phi$, i.e. when $\psi^* = 135 + \phi/2$. The difference in sign from the corresponding criterion in Cartesian co-ordinates is simply an artefact of using ψ^* instead of ψ, and has no physical significance.

Equations (7.4.12) and (7.4.13) can be solved by the techniques illustrated in examples (7.3.1) and (7.3.2) though again Runge–Kutta integration is more accurate. As explained in §7.2, it is unlikely that we will have an active stress state in a conical hopper so that normally we will have $\kappa = +1$. At the centre-line the major principal stress will be horizontal and therefore $\psi^* = 90°$. At the left-hand wall, $\theta = \alpha$, where α is the hopper half-angle, the wall shear will be negative and since we are concerned with passive failure, the wall stresses will be given by the greater intersection of the lower wall yield locus with Mohr's circle as shown by point Θ in figure 7.4. Since ψ^* is the angle between the major principal stress direction and the r co-ordinate, its value at the wall can be found from this figure to be $90 + \frac{1}{2}(\omega+\phi_w)$. The usual boundary conditions are therefore

$$\psi^* = 90° \text{ on } \theta = 0° \qquad (7.4.15)$$

and

$$\psi^* = \psi_w^* = 90 + \tfrac{1}{2}(\omega + \phi_w) \text{ on } \theta = \alpha \qquad (7.4.16)$$

A problem, however, arises on the centre-line as in the cylindrical case of §7.3. Here $\psi^* = 90°$ and $\theta = 0°$ and the quantities $\sin 2\psi^* \cot \theta$ and $(1+\cos 2\psi^*) \cot \theta$ appearing in equations (7.4.8) and (7.4.11) are indeterminate in their present form. Expressing the former term in the form, $\sin 2\psi^*/\tan \theta$, we can use l'Hopital's rule to give

$$\frac{\sin 2\psi^*}{\tan \theta} = \frac{2 \cos 2\psi^*}{\sec^2 \theta} \frac{d\psi^*}{d\theta} = -2 \frac{d\psi^*}{d\theta} \qquad (7.4.17)$$

and similarly

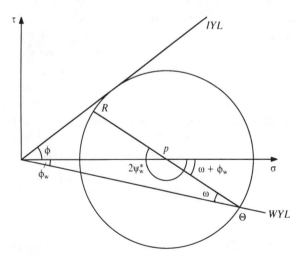

Figure 7.4 Mohr's circle for the evaluation of the stress parameter ψ^* at the wall.

$$\frac{1 + \cos 2\psi^*}{\tan \theta} = -\frac{2 \sin 2\psi^*}{\sec^2 \psi^*} \frac{d\psi^*}{d\theta} = 0 \tag{7.4.18}$$

Thus on the centre-line the quantities A to F become

$$A = 0 \tag{7.4.19}$$

$$B = -2q \sin \phi \tag{7.4.20}$$

$$C = 1 + q\left[1 - \sin \phi \left(5 + 2\frac{d\psi^*}{d\theta}\right)\right] \tag{7.4.21}$$

$$D = 1 + \sin \phi \tag{7.4.22}$$

$$E = 0 \tag{7.4.23}$$

$$F = 0 \tag{7.4.24}$$

Thus from equation (7.4.12),

$$\frac{dq}{d\theta} = 0 \tag{7.4.25}$$

and from equation (7.4.13),

$$\frac{d\psi^*}{d\theta} = \frac{1 + q(1 - 5 \sin \phi)}{4 \, q \sin \phi} \tag{7.4.26}$$

Equations (7.4.25) and (7.4.26) must be used instead of equations (7.4.12) and (7.4.13) for the first step of any integration starting from $\theta = 0$. As in §7.3 we must assume a value for q on $\theta = 0$, i.e. q_0, in order to start our calculations and check whether the resulting value of ψ^* on $\theta = \alpha$ satisfies the wall boundary condition.

Figures 7.5 and 7.6 shows the results of such calculations for the case of $\phi = 30°$ and values of θ up to $40°$ using various values of q_0. If we take as an example a hopper of half-angle $\alpha = 25°$, we see from figure 7.5 that the value of $q_0 = 0.2$ corresponds to $\psi_w^* = 111.23°$ and hence from equations (7.2.15) and (7.2.16), recalling that $c = 0$,

$$\tan \phi_w = -\frac{(\tau_{\theta r})_w}{(\sigma_{\theta\theta})_w} = -\frac{\sin \phi \sin 2\psi_w^*}{1 - \sin \phi \cos 2\psi_w^*} \tag{7.4.27}$$

giving $\phi_w = 13.85°$. Thus for each selected value of q_0 we can evaluate ϕ_w and hence by using the shooting technique described in §7.3 we can find the value of q_0 corresponding to any specified ϕ_w.

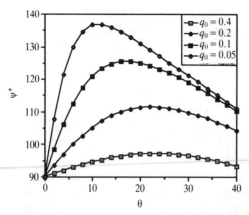

Figure 7.5 Variation of the stress parameter ψ^* with angular position θ.

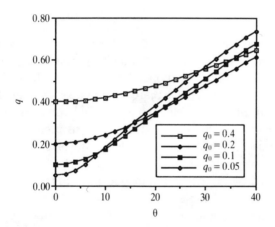

Figure 7.6 Variation of the stress parameter q with angular position θ.

From figure 7.6 at $q_0 = 0.2$ and $\theta = 25°$ we find that $q = 0.410$. Thus for $\phi = 30°$, $\phi_w = 13.85°$ and $\alpha = 25°$ we have that $q_w = 0.410$ and, as above, $\psi_w^* = 111.23°$. Thus from equations (7.2.14) to (7.2.16)

$$(\sigma_{rr})_w = 0.410\,\gamma r\,(1 + \sin 30 \cos 222.46) = 0.259\,\gamma r$$
$$(\sigma_{\theta\theta})_w = 0.410\,\gamma r\,(1 - \sin 30 \cos 222.46) = 0.561\,\gamma r$$
$$(\tau_{\theta r})_w = -\,(\tau_{r\theta})_w = -\,0.410\,\gamma r \sin 30 \sin 222.46 = -\,0.138\,\gamma r$$

The corresponding stresses on the centre-line can be obtained by putting $q = 0.2$ and $\psi^* = 90°$ giving,

$$\sigma_{rr} = 0.2 \, \gamma r \, (1 + \sin 30 \cos 180) = 0.1 \, \gamma r$$
$$\sigma_{\theta\theta} = 0.2 \, \gamma r \, (1 - \sin 30 \cos 180) = 0.3 \, \gamma r$$
$$\tau_{r\theta} = - \, \tau_{\theta r} = 0.2 \, \gamma r \sin 30 \sin 180 = 0$$

The stresses at any intermediate value of θ can be found by the same technique, and hence the stress profile can be evaluated.

Jenike (1961) presents graphs of solutions for the radial stress field for a wide range of parameters. These are very convenient for cases when the accuracy to which a graph can be read is sufficient and also provide a useful check of any numerical solution.

It can be seen from figure 7.5 that the curves for low values of q_0 become crowded together at large θ. For example at $\theta = 40°$, $q_0 = 0.1$ gives $\psi_w^* = 110.31°$ which corresponds to $\phi_w = 13.27°$. The value of $q_0 = 0.05$ gives $\psi_w^* = 110.93°$ and $\phi_w = 13.66°$. Thus it becomes very difficult to obtain an accurate value of q_0 from a specified value of ϕ_w and any solution obtained in this region must be used with caution. Furthermore, at very low values of q_0 the curves can overlap slightly and multiple roots can therefore occur. Because of this overlap, there is an upper limit to the value of ψ^* at a given value of α. For example at $\phi = 30°$ and $\alpha = 40°$, the upper limit of ψ^* is about $112°$ which corresponds to $\phi_w \approx 14°$. Hence there are no radial stress field solutions for $\phi_w >$ about $14°$. This phenomenon should come as no surprise, since we concluded in §5.8 that Walker's analysis predicts that a linear asymptote exists only when $m > 1$, which also places an upper limit on the permissible value of ϕ_w.

The curves of figure 7.5 differ in one respect from those for the Cartesian case given in table 7.1, namely that ψ^* can pass through a maximum within the range $0 < \theta < \alpha$. Thus the critical condition of $\psi^* = 135 + \phi/2$ can occur within the material for a partially rough wall. Great care must be taken in any numerical solution in which ψ^* approaches this critical value since round-off errors can become dominant and meaningless solutions generated. Any calculation which gives values of $\psi^* > 135 + \phi/2$ must be regarded as suspect and should be repeated using a smaller step length.

The radial stress field solutions are compared with the analyses of Walker and Enstad in §7.5.

Wedge-shaped hoppers

The analysis for the wedge-shaped hopper is very similar to that for a conical hopper. We now start with the stress equation in polar co-ordinates (equations (7.2.15) and (7.2.16)) and find that equations of the form of (7.4.4) and (7.4.5) occur. The parameters A, B, D and E are unchanged but C and F take the somewhat simpler form

$$C = q\,(1 + 3\sin\phi\cos 2\psi^*) + \cos\theta \qquad (7.4.28)$$

$$F = 3q\sin\phi\sin 2\psi^* - \sin\theta \qquad (7.4.29)$$

The method of solution is identical to the conical case except that we get no problems on the centre-line due to the absence of the terms in cot θ. We can therefore use the basic form of the equations throughout the calculation. Equations (7.4.28) and (7.4.29) can also be used to predict the stress distribution near the top of a retaining wall as will be illustrated in §7.8.

7.5 Assessment of the method of differential slices

In the last two sections we have analysed the asymptotic stress distributions at great depth in cylindrical and conical bins. Apart from round-off errors which can be minimised by careful programming, these results are exact for a cohesionless Coulomb material at incipient failure. These results can therefore be used to assess the approximate analyses of chapter 5 which contain additional assumptions. First, we will compare the values of the distribution factor in a cylindrical bin predicted by the two methods and then we will compare Walker's and Enstad's analyses of the conical hopper with the results of the radial stress field. Most of these comparisons will be made for a material with $\phi = 30°$ and $\phi_w = 20°$ which may be considered typical of a free-flow material such as a rounded sand, sugar or many agricultural products. In addition, a few comparisons will be made for materials with greater angles of friction.

The evaluation of the distribution factor $\mathcal{D}$ given in §5.4 is exact for a cohesionless Coulomb material except for the assumption that the stress σ_{rr} is constant across the bunker. Considering our typical material with $\phi = 30°$ and $\phi_w = 20°$, we see from the results of table 7.2 that σ_{rr} varies from 1.421 γa on the centre-line to 1.374 γa at the wall, a 3.4% variation, and we must expect a consequential error in the evaluation of $\mathcal{D}$ since the assumption of constant σ_{rr} is the only error in Walker's analysis. From table 7.2, we see that the axial stress σ_{zz} varies from 4.263 γa at the centre line to 3.711 γa at the wall. The mean value is given by

$$\bar{\sigma}_{zz} = \frac{1}{\pi a^2} \int_0^a 2\pi r \, \sigma_{zz} \, dr \qquad (7.5.1)$$

The results of table 7.2, together with the values at $R = 0.1, 0.3$, etc., have been used to evaluate this quantity using Simpson's rule integration. The resulting value of $\bar{\sigma}_{zz}$ is 4.000 γa from which we find that $\mathcal{D} = 3.711/4.000 = 0.928$. This value can be compared with the value of 0.946 calculated from equation (5.4.23). Thus Walker's approximate results differs from the exact value by only 2%.

Similarly, in the passive state for the same material we find σ_{rr} varies from 1.220 γa at the centre to 1.374 γa at the wall, a 12.6% variation, and σ_{zz} varies from 0.407 γa to 0.868 γa with a mean of 0.637 γa giving the exact value of $\mathcal{D}$ as 1.362. This differs from Walker's value of 1.326 by 2.7%. Thus under normal circumstances, Walker's approximate analysis is quite accurate enough for general use, though it must be remembered that we have only confirmed the accuracy of the value of $\mathcal{D}$ at great depth. The analysis of §7.3 can give no information about the variation of $\mathcal{D}$ with depth. It may be noted that in both cases the exact value of $\mathcal{D}$ is further from unity than Walker's estimate.

As so often, the difference between the exact and approximate analyses increases with increasing ϕ and as ϕ_w approaches ϕ. For the extreme case of the passive failure of a material with $\phi = 50°$ and $\phi_w = 40°$, there is a 3.7-fold variation in σ_{rr} from 0.161 γa to 0.596 γa and σ_{zz} varies from 0.021 γa to 0.719 γa, a 34-fold variation. Even so the exact value of $\mathcal{D}$ of 2.04 compares very favourably with the value of 2.03 predicted from equation (5.4.23) on the assumption that σ_{rr} is constant. These results show that Walker's method for evaluating the distribution factor is insensitive to the rather questionable assumptions implicit in that method.

The exact solution for the asymptotic stresses in a 15° hopper containing the typical material with $\phi = 30°$ and $\phi_w = 20°$ has been found from the radial stress field analysis of §7.4. The required value of ψ^* at the wall is $90 + \frac{1}{2}(\omega - \phi_w) = 121.58°$ and this is obtained by putting $q_0 = 0.1179$ which gives $q_w = 0.2526$. Hence from equation (7.2.15) we find that $\sigma_w = 0.2526 \, (1 - \sin 30 \cos 243.36) \, \gamma r = 0.309 \, \gamma r$.

This result can be compared with the predictions of Walker's method given in §5.8. Using the simplest version of that method, we find from equation (5.8.18) that $m = 3.626$, from which using equations (5.8.3), (5.8.7) and (5.8.19) gives $\sigma_w = 0.417 \, \gamma r$.

If we use the first order improvement to Walker's method, resulting in equation (5.8.21), we have from equation (5.8.22) that $\eta = 25.896°$ and hence from equation (5.8.23), $\mathcal{D} = 1.5522$. Thus $m^* = 5.628$ and $\sigma_w = 0.347 \, \gamma r$. Taking Walter's modification, we have from equation (5.8.7) $m^{**} = 6.7324$ and hence $\sigma_w = 0.281 \, \gamma r$.

It can be seen that the simplest version of Walker's method gives a poor estimate of the wall stress and that the exact value lies almost exactly half-way between the values obtained using m^* and m^{**}.

We can investigate the reasons for these results by calculating the variation of σ_{hh} across the hopper. In doing so we must recall that Walker's method considers a horizontal slice across the hopper. If this intersects the wall at the radial co-ordinate r_1, the radial distance to the centre of this slice will be $r_1 \cos 15$. Thus σ_{hh} at the wall will be given by $\gamma r_1 [1 + \sin 30 \cos 2 \times (121.58+15)] = 0.259 \, \gamma r_1$ and at the centre-line σ_{hh} is given by $\gamma r_1 \cos 15(1 - \sin 30) = 0.057 \, \gamma r_1$. The simplest version of Walker's analysis assumes that σ_{hh} is constant and this is clearly far from the case. We can obtain the exact values of the distribution factor by evaluating the mean of σ_{hh} using equation (7.5.1) from which we obtain $\mathcal{D} = 1.57$ which compares favourably with the value of 1.55 predicted by equation (5.8.23) but not with the value of 1.326 calculated from equation (5.4.23). Thus Walters' modification of Walker's method of evaluating the distribution factor in a conical hopper seems to be preferable. It seems that the uncertainty in the prediction of the wall stress results occurs because of the choice of the element used in the force balance. Neither the cylindrical element, as used by Walker, nor the conical element used by Walters is exact.

We can also consider Enstad's prediction of the stress distribution in the same situation. From equation (5.10.1) we have $\beta = 31.58°$. Thus if we take the distance OA of figure 5.20 to be r_1 the radius $R = r_1 \sin 15 \sin (15 + 31.58) = 0.356 \, r_1$ and the distance OC is $r_1 \sin 31.58 \sin (15 + 31.58) = 0.721 \, r_1$. Thus the distance OC is $1.077 \, r_1$. Recalling that $q_w = 0.2526$ and $q_0 = 0.1179$ we have that $(\sigma_3)_w = 0.2526 \, \gamma r_1 (1 - \sin 30) = 0.126 \, \gamma r_1$ and that $(\sigma_3)_0 = 0.1179 \times 1.077 \, \gamma r_1 (1 - \sin 30) = 0.063 \, \gamma r_1$. Thus Enstad's assumption that σ_3 is constant along the surface of his element is not exact. None-the-less, we can evaluate Enstad's parameters from equations (5.10.17) and (5.10.18) giving $X = 9.563$ and $Y = 2.287$. The linear asymptote of Enstad's result can be expressed in the form, $q = Y/(X-1) = 0.2670$. This is in excellent agreement with the value of q_w quoted above, though not with q_0. Thus despite the erroneous assumptions, Enstad's method, like Walker's and Walters' methods, works well for the prediction of the wall stress. The reasons for this are obscure and these examples must therefore not be taken to suggest that the approximate methods are adequate for all combinations of parameters. Care should therefore be taken when using these methods especially with extreme combinations of parameters.

7.6 The method of characteristics

In §7.3 and §7.4 we considered the exact solutions for the asymptotes to which the stresses tend in cylindrical and conical bunkers. However, these analyses give no information about the rate of approach to the asymptotes nor do they give any indication about the stresses in the upper parts of the bunker. For these we require a complete solution to the stress equations which satisfies the boundary conditions specified by the imposed stresses on the top surface of the material. This is best done by the method of characteristics as described below. Other mathematical techniques have been used, e.g. by Pitman (1986), but the method of characteristics is recommended since, as we will see, the calculational procedure is closely linked to the structure of the solution and therefore copes better than other methods with certain special features of this topic.

The method of characteristics is a convenient way of solving certain types of first and second order partial differential equations, and utilises the property that these equations reduce to ordinary differential equations along particular lines, which are known as the characteristics, or, more strictly, the characteristic ground curves. Since ordinary differential equations are usually much easier to solve than partial differential equations, this phenomenon greatly simplifies the method of solution. The method of characteristics is described in many standard texts, such as Abbott (1966) and its use for stress analysis in granular materials is described in detail by Sokolovskii (1965). The derivation given below is not the most elegant or complete but is presented in this form in the hope that it may be more comprehensible than that found in more mathematical texts.

In Cartesian co-ordinates, we must start with equations (7.2.6) and (7.2.7),

$$(1 + \sin \phi \cos 2 \psi) \frac{\partial p^*}{\partial x} - 2p^* \sin \phi \sin 2\psi \frac{\partial \psi}{\partial x} + \sin \phi \sin 2\psi \frac{\partial p^*}{\partial y}$$

$$+ 2p^* \sin \phi \cos 2\psi \frac{\partial \psi}{\partial y} = 0 \tag{7.6.1}$$

and

$$\sin \phi \sin 2\psi \frac{\partial p^*}{\partial x} + 2p^* \sin \phi \cos 2\psi \frac{\partial \psi}{\partial x} + (1 - \sin \phi \cos 2\psi) \frac{\partial p^*}{\partial y}$$

$$+ 2p^* \sin \phi \sin 2\psi \frac{\partial \psi}{\partial y} = \gamma \qquad (7.6.2)$$

It will be noted that the cohesion c does not enter these equations explicitly, so that these equations are equally valid for cohesive and cohesionless materials. In the next few sections we will be concerned solely with cohesionless materials for which p^* and p are identical. Thus for convenience we will replace p^* in equations (7.6.1) and (7.6.2) by p. Cohesive materials are considered in §7.9.

We can eliminate $\partial p/\partial y$ from these equations by multiplying equation (7.6.1) by $(1 - \sin \phi \cos 2\psi)$ and subtracting equation (7.6.2) multiplied by $\sin \phi \sin 2\psi$, giving

$$\cos^2\phi \frac{\partial p}{\partial x} = -2p \sin \phi \left[\frac{\partial \psi}{\partial x} \sin 2\psi + \frac{\partial \psi}{\partial y} (\sin \phi - \cos 2\psi) \right] - \gamma \sin \phi \sin 2\psi$$

$$(7.6.3)$$

Similarly, we can eliminate $\partial p/\partial x$ and show that

$$\cos^2\phi \frac{\partial p}{\partial y} = -2p \sin \phi \left[\frac{\partial \psi}{\partial x} (\sin \phi + \cos 2\psi) + \frac{\partial \psi}{\partial y} \sin 2\psi \right]$$

$$+ \gamma (1 + \sin \phi \cos 2\psi) \qquad (7.6.4)$$

We can now evaluate the gradient of p along a line inclined at angle ζ anticlockwise from the x axis, as shown in figure 7.7, since from the laws of vector resolution

$$\frac{\mathrm{d}p}{\mathrm{d}s} = \frac{\partial p}{\partial x} \cos \zeta + \frac{\partial p}{\partial y} \sin \zeta \qquad (7.6.5)$$

where s is distance along this line. Substituting from equation (7.6.3) and (7.6.4) gives

$$\frac{\mathrm{d}p}{\mathrm{d}s} \cos^2\phi = 2p \sin \phi \, [\sin(2\psi - \zeta) - \sin \phi \sin \zeta] \frac{\partial \psi}{\partial x}$$

$$+ 2p \sin \phi \, [\sin \phi \cos \zeta - \cos (2\psi - \zeta)] \frac{\partial \psi}{\partial y} \qquad (7.6.6)$$

$$+ \gamma [\sin \zeta - \sin \phi \sin(2\psi - \zeta)]$$

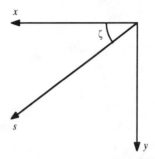

Figure 7.7 Definition of the parameters ζ and s.

We can eliminate $\partial\psi/\partial x$ from this equation using

$$\frac{d\psi}{ds} = \frac{d\psi}{dx} \cos \zeta + \frac{\partial\psi}{\partial y} \sin \zeta \qquad (7.6.7)$$

which is analogous to equation (7.6.5). This gives

$$\frac{dp}{ds} \cos^2\phi = 2p \sin \phi \, [\sin(2\psi - \zeta) - \sin \phi \sin \zeta] \sec \zeta \, \frac{d\psi}{ds}$$

$$+ 2p \sin \phi \, \{\sin \phi \cos \zeta - \cos(2\psi - \zeta)$$

$$- [\sin(2\psi - \zeta) - \sin \phi \sin \zeta] \tan \zeta\} \frac{\partial\psi}{\partial y}$$

$$+ \gamma[\sin \zeta - \sin \phi \sin(2\psi - \zeta)] \qquad (7.6.8)$$

Equation (7.6.8) is seen to contain both partial and ordinary derivatives, but reduces to an ordinary differential equation when the coefficient of the $\partial\psi/\partial y$ term happens to equal zero, i.e. when

$$\sin \phi \cos \zeta - \cos(2\psi - \zeta) - [\sin(2\psi - \zeta) \qquad (7.6.9)$$
$$- \sin \phi \sin \zeta] \tan \zeta = 0$$

Rearranging this equation, gives

$$\cos(2\psi - 2\zeta) = \sin \phi = \cos 2\varepsilon \qquad (7.6.10)$$

where ϵ is defined by

$$2\varepsilon = 90 - \phi \qquad (7.6.11)$$

as in equation (3.3.4). Hence, from equation (7.6.10),

$$\zeta = \psi \pm \varepsilon \qquad (7.6.12)$$

and we see that there are two directions along which equation (7.6.8) reduces to the ordinary differential equation

$$\frac{dp}{ds} \cos^2 \phi = 2p \sin \phi \, [\sin(2\psi-\zeta) - \sin \phi \sin \zeta] \sec \zeta \frac{d\psi}{ds}$$

$$+ \gamma[\sin \zeta - \sin \phi \sin(2\psi-\zeta)] \qquad (7.6.13)$$

We can call these directions the α-direction, when $\zeta = \psi-\varepsilon$, and the β-direction, when $\zeta = \psi+\varepsilon$ and lines along these directions are called the α- and β-characteristics respectively. The equation of an α-characteristic is given by

$$\frac{dy}{dx} = \tan(\psi-\varepsilon) \qquad (7.6.14)$$

and that of a β-characteristic is given by

$$\frac{dy}{dx} = \tan(\psi+\varepsilon) \qquad (7.6.15)$$

Putting $\zeta = \psi-\varepsilon$ into equation (7.6.13) gives

$$\frac{dp}{ds} = 2p \tan \phi \frac{d\psi}{ds} + \gamma[\sin(\psi-\varepsilon) - \tan \phi \cos(\psi-\varepsilon)] \qquad (7.6.16)$$

or, since $\dfrac{dx}{ds} = \cos \zeta = \cos(\psi-\varepsilon)$ and $\dfrac{dy}{ds} = \sin(\psi-\varepsilon)$,

$$\frac{dp}{ds} - 2p \tan \phi \frac{d\psi}{ds} = \gamma \left[\frac{dy}{ds} - \tan \phi \frac{dx}{ds}\right] \qquad (7.6.17)$$

Similarly, along the β-characteristic, where $\zeta = \psi+\varepsilon$,

$$\frac{dp}{ds} + 2p \tan \phi \frac{d\psi}{ds} = \gamma \left[\frac{dy}{ds} + \tan \phi \frac{dx}{ds}\right] \qquad (7.6.18)$$

Figure 7.8 shows the inclination of the relevant planes and lines. Figure 7.8(a) shows the major principal stress direction, which is inclined at angle ψ anticlockwise from the x-direction and the characteristic directions which are inclined at $\pm\varepsilon$ to the major principal stress direction. Figure 7.8(b) shows the major and minor principal planes, which are inclined at ψ and $90 + \psi$ to the x-plane and also the slip planes which we showed in §3.3 are inclined at $\pm\varepsilon$ to the minor principal plane. Comparison of the two parts of this figure shows that the characteristic directions lie along the slip planes and thus have

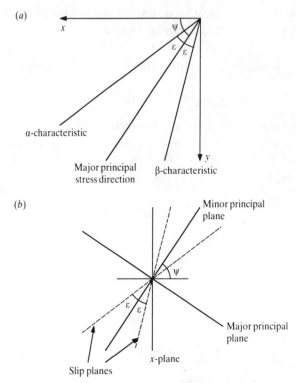

Figure 7.8 Relationship between the characteristic directions and the slip planes.

physical as well as mathematical significance. We will see below that information about the boundary conditions propagates along the characteristics and this accounts for the greater success of this method than those based on arbitrarily selected grid points.

The use of the equations derived above can be illustrated as follows. Suppose we know the stress parameters p and ψ at two nearby points 1 and 2, i.e. (x_1, y_1, p_1, ψ_1) and (x_2, y_2, p_2, ψ_2) are specified. We can then construct the α-characteristic through point 1, since its slope, $\tan(\psi_1 - \varepsilon)$, is known and we can also construct the β-characteristic through point 2 as shown in figure 7.9. These characteristics will intersect at point 3 with the, as yet unknown, co-ordinates (x,y) and stress parameters (p,ψ).

Equation (7.6.14), which becomes

$$dy = \tan(\psi_1 - \varepsilon)\, dx \qquad (7.6.19)$$

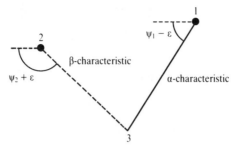

Figure 7.9 The intersection of an α- and β- characteristic.

applies along the α-characteristic, i.e. along the line 1–3, and hence, if the distance 1 to 3 is small, we can say that

$$y - y_1 = (x-x_1) \tan(\psi_1 - \varepsilon) \tag{7.6.20}$$

Similarly, we can use equation (7.6.15) along the β-characteristic giving

$$y - y_2 = (x-x_2) \tan(\psi_2 + \varepsilon) \tag{7.6.21}$$

These equations can be solved to give x and y.

We can note that the term ds cancels from equation (7.6.17) so that this equation can be written

$$dp - 2\mu p \, d\psi = \gamma(dy - \mu \, dx) \tag{7.6.22}$$

or, in finite difference form

$$p - p_1 - 2\mu p_1(\psi - \psi_1) = \gamma[y - y_1 - \mu(x-x_1)] \tag{7.6.23}$$

Similarly, from equation (7.6.18), we have

$$p - p_2 + 2\mu p_2(\psi - \psi_2) = \gamma[y - y_2 + \mu(x-x_2)] \tag{7.6.24}$$

The values of p and ψ can therefore be obtained from the simultaneous solution of these equations.

More accurately, we can say that the value of ψ in equation (7.6.20) should be the average over the interval 1 to 3 so that ψ_1 should be replaced by $\bar{\psi}_1$, defined by

$$\bar{\psi}_1 = \tfrac{1}{2}(\psi + \psi_1) \tag{7.6.25}$$

We can, similarly, define

$$\bar{\psi}_2 = \tfrac{1}{2}(\psi + \psi_2) \tag{7.6.26}$$

$$\bar{p}_1 = \tfrac{1}{2}(p + p_1) \tag{7.6.27}$$

$$\bar{p}_2 = \tfrac{1}{2}(p+p_2) \qquad\qquad (7.6.28)$$

and write our equations in the following form:
along the α-characteristics

$$y - y_1 = (x-x_1)\tan(\bar{\psi}_1-\varepsilon) \qquad (7.6.29)$$

$$p - p_1 - 2\mu\bar{p}_1(\psi-\psi_1) = \gamma[y-y_1-\mu(x-x_1)] \qquad (7.6.30)$$

and along the β-characteristics

$$y - y_2 = (x-x_2)\tan(\bar{\psi}_2+\varepsilon) \qquad (7.6.31)$$

$$p - p_2 + 2\mu\bar{p}_2(\psi-\psi_2) = \gamma[y-y_2+\mu(x-x_2)] \qquad (7.6.32)$$

Example 7.6.1 We will illustrate the use of these equations by considering a numerical example involving a material of weight density $\gamma = 15.0$ kN m^{-3} and angle of friction 40°, i.e. $\mu = 0.837$. Let us suppose that at $x_1 = 0.04$ m, $y_1 = 0.05$ m, the stress parameters are $p_1 = 1.7$ kN m^{-2} and $\psi_1 = 80°$ and, at $x_2 = 0.55$ m, $y_2 = 0.045$ m, $p_2 = 1.6$ kN m^{-2} and $\psi_2 = 85°$.

From equation (7.6.11) we have that

$$\varepsilon = \tfrac{1}{2}(90-\phi) = 25°$$

Taking as our first approximation that $\bar{\psi}_1 = \psi_1$, $\bar{\psi}_2 = \psi_2$ etc., we have, from equation (7.6.20),

$$y - 0.05 = (x-0.04)\tan(80-25) = 1.428\,(x-0.04)$$

and, from equation (7.6.21),

$$y - 0.045 = (x-0.055)\tan(85+25) = -2.747(x-0.055)$$

These equations are linear and can be solved directly to give $x = 0.0487$ m, $y = 0.0624$ m.

From equations (7.6.23) and (7.6.24) we have

$$(p-1.7) - 2\times0.837\times1.7(\psi-80)\frac{\pi}{180}$$
$$= 15[(0.0624-0.05) - 0.837\times(0.0487-0.04)]$$

and

$$(p-1.6) + 2\times0.837\times1.6(\psi-85)\frac{\pi}{180}$$
$$= 15[(0.0624-0.045) + 0.837\times(0.0487-0.0055)]$$

where the factor of $\pi/180$ appears since ψ must be measured in radians. These equations are also linear and have solution $\psi = 82.48°$, $p = 1.90$ kN m^{-2}.

We can now evaluate the mean values of ψ and p from equations (7.6.25) to (7.6.28) giving $\bar{\psi}_1 = 81.24°$, $\bar{\psi}_2 = 83.74°$, $\bar{p}_1 = 1.80$ kN m^{-2} and $\bar{p}_2 = 1.75$ kN m^{-2} and repeat the calculations as follows

$$y - 0.05 = (x-0.04)\tan(81.24-25) = 1.496(x-0.04)$$

$$y - 0.045 = (x-0.055)\tan(83.75+25) = -2.946(x-0.055)$$

from which we obtain $x = 0.0488$ m, $y = 0.0632$ m. Hence,

$$(p-1.7) - 2 \times 0.837 \times 1.80(\theta-80)\frac{\pi}{180}$$
$$= 15[(0.0632-0.05) - 0.837 \times (0.0488-0.04)]$$

and

$$(p-1.6) + 2 \times 0.837 \times 1.75(\theta-85)\frac{\pi}{180}$$
$$= 15[(0.0632-0.045) + 0.837 \times (0.0488-0.055)]$$

which have solution $\psi = 82.54°$, $p = 1.92$ kN m^{-2}. It can be seen that there have been only minor changes from our preliminary solution. A third iteration gives the same answer to at least three significant figures. Clearly, the iteration converges rapidly in this example.

We can now consider the implementation of this method when the stress parameters p and ψ are known along a line such as the line ABC of figure 7.10. We can select an arbitrary set of points along this line such as the points marked 1 to 6. These points need not be equally spaced though this may be convenient under certain circumstances.

Using the method described above we can consider the α-characteristic through point 1 and the β-characteristic through point 2 to evaluate the co-ordinates and stress parameters of point a. Repeating this procedure for all five pairs of adjacent points gives us the results for the five points a to e of the second row. This procedure can be repeated again to give four points on the third row and eventually we can obtain the stress distribution throughout the region ACD bounded by the original line AC, the α-characteristic AD and the β-characteristic CD. Clearly, given sufficient boundary conditions this procedure will give us the stress distribution throughout the material.

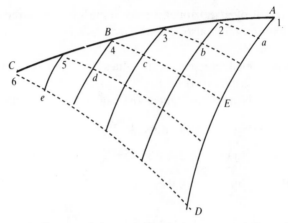

Figure 7.10 Characteristic mesh illustrating zones of dependency.

It is, however, worth noting that within the region *BAE*, where *BE* is the β-characteristic through *B*, we have made no use of the specified values at the points 5 and 6. Within this region the stresses are *totally independent* of the boundary conditions along *BC* and the region *BAE* is said to be the domain of influence of the boundary conditions along *AB*. Perhaps more importantly, we can say that the region *BAE* is outside the domain of influence of *BC*. It is seen that changing the boundary conditions at a point, only affects the solution in the region bounded by the two characteristics emanating from that point. Thus we can regard the characteristics as the lines along which information about the stresses is propagated and we must therefore expect that sudden changes can occur across characteristics. Furthermore, the method of characteristics gives us the stresses at important points within the solution, unlike methods based on a regular mesh of points which give information at arbitrarily selected points.

The fact that the stresses in some part of the material are independent of some of the boundary conditions, may come as a surprise to those whose experience of partial differential equations is based on the heat conduction or potential flow equations. In these cases the differential equations are of the type known as elliptic equations and for such equations the solution at *all* points depends on *all* the boundary conditions. The stress equations are of the type known as hyperbolic equations and the stress at a particular point depends on *some* of the boundary conditions. The first equations of the hyperbolic type met by most students are those of supersonic flow. Here the velocities depend only on upstream conditions, unlike subsonic flow for which the equations are elliptic and the velocities depend on both upstream and downstream conditions. As such students will know, supersonic flow can contain shock waves, which are discontinuous changes in the velocity. Hyperbolic equations often have discontinuous solutions and such solutions can also occur in stress distributions. This topic is considered in more detail in §7.12.

Finally, we must consider the accuracy of this method. Equations (7.6.14) to (7.6.18) are exact for a Coulomb material at incipient yield. However, the solutions may contain errors resulting from two causes. First, there will be round-off errors unless sufficient significant figures are maintained in the calculation. This used to be a significant problem but is now less so with the advent of modern computers. More importantly, we must consider the accuracy of the integration method. Without iteration, i.e. when using equations (7.6.20) to (7.6.24), we are performing an Euler integration which is only accurate if the spacing of the characteristics is very small. The use of iteration improves the accuracy but there seems to be no way of implementing an integration scheme of the Runge–Kutta type. It is therefore wise to use a narrow characteristic spacing and, ideally, the calculation should be repeated with a closer spacing. If there is a significant difference between the two results, an even closer spacing should be used.

7.7 Single retaining wall with surcharge

We can illustrate the use of the equations derived in the previous section by considering the active failure of a vertical retaining wall as shown in figure 7.11. We will consider the case of a cohesionless material with an angle of internal friction $\phi = 30°$ and an angle of wall friction of $\phi_w = 20°$ subject to a uniform surcharge Q_0.

To maintain consistency with subsequent examples, we will take an arbitrary length a of the top surface and work in terms of the dimensionless variables, X, Y and P defined by

$$X = x/a \tag{7.7.1}$$

$$Y = y/a \tag{7.7.2}$$

and

$$P = p/\gamma a \tag{7.7.3}$$

Exact stress analyses

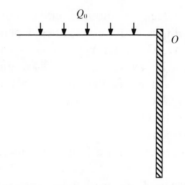

Figure 7.11 Single retaining wall with uniform surcharge.

where (x,y) is a set of Cartesian co-ordinates with origin at the top corner O.

The characteristic equations, equations (7.6.29) to (7.6.32), can now be written: along the α-characteristic

$$Y - Y_1 = (X-X_1)\tan(\psi_1-\epsilon) \qquad (7.7.4)$$

$$P - P_1 - 2\mu\bar{P}_1(\psi-\psi_1) = Y - Y_1 - \mu(X-X_1) \qquad (7.7.5)$$

and along the β-characteristic

$$Y - Y_2 = (X-X_2)\tan(\psi_2+\epsilon) \qquad (7.7.6)$$

$$P - P_2 + \mu\bar{P}_2(\psi-\psi_2) = Y - Y_2 + \mu(X-X_2) \qquad (7.7.7)$$

The boundary conditions are the uniform applied stress on the top surface, Q_0, which we will take to be $0.3\,\gamma a$, and the major principal stress direction at the wall ψ_w. Since we are considering active failure against a right-hand wall, the wall stresses are given by the lesser intersection of the upper yield locus with Mohr's circle. Thus from figure 3.16

$$\psi_w = 90 + \tfrac{1}{2}(\omega-\phi_w) \qquad (7.7.8)$$

where from equation (3.7.8) $\sin\omega = \sin 20/\sin 30$ i.e. $\omega = 43.160°$ and therefore $\psi_w = 101.580° = 1.7729$ radians.

To start the calculations we must select an arbitrary set of points on the top surface. As mentioned in the previous section, these need not be equally spaced but in this case it is convenient to take six points at equal increments of 0.2 in X as shown in figure 7.12. The naming of these points is also arbitrary but it is convenient if we denote the α-

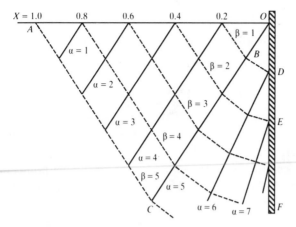

Figure 7.12 Characteristic mesh for the single retaining wall problem.

characteristic through the point (1,0) as $\alpha = 0$ and the β-characteristic through the origin as $\beta = 0$. The characteristics will be numbered consecutively and their points of intersection will be denoted by the values of α and β. Thus the co-ordinates of our selected points are $X_{0,5} = 1$, $X_{1,4} = 0.8$, ..., $X_{5,0} = 0$ and $Y_{0,5} = Y_{1,4} = ... = Y_{5,0} = 0$. This is illustrated in figure 7.12 in which the α-characteristics are shown by the full lines and the β-characteristics by the dashed lines.

Since we are considering the active state, the major principal stress is vertical and hence the boundary condition on the top surface is $\psi = 90°$ and $\sigma_{yy} = Q_0 = \gamma a q_0$. From equation (7.2.2), we see that this is equivalent to

$$P = \frac{q_0}{1 + \sin \phi} = 0.2 \qquad (7.7.9)$$

We can now find the co-ordinates and stress parameters of the points **(1,5)**, **(2,4)**, ..., **(5,1)** using the procedure of the previous section. Here, the use of bold numbers indicates that we are naming a point by its values of α and β; the Cartesian co-ordinates of a point will be given in plain numbers. From equations (7.7.4) to (7.7.7) we have

$$Y_{1,5} - 0 = (X_{1,5} - 0.8) \tan(90-30) \qquad (7.7.10)$$

$$Y_{1,5} - 0 = (X_{1,5} - 1.0) \tan(90+30) \qquad (7.7.11)$$

which have solution, $X_{1,5} = 0.9$, $Y_{1,5} = 0.1732$, and

$$(P_{1,5}-0.2) - 2 \times 0.577 \times 0.2(\psi_{1,5}-\pi/2)$$

$$= (0.1732-0) - 0.577 \times (0.9-0.8) \qquad (7.7.12)$$

$$(P_{1,5}-0.2) + 2 \times 0.577 \times 0.2(\psi_{1,5}-\pi/2)$$

$$= (0.1732-0) + 0.577 \times (0.9-1.0) \qquad (7.7.13)$$

which have solution $P_{1,5} = 0.3155$, $\psi_{1,5} = \pi/2$ radians $= 90°$. Since there has been no change in the value of ψ, these results are exact and no iteration is required.

This procedure can be repeated for the remaining points in the row, i.e. points (2,4) to (5,1) and we find that $P = 0.3155$, $\psi = \pi/2$ and $Y = 0.1732$ for all these points with X taking the values 0.9, 0.7, 0.5, 0.3 and 0.1.

Exactly the same procedure can be used for the next four rows ultimately giving for point C, i.e. point (5,5), $X_{5,5} = 0.5$, $Y_{5,5} = 0.866$, $\psi_{5,5} = \pi/2$ and $P_{5,5} = 0.777$. At this stage we have run out of characteristics and can proceed no further using the basic method. We have evaluated the stress distribution within the domain of influence of the top surface OA. It can be seen that $\psi = \pi/2$ throughout this region just as in the Rankine active state below an infinite horizontal fill. For this reason such a zone is often called a Rankine zone. Using the method of §3.4, we see that

$$\psi = \pi/2 \qquad (7.7.14)$$

and that

$$\sigma_{yy} = q_0 + \gamma y \qquad (7.7.15)$$

which can be rearranged to give

$$P = \frac{q_0 + Y}{1 + \sin\phi} \qquad (7.7.16)$$

Because of the existence of a simple analytic solution for the Rankine zone, most workers start their calculations on the line OC, using equations (7.7.14) and (7.7.16) as the boundary conditions. As we will see in the next section, this is particularly convenient when $Q_0 = 0$ since in this case the solution using equations (7.7.4) to (7.7.7) breaks down on the top row due to the necessity of division by zero.

We can break out of the Rankine zone OAC across the line OC by considering the β-characteristic through the point (5,1) which is marked as B on figure 7.12. This characteristic (i.e. β = 1) will cut the wall at the point we will denote by D whose co-ordinates (X,Y) and stress parameters (P,ψ) can be found from the two equations along the β-characteristic, equations (7.7.6) and (7.7.7), and the conditions

$$X = 0 \text{ and } \psi = \psi_w = 1.7729 \text{ radians} \qquad (7.7.17)$$

Since both ψ_2 and ψ_w are known we can evaluate $\overline{\psi}_2$ directly as 1.6227 so that from equation (7.7.6) we have $Y = 0.3273$. Similarly, we can put $\bar{P}_2 = \frac{1}{2}(P+P_2)$ and from equation (7.7.7) we find that $P = 0.3212$. We will call calculations of this sort 'wall calculations' and it is seen that no iteration is required in a wall calculation, unlike the basic calculation which requires iteration except within a Rankine zone. These values can be taken as the co-ordinates (0,0.3273) and stress parameters of the point **(6,1)** and we can start a new α-characteristic ($\alpha = 6$) from this point. We can now use the basic method to find the co-ordinates and stress parameters along the $\alpha = 6$ characteristic and then from the point **(6,2)** we can use the wall calculation to give the values at point E, which we will call **(7,2)**. By this means we can evaluate the stress distribution throughout the material, using the basic calculation whenever characteristics of the different families meet and using the wall calculation when a β-characteristic cuts the wall. The β-characteristics terminate when they reach the wall and a new α-characteristic can be started.

However, if the angle of wall friction is not small, the evaluation of the point D may not be accurate since the $\beta = 1$ characteristic is significantly curved and the assumption that the slope of the chord is the average of the slopes of the two end tangents may not be good enough. We can obtain a more accurate solution by starting several α-characteristics from the point O, thereby taking the sudden change in ψ from $90°$ to ψ_w in small steps. All these α-characteristics start from the degenerate $\beta = 0$ characteristic and we can construct a schematic characteristic grid as in figure 7.13. As an example, we will take four further α-characteristics through the point O in addition to the $\alpha = 5$ characteristic already considered.

Along the degenerate $\beta = 0$ characteristic, we have that dX and dY are both zero and so equation (7.7.7) becomes

$$dP + 2\mu P \, d\psi = 0 \qquad (7.7.18)$$

or, on integration,

$$P = A \, e^{-2\mu\psi} \qquad (7.7.19)$$

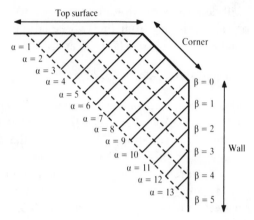

Figure 7.13 Schematic representation of the characteristic mesh for the single retaining wall problem. Note that the corner is represented by the line $\beta = 0$ instead of a point.

where we can evaluate the arbitrary constant A if we know the value of P, say P_0, at $\psi = \psi_0$, giving

$$\frac{P}{P_0} = \exp[-2\mu(\psi-\psi_0)] \qquad (7.7.20)$$

In our case we have at the point $(5,0)$, $P_0 = 0.2$ and $\psi_0 = 90°$, and arbitrarily selecting equal increments in ψ we have, $\psi_{6,0} = 92.895°$, $\psi_{7,0} = 95.790°$, $\psi_{8,0} = 98.865°$ and $\psi_{9,0} = \psi_w = 101.580°$. Thus from equation (7.7.20) we have $P_{6,0} = 0.1887$, $P_{7,0} = 0.1780$, $P_{8,0} = 0.1679$ and $P_{9,0} = 0.1584$. We can therefore tabulate the solutions as the first row of table 7.3. The values for $(5,0)$, $(5,1)$ and $(4,1)$ have been found previously and are also listed in the table. We can now use the basic calculations to obtain the points $(6,1)$ to $(9,1)$ and then use the wall calculation to give the point $(10,1)$. The procedure can then be repeated along the $\beta = 2$ characteristic and hence throughout the rest of the material.

Comparison of $Y_{10,1}$ and $P_{10,1}$ with the preliminary estimate obtained previously $(Y = 0.3273, P = 0.3212)$ shows a small but significant difference. In fact the difference is less important than it seems at first sight. It can be shown that in this region the stress increases linearly with depth so that extrapolating our values at $Y = 0$ and $Y = 0.3025$ to $Y = 0.3273$ gives $P = 0.3287$ which compares favourably with the preliminary estimate of 0.3212. Thus we see that, though the preliminary

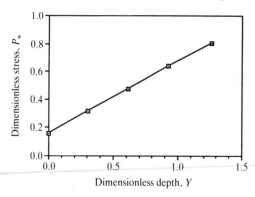

Figure 7.14 Variation of the dimensionless stress parameter P on the dimensionless depth Y.

calculation gave a poor estimate of the position at which the $\beta = 1$ characteristic cuts the wall, it gives the stress at a particular point to within 2% of that obtained by the more accurate method.

Similar calculations along the $\beta = 2$ characteristic give $Y_{11,2} = 0.6120$ and $P_{11,2} = 0.4775$. Comparison of the stress gradients over the intervals $(9,0)$ to $(10,1)$ and $(10,1)$ to $(11,2)$ shows that, as expected, the wall stress varies linearly with depth to an accuracy of about 0.4%, an error which is due partly to retaining too few significant figures and partly to using too coarse a characteristic spacing. This accuracy is, however, quite adequate for most practical purposes, particularly in view of the precision to which the physical parameters can be measured. Calculations along the remaining β-characteristics show that P_w continues to increase linearly without limit as shown in figure 7.14.

A best-fit straight line to the results gives

$$P_w = 0.1581 + 0.5223\, Y \qquad (7.7.21)$$

or, in dimensional terms

$$p_w = 0.1581\,\gamma a + 0.5223\,\gamma y \qquad (7.7.22)$$

Substituting this result into equation (7.2.1) gives

$$\sigma_w = p_w(1 + \sin\phi\cos 2\psi_w) = 0.08542\,\gamma a + 0.2822\,\gamma y$$

$$= 0.2854\, Q_0 + 0.2822\,\gamma y \qquad (7.7.23)$$

This result can be compared with the prediction of the method of wedges, equation (4.7.8), which gives

Table 7.3

$\alpha =$		4	5	6	7	8	9	10
$\beta = 0$	$X =$		0	0	0	0	0	
	$Y =$		0	0	0	0	0	
	$\psi =$		90.000	92.895	95.790	98.685	101.580	
	$P =$		0.2000	0.1887	0.1780	0.1679	0.1584	
$\beta = 1$	$X =$	0.2000	0.1000	0.0951	0.0899	0.0844	0.0787	0.0000
	$Y =$	0	0.1732	0.1814	0.1895	0.1976	0.2055	0.3025
	$\psi =$	90.000	90.000	91.785	93.466	95.045	96.5423	101.580
	$P =$	0.2000	0.3155	0.3096	0.3043	0.2996	0.2954	0.3158

$$\sigma_w = 0.279 \, (Q_0 + \gamma y) \qquad (7.7.24)$$

It is seen that the method of characteristics has given a slightly higher wall stress than the method of wedges. The latter method finds the largest wall stress that can occur as the result of slip along any planar slip surface. The method of characteristics has found a curved slip surface that gives an even greater stress. The difference is, however, slight as we deduced in §4.6.

7.8 Single retaining wall with zero or very large surcharge

Zero surcharge

If we attempt to repeat the analysis of the previous section for the case of zero surcharge, i.e. $Q_0 = 0$, we run into two problems.

First, on the top surface, where $P = 0$ at all points, the basic calculation breaks down since we find that we have to divide by zero during the first iteration. This causes no great difficulty since we can use the analytic solution for the Rankine zone, equations (7.7.14) and (7.7.16), i.e.

$$\psi = 90° \qquad (7.8.1)$$

and

$$P = \frac{Y}{1 + \sin \phi} \qquad (7.8.2)$$

and start our calculations from the line OC of figure 7.12. Alternatively, we could use the method of characteristics but during the first iteration put $\bar{P}_1 = P_1 + \delta$ where δ is some small quantity such as 0.1. Since the method converges to the correct solution, whatever the starting point, this device is always successful.

Secondly, there are difficulties with the fan of characteristics emanating from the point O. Since $P = 0$ at the point $(\mathbf{5,0})$, we have from equation (7.7.20) that $P = 0$ along the whole of the $\beta = 0$ characteristic. If we were to proceed with the basic method along the $\beta = 1$ characteristic, as in the example presented in the previous section, we would obtain the results shown in table 7.4.

It can be seen that along $\beta = 1$, the α-characteristics have become closely spaced with all the values of ψ close to 90°. The final step to the wall involves the large change in ψ from 90.140° to 101.58° and is

Table 7.4

$\alpha =$	5	6	7	8	9
$\beta = 0$					
$X =$	0	0	0	0	0
$Y =$	0	0	0	0	0
$P =$	0	0	0	0	0
$\psi =$	90.000	92.895	95.790	98.685	101.580
$\beta = 1$					
$X =$	0.1000	0.0969	0.0937	0.0904	0.0874
$Y =$	0.1732	0.1786	0.1841	0.1898	0.1950
$P =$	0.1155	0.1190	0.1226	0.1263	0.1296
$\psi =$	90.000	90.016	90.039	90.085	90.140

therefore inaccurate. Furthermore, the $\alpha = 9$ characteristic is also markedly curved, being bent through the same angle, and this casts doubts on the accuracy of the values in this table. Thus the method of the previous section becomes unreliable and we must look for an alternative.

We can overcome this difficulty by noting that the boundary conditions for the stress state in region OCF of figure 7.12 are:

along the Rankine zone boundary, OC,

$$\psi = 90° \text{ and } P = \frac{Y}{1 + \sin \phi} \qquad (7.8.3)$$

and along the wall, OF, $\qquad \psi = \psi_w$ $\qquad (7.8.4)$

These are the boundary conditions for the radial stress field, the equations for which are given in polar co-ordinates (r, θ) in §7.4 with the axis $\theta = 0$ directed vertically upwards. The α-characteristic OC is inclined to the horizontal by the angle $\psi - \epsilon = 90 - \epsilon$ and hence $\theta = 180 - \epsilon$. Substituting these results into equation (7.2.11) gives

$$\psi^* = \psi + 90 - \theta = \epsilon \qquad (7.8.5)$$

and equations (7.4.1) and (7.8.3) give

$$q = \frac{p}{\gamma r} = \frac{y}{r(1 + \sin \phi)} = -\frac{\cos \theta}{1 + \sin \phi} = \frac{\cos \epsilon}{1 + \sin \phi} \qquad (7.8.6)$$

On the wall ($\theta = 180°$)

$$\psi_w^* = \psi_w + 90 - \theta = \psi_w - 90 = \tfrac{1}{2}(\omega - \phi_w) \qquad (7.8.7)$$

We can now evaluate the radial stress field for the case of a vertical wall with $\phi = 30°$ and $\phi_w = 20°$ as in the previous section, using the equations presented in §7.4. From equations (7.8.5) to (7.8.7), the boundary conditions are:

on $\theta = 180°$, $\quad \psi_w^* = 11.580°$,

and on $\theta = 150°$, $\quad \psi^* = 30°$ and $q = \dfrac{\cos 30}{1 + \sin 30} = \dfrac{1}{\sqrt{3}} = 0.57735.$

It is inconvenient to start the calculations at $\theta = 150°$ since here both the numerator and denominator of equation (7.4.12) are zero and $dq/d\theta$ becomes indeterminate. Instead, we can use a shooting technique, starting with assumed values of q at the wall and integrating from 180° to 150°. The values of ψ^* and q at $\theta = 150°$ have been found for various values of q on the wall using Runge–Kutta integration with a step length of 1°. The results of these calculations are given in table 7.5.

Plotting these results, as in figure 7.15 shows that the calculations become unreliable when ψ_{150}^* approaches the critical value of $\epsilon = 30°$ since the evaluation of both $d\psi^*/d\theta$ and $dq/d\theta$ become subject to large round-off errors as the numerators and denominators of equations (7.4.12) and (7.4.13) fall to zero. Thus the results of the last two columns of this table are suspect. The markedly anomalous value for $q_w = 0.524$ occurs because the predicted value at some intermediate stage of the calculation became almost exactly 30° with the result that an erroneous slope was predicted, throwing out all the subsequent calculations. However, the value of q_w that would give $\psi_{150}^* = 30°$, if the calculations had been perfect, can readily be found by extrapolating the results of the first three columns, giving $q_w = 0.5228$.

It will be noted that we have not used the result that $q_{150} = 0.577\,35$. Examination of the table shows that this has been achieved automatically. If q had not equalled this value on OC the numerator of equation (7.4.12) would not have been zero, giving $dq/d\theta = \infty$, and hence q

Table 7.5

q_w	0.516	0.518	0.520	0.522	0.524
ψ_{150}	29.448	29.604	29.764	29.949	28.265
q_{150}	0.5689	0.5713	0.5737	0.5755	0.5993

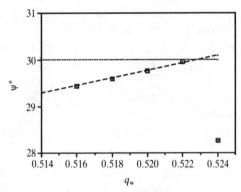

Figure 7.15 Variation of ψ^* on the bounding characteristic of the fan with choice of wall stress parameter q_w. Note that the calculations become unreliable as ψ^* approaches the critical value of 30°.

would have rapidly adjusted to the desired value. Thus the apparent over-specification of the problem, by having three boundary conditions for two first order differential equations, does not cause any problems since the values of q and ψ on OC are not independent. However, the fact that q is a constant along the whole of OC shows that the radial stress field is exact for the whole of the region OCF. This may be contrasted with our use of the radial stress field equations in §7.4 where we only had two boundary conditions, namely the values of ψ^* at two values of θ. Under these circumstances the radial stress field gives the asymptote to which the full solution tends near the apex.

We can compare the result of this calculation,

$$q_w = 0.5228$$

i.e.

$$\sigma_w = \gamma r\, q_w(1 - \sin\phi \cos 2\psi^*) = 0.2825\,\gamma y$$

with the stress gradient of the previous example (0.2822) and with the result of the method of wedges (0.279). It is seen that the two former results are in excellent agreement and are somewhat greater than those found by the method of wedges. This result is self-evident; in the method of wedges we were looking for the slip surface which gave the largest force on the wall, but we confined our search to planar surfaces. It is therefore not surprising that we can find a surface giving an even greater wall force. Thus the method of wedges inevitably under-estimates the wall force. This is a well-known result in the field of soil mechanics where the problem of prime interest is the determination of the surcharge that a wall of given strength can support. Here the method of wedges will over-estimate the surcharge that can be supported by a wall of specified strength and is said to give an *upper bound* solution to the problem. The method of characteristics is similarly said to give a *lower bound* solution. If, on the other hand, the objective is the prediction of the wall force for a given surcharge, the upper bound solution will, paradoxically, give the lower prediction.

The lower bound solution is correct for a material in incipient failure against a perfectly rigid wall. If, on the other hand, the wall were to move outwards slightly, slip would occur along the appropriate β-characteristic. However, slip could not be sustained since the curvature of the β-characteristic increases in the direction of motion. The material would then cease to be in a state of incipient yield. It can be shown that if the material obeys the associated flow rule, discussed in §8.3 below, the stresses exerted by a static fill on an elastic retaining wall would lie between the upper and lower bound solutions. In steady flow, however, when slip can occur along both sets of characteristics, the solution by the method of characteristics, i.e. the lower bound solution, is correct.

In our example there is very little difference between the upper and lower bound solutions. As discussed in §4.6 this is because the β-characteristics are bent through the small angle $\frac{1}{2}(\omega - \phi_w) = 11.58°$ and therefore the prediction based on the assumption that the slip surfaces are planar is very nearly correct.

Finally, we can note that the Rankine zone is also part of the radial stress field. This can be checked by putting $\psi^* = 180 - \theta$ and $q = -\cos \theta / (1 + \sin \phi)$ into the equations for the radial stress field in polar co-ordinates. Thus we can plot the dependence of ψ^* on θ for this example as in figure 7.16. The sudden change in behaviour at $\theta = 150°$ is related to the observation noted above that the equations of the radial stress field become indeterminate when $\cos \psi^* = \sin \phi$.

Very large surcharges

The case of a very large surcharge is particularly simple. If $Q_0 \gg \gamma y_{max}$ where y_{max} is the greatest depth in which we are interested, we can neglect the weight of the material. Under these circumstances all the α-characteristics become straight as shown in figure 7.17, and the material can be divided into three regions:

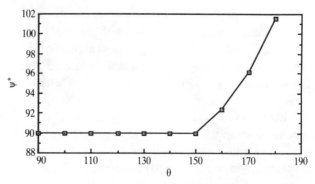

Figure 7.16 Variation of the stress parameter ψ^* with angular position θ, showing a discontinuity of slope at the edge of the fan.

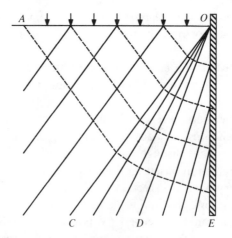

Figure 7.17 Characteristic mesh for the single retaining wall with zero surcharge.

For $\theta < 180 - \epsilon$, we have a Rankine zone AOC in which $\psi = \psi_0 = 90°$ and, from equation (7.7.9),

$$p = p_0 = \frac{q_0}{1 + \sin \phi} \qquad (7.8.8)$$

Below the Rankine zone there is a fan COD which subtends the angle $\psi_w - \psi_0$ in which $\psi^* = \epsilon$ or, from equation (7.2.11),

$$\psi = \theta + \epsilon - 90 \qquad (7.8.9)$$

and in which, from equation (7.7.20),

$$p = p_0 \exp[-2\mu(\psi - \psi_0)] \qquad (7.8.10)$$

or

$$p = p_0 \exp[-2\mu(\theta + \epsilon - 180)] \qquad (7.8.11)$$

The lowest characteristic of the fan is the line $\theta = \psi_w + 90 - \epsilon$ and below this we have a second Rankine zone DOE in which $\psi = \psi_w$ and p has the constant value p_w given by

$$p_w = p_0 \exp[-2\mu(\psi_w - \psi_0)] \qquad (7.8.12)$$

Within both the Rankine zones, all the stress components are constant and therefore clearly satisfy the equations of static equilibrium, equations (7.2.4) and (7.2.5) for a weightless material, i.e. $\gamma = 0$. It is left as an exercise for the reader to confirm that equations (7.8.9) and (7.8.11) satisfy the equilibrium equations in polar co-ordinates, equations (7.2.12) and (7.2.13).

7.9 Inclined top surfaces, inclined walls and cohesive materials

Inclined top surfaces

The case of a cohesionless material with an inclined top surface, such as that shown in figure 7.18, can be treated by the same techniques as the previous problem. Here the only difficulty is formulating the boundary condition on the top surface.

Since the top surface is outside the zone of influence of the wall, the stresses are identical to those beneath an infinite surface of the same slope. We can take axes (n,s) normal and along the surface and by analogy with equations (7.2.4) and (7.2.5), we have

$$\frac{\partial \sigma_{ss}}{\partial s} + \frac{\partial \tau_{ns}}{\partial n} = -\gamma \sin \alpha \qquad (7.9.1)$$

$$\frac{\partial \sigma_{nn}}{\partial n} + \frac{\partial \tau_{ns}}{\partial s} = \gamma \cos \alpha \qquad (7.9.2)$$

where α is the angle between the top surface and the horizontal. By symmetry, all derivatives with respect to s must be zero and hence at a small distance δn below the surface

$$\tau_{ns} = -\gamma \sin \alpha \, \delta n \qquad (7.9.3)$$

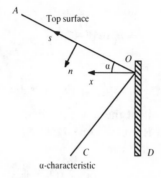

Figure 7.18 Definition of the co-ordinates (n,s) for a fill with an inclined upper surface.

and

$$\sigma_{nn} = \gamma \cos \alpha \, \delta n \qquad (7.9.4)$$

Assuming that we have active failure, $\sigma_{nn} > \sigma_{ss}$, and we can identify the point N representing the n-plane, on the Mohr–Coulomb diagram of figure 7.19. The angle NOC is clearly α and by analogy with the analysis of §3.7 the angle A is given by

$$\sin A = \frac{\sin \alpha}{\sin \phi} \qquad (7.9.5)$$

The s-plane lies at the opposite end of the diameter from N and is denoted by S. The point X representing the x-plane is inclined at 2α anticlockwise from S and therefore

$$2\psi = 180 + A + \alpha - 2\alpha \qquad (7.9.6)$$

i.e.

$$\psi = 90 + \tfrac{1}{2}(A - \alpha) \qquad (7.9.7)$$

Within the region AOC of figure 7.18, where OC is the α-characteristic through O, having slope $\tan [90 + \tfrac{1}{2}(A-\alpha) - \varepsilon]$, we have a Rankine-type zone in which the characteristics are straight. It can be seen from figure 7.19 and equation (7.9.4) that

$$p = \frac{\sigma_{nn}}{1 + \sin \phi \cos(A + \alpha)} = \frac{\gamma n \cos \alpha}{1 + \sin \phi \cos(A + \alpha)} \qquad (7.9.8)$$

and ψ is given by equation (7.9.7). Within the region COD the stress

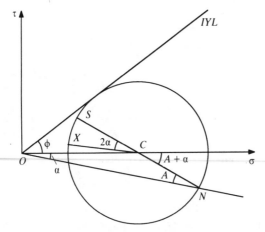

Figure 7.19 Mohr's circle for the stresses below an inclined surface.

distribution can be found from the radial stress field as in the previous section.

It may be noted that there is no solution to equation (7.9.5) for $\alpha > \phi$. This proves that for a cohesionless material the angles of repose and internal friction are equal, a result we derived by a less convincing method in §3.5.

Inclined retaining walls

When considering a right-hand wall inclined to the vertical by η as shown in figure 7.20, we can note that the major principal stress plane is inclined to the wall plane by $90 + \frac{1}{2}(\omega - \phi_w)$ in active failure and by $-\frac{1}{2}(\omega + \phi_w)$ in passive failure. Thus in active failure

$$\psi = 90 + \tfrac{1}{2}(\omega - \phi_w) - \eta \qquad (7.9.9)$$

and in passive failure

$$\psi = -\tfrac{1}{2}(\omega + \phi_w) - \eta \qquad (7.9.10)$$

which is entirely equivalent to

$$\psi = 180 - \tfrac{1}{2}(\omega + \phi_w) - \eta \qquad (7.9.11)$$

This latter form is more convenient since, as we will see in §7.12, it is helpful to define ψ so that it lies within the range $0 < \psi < 180$.

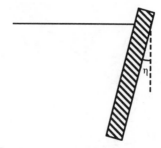

Figure 7.20 Inclined retaining wall.

In the case of active failure in a horizontal fill, ψ changes in the fan emanating from the corner O from its value of 90° in the Rankine zone to the wall value of $[90+\frac{1}{2}(\omega-\phi_w)-\eta]$. As η is increased, the width of the fan narrows and when $\eta = \frac{1}{2}(\omega-\phi_w)$, the fan disappears and the Rankine zone extends up to the wall.

For values of $\eta < \frac{1}{2}(\omega-\phi_w)$, a radial stress field solution can be obtained as in §7.8. For this it is convenient to express the principal stress direction by ψ^* as defined by equation (7.2.11). On the Rankine zone boundary, $\theta = 180-\varepsilon$, we have

$$\psi^* = \varepsilon \qquad (7.9.12)$$

as in equation (7.8.5), and on the wall, $\theta = 180-\eta$,

$$\psi^* = 90 + [90+\tfrac{1}{2}(\omega-\phi_w)-\eta] - (180-\eta) = \tfrac{1}{2}(\omega-\phi_w) \quad (7.9.13)$$

as in equation (7.8.7). Thus the change in ψ^* is independent of the value of η so that as η is decreased the gradients of ψ^* increase in magnitude, reaching the Rankine value of -1 when $\eta = \frac{1}{2}(\omega-\phi_w)$.

For values of $\eta > \frac{1}{2}(\omega-\phi_w)$, the α-characteristics leaving the top surface are steeper than those leaving the wall and therefore these characteristics appear to cross. Under these circumstances a stress discontinuity is formed. Stress discontinuities are discussed in §7.12 and the rest of the discussion of inclined retaining walls is therefore reserved until that section.

Cohesive materials

As noted in §7.2, the differential equations governing stress distributions in granular materials are independent of the cohesion. The methods described in the preceding sections can therefore be used for cohesive materials if p is replaced by p^*. On the other hand, the boundary conditions do depend on the value of the cohesion.

On a stress-free horizontal surface in active failure, we have that $\sigma_{yy} = 0$ and $\psi = 90°$. Therefore from equation (7.2.2) we have

$$p^* = \frac{c \cot \phi}{1 + \sin \phi} \qquad (7.9.14)$$

and on the wall we must satisfy the wall failure criterion,

$$\tau_w = c_w + \sigma_w \tan \phi_w \qquad (7.9.15)$$

In general, a right-hand wall calculation consists of the simultaneous solution of the two equations along the β-characteristic, the equation of the wall ($x = 0$) and equation (7.9.15). However, for the special case when $c_w \cot \phi_w = c \cot \phi$, equation (7.9.15) reduces to

$$\psi_w = 90 + \tfrac{1}{2}(\omega - \phi_w) \qquad (7.9.16)$$

in the active case and the wall calculation is just the same as in a cohesionless material.

Cohesive materials with non-linear yield loci can be treated by the methods described in the preceding sections if we take ϕ and c to be the equivalent angle of friction ϕ_e and equivalent cohesion c_e as defined by equation (6.5.9). Since the values of ϕ_e and c_e are not constant but depend on the current stress state, they must be up-dated at every step of the integration.

7.10 Twin retaining walls

Once the analysis of the single retaining wall, as presented in §7.7 and §7.8, has been mastered, the analysis of the twin retaining wall problem becomes straight-forward. Similarly the wedge-shaped hopper can be solved by combining the techniques of this section and those of §7.9.

Let us consider twin vertical walls a distance $2a$ apart as in figure 7.21 and again pay attention to the active case with $\phi = 30°$, $\phi_w = 20°$ and $Q_0 = 0.3 \, \gamma a$ as in §7.7. We can solve the stress distribution throughout the region between the walls or save effort by noting that the system is symmetrical about the mid-plane and pay attention only to the half-problem. To preserve compatibility with the problem of §7.7, we will work in terms of the dimensionless parameters X, Y and P and again start with points along the top surface at intervals of 0.2 in X. Since the region OAC of figure 7.21 lies outside the zones of

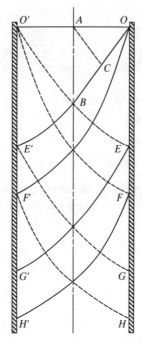

Figure 7.21 Characteristic mesh for twin retaining walls.

influence of either the left-hand wall or the left-hand half of the top surface, the stresses in this region are identical to those beneath the single retaining wall, and have already been evaluated in §7.7.

We can break out of this region across the line AC in the same way as we broke out across OC in §7.7. Figure 7.22 shows the region in the vicinity of point A. The $\alpha = 1$ characteristic cuts the centre-line at point D which has Cartesian co-ordinates $(1,Y)$. By symmetry $\psi = 90°$ along the centre-line but the value of P at point D is as yet unknown. The equations along the α-characteristic plus the conditions

$$X = 1 \tag{7.10.1}$$

and

$$\psi = 90° \tag{7.10.2}$$

are sufficient for solution and we will call a calculation of this sort a 'centre-line calculation'.

Using this method we find that point D has co-ordinates and stress parameters $X = 1$, $Y = 0.3464$, $P = 0.4310$ and $\psi = 90°$. Since there

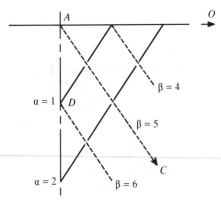

Figure 7.22 Characteristic mesh near the centre-line.

has been no change in the value of ψ no iteration is required, nor is it necessary to start a fan of characteristics from the point A. Indeed, if we were to attempt to do so we would find that all the characteristics were exactly co-incident with the $\beta = 5$ characteristic since ψ is single-valued at point A. We can denote point D as **(1,6)** and start the $\beta = 6$ characteristic from it. The calculations can then proceed as in §7.7.

It will be noted that the co-ordinates and stress parameters of point D are typical of those for the second row of points in §7.7 i.e. points **(2,5)**, **(3,4)**, **(4,3)** and **(5,2)**. Indeed, it is self-evident that we have a Rankine zone throughout the region $OO'B$ of figure 7.21, since this region is outside the zones of influence of either wall.

Within the region OBE, we can use the method of characteristics, or the radial stress field solution as appropriate and hence we see that the wall stresses are identical to those on a single retaining wall and therefore increase linearly with depth up to the point E. Thus the examples of §7.7 and §7.8 have given us the stress distribution throughout the region $OABE$ and we can use the values along the β-characteristic BE as the starting point for the rest of this problem. The solution for the region below BE is straight-forward. We can use the basic calculation whenever an α- and a β-characteristic intersect, the wall calculation when a β-characteristic meets the wall and the centre-line calculation when an α-characteristic meets the centre-line. In both these last two cases the characteristic terminates and a new characteristic of the other family can be started.

At point E the wall first encounters the influence of the far corner O'. This can alternatively be regarded as the reflection of the influence of O in the centre-line. The section EF is influenced both by the top surface and the corner and below F the influence of the far wall is felt. Thus we must expect to find sudden changes in the stress gradients at both E and F. Furthermore we must expect to find echos of these changes as the characteristics from E' and F' meet the wall at G and H. It can be seen that figure 7.21 shows striking similarity with figure 4.25 which shows the results for the same problem solved by the method of wedges. The main difference is that whereas in figure 4.25 the slip surfaces are straight by definition, the characteristics in figure 7.21 are gently curved except in the region $OO'B$. The wall stresses obtained by the two methods are compared in figure 7.23 and it is seen that there is substantial agreement between the two predictions. As discussed in §7.8, the solution by the method of characteristics will be correct for a rigid wall but for an elastic wall the stresses will lie between those predicted by these upper and lower bound solutions.

Hancock (1970) has used the method of characteristics to assess the accuracy of Janssen's analysis. For a cohesionless material with zero surcharge, the exact stress analysis gives a wall stress profile which consists of a series of almost straight sections, the first being exactly straight. None-the-less, the numerical difference from the exponential curve predicted by Janssen is very small, typical errors being only 1 or 2%. However, greater discrepancies occur in the presence of a surcharge as can be seen in figure 7.24. Indeed if Q_0 is greater than the asymptotic value of σ_{zz} the initial stress gradients are of opposite sign. Janssen's analysis predicts that the stresses will decrease monotonically to the asymptotic values, but in reality there is initially

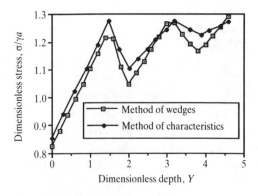

Figure 7.23 Comparison of the stresses predicted by the method of wedges and by the method of characteristics.

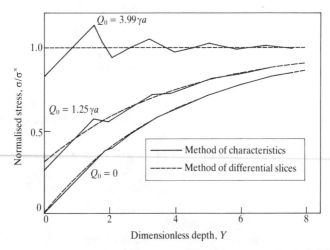

Figure 7.24 Comparison of the stresses predicted by the method of differential slices and by the method of characteristics.

an increase in the wall stresses throughout the region OE of figure 7.21 before the influence of the far wall is felt and the stresses begin to decrease. Thus while Janssen's method gives an excellent prediction in the absence of a surcharge, it should not be used when there is a significant surcharge. Furthermore, this result casts considerable doubt on the accuracy of Walters' switch stress analysis, which uses the Janssen's result in the presence of a surcharge.

Very similar calculations can be performed for a cylindrical bunker using the characteristic equations in cylindrical co-ordinates as derived in §7.11 below. However, in this case we cannot use the radial stress field solution throughout the region OBE of figure 7.21, since this is a two-dimensional analysis which cannot strictly be applied in an axi-symmetric situation. The radial stress field solution will be sufficiently accurate for some small region in the neighbourhood of the corner O over which there is negligible change in the distance from the axis of symmetry. We can therefore select an arc of small radius, say $0.1\,a$, about O, calculate the values of P and ψ on this arc from the radial stress field and use these values as the boundary conditions for a solution by the method of characteristics. The initial stress gradient is given by the radial stress field, as before, but the wall stress does not increase linearly throughout the section OE of figure 7.21. The information that the system is axi-symmetric is in the differential equations and the influence of the wall curvature is therefore felt immediately, unlike the case of the parallel walls in which information about the presence of the far wall enters the problem only through the boundary conditions. Figure 7.25 compares the results for the twin-wall case and for a cylindrical bunker for the case of $\phi = 30°$, $\phi_w = 20°$ and $Q_0 = 0$.

It is appropriate at this stage to discuss whether the calculations must proceed downwards as in the examples presented above or whether we could start at the base of the bunker and work upwards.

In principle, the calculations can proceed in either direction and we could start at the base to find the surcharge, or shape of the top surface, that gives a specified stress distribution at the base. In practice this is never done. Not only is this a most unrealistic problem but such calculations are found to be unstable.

The reason for this can be seen by analogy with Janssen's analysis which gives rise to the differential equation (5.2.7), which can be written

$$\frac{d\sigma_{zz}}{dz} + k\,\sigma_{zz} = \gamma \qquad (7.10.3)$$

where

$$k = \frac{4K\mu_w}{D} \qquad (7.10.4)$$

and which has the analytic solution

$$\sigma_{zz} = \frac{\gamma}{k}(1 - e^{-kz}) \qquad (7.10.5)$$

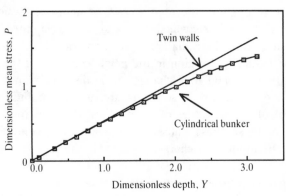

Figure 7.25 Comparison of the wall stresses in a cylindrical bunker with those on twin retaining walls.

If we attempted a numerical solution to equation (7.10.3) starting at the top surface, and at some point made an arithmetical error of magnitude δ, we would find that this error died away exponentially according to $\delta\,e^{-kz}$. On the other hand if we were to work upwards, the inevitable round-off errors would increase exponentially. Thus we see that under all realistic circumstances we must start our calculation at the top surface and work downwards.

The same conclusion can be reached qualitatively by noting that the stresses at a particular section are determined primarily by the weight of the material above that section. The situation is analogous to that in differential equations involving time. We must normally work forward in time since conditions are determined by past events and are unaffected by future events. Calculations working backwards in time to determine what must have happened to cause the current situation rarely give unambiguous results. Thus in situations in which the stresses are primarily determined by the weight of the material, the calculation must proceed downwards. Similarly, in problems concerned with extrusion, the calculations must proceed from the point of origin of the extruding force.

7.11 Characteristic equations in non-Cartesian co-ordinate systems

Polar co-ordinates

The characteristic equations can be derived for polar co-ordinates using the method of §7.6 but starting with the basic stress equations in polar co-ordinates, equations (7.2.17) and (7.2.18). In fact, it is easier to start with the characteristic equations in Cartesian co-ordinates, equations (7.6.29) to (7.6.32) and convert to polar co-ordinates using the simple transformations

$$x = r \sin \theta \qquad (7.11.1)$$

and

$$y = r \cos \theta \qquad (7.11.2)$$

However, it is very rare indeed that it is necessary to use polar co-ordinates. Under almost all circumstances it is more convenient to perform the calculations in Cartesian form and then convert to polar co-ordinates using equations (7.11.1) and (7.11.2). Perhaps the only case in which the use of the characteristic equations in polar form might be more convenient is when considering the wall calculation in a wedge-shaped hopper, as here the equation of the wall takes the simple form

$$\theta = \theta_w \qquad (7.11.3)$$

The calculations using Cartesian co-ordinates is almost as simple. The wall calculation involves the solution of equation (7.6.29) (in the case of an α-characteristic) and the equation of the wall, which, if straight, can be written in the form

$$y = mx + c \qquad (7.11.4)$$

Since both equations (7.6.29) and (7.11.4) are linear, their solution is straight-forward. For curved walls, the equation of the wall, $y = f(x)$ is non-linear and the advantage of using polar co-ordinates disappears.

Cylindrical co-ordinates

The characteristic equations for cylindrically symmetrical systems can be derived by starting from the stress equations in cylindrical co-ordinates, equations (7.2.25) and (7.2.26). Following the procedure of §7.6, we find that the characteristic directions are unchanged so that the equations of the characteristics are:

for the α-characteristic

$$\frac{dz}{dr} = \tan(\psi - \varepsilon) \qquad (7.11.5)$$

and for the β-characteristic

$$\frac{dz}{dr} = \tan(\psi + \varepsilon) \qquad (7.11.6)$$

However, the terms $(\sigma_{rr} - \sigma_{xx})/r$ and τ_{zr}/r appearing in equations (7.2.19) and (7.2.20) give rise to extra terms in the equations along the characteristics, so that equations (7.6.17) and (7.6.18) become:

along the α-characteristic

$$\frac{dp}{ds} - 2p \tan \phi \, \frac{d\psi}{ds} = \gamma \left(\frac{dz}{ds} - \tan \phi \, \frac{dr}{ds} \right) + E_\alpha \qquad (7.11.7)$$

and along the β-characteristic

$$\frac{dp}{ds} + 2p \tan\phi \frac{d\psi}{ds} = \gamma\left(\frac{dz}{ds} + \tan\phi \frac{dr}{ds}\right) + E_\beta \qquad (7.11.8)$$

where

$$E_\alpha = \frac{p \tan\phi}{r}\left(\kappa \cos\phi \frac{dr}{ds} + (1 + \kappa \sin\phi)\frac{dz}{ds}\right) \qquad (7.11.9)$$

and

$$E_\beta = \frac{p \tan\phi}{r}\left(\kappa \cos\phi \frac{dr}{ds} - (1 + \kappa \sin\phi)\frac{dz}{ds}\right) \qquad (7.11.10)$$

Equations (7.11.5) to (7.11.8) can be used in exactly the same way as the corresponding equations in Cartesian co-ordinates, with only a slight increase in the amount of arithmetic resulting from the extra terms E. Difficulties, however, occur on the centre-line due to the presence of the factor $1/r$ in E. By substituting from equation (7.11.5) into (7.11.9) we have

$$E_\alpha = \frac{p \tan\phi}{r}[\kappa \cos\phi + (1 + \kappa \sin\phi)\tan(\psi - \varepsilon)]\frac{dr}{ds} \qquad (7.11.11)$$

which, on rearranging, gives

$$E_\alpha = \frac{p \tan\phi}{r}\left[\frac{\kappa \sin(\psi + \varepsilon) + \sin(\psi + \varepsilon)}{\cos(\psi - \varepsilon)}\right]\frac{dr}{ds} \qquad (7.11.12)$$

When $\kappa = 1$, equation (7.11.12) becomes

$$E_\alpha = \frac{2p \tan\phi \sin\psi \cos\varepsilon}{r \cos(\psi - \varepsilon)}\frac{dr}{ds} \qquad (7.11.13)$$

and when $\kappa = -1$,

$$E_\alpha = -\frac{2p \tan\phi \cos\psi \sin\varepsilon}{r \cos(\psi - \varepsilon)}\frac{dr}{ds} \qquad (7.11.14)$$

This last pair of equations provides what is probably the most convincing explanation that $\kappa = +1$ in the passive case and -1 in the active case. In the passive case $\psi = 0$ on $r = 0$ and it is seen that equation (7.11.14) gives $E_\alpha = -\infty$ and thus equation (7.11.13) must be used. Similarly in the active case $\psi = 90°$ on $r = 0$ and (7.11.14) is the appropriate form.

However, on the centre-line the values of E are indeterminate from the above forms of the equations and we must evaluate the awkward fractions $(\sin\psi)/r$ and $(\cos\psi)/r$ using l'Hopital's rule.

For the passive case, where $\psi = 0$ on $r = 0$,

$$\frac{\sin \psi}{r} = \frac{\cos \psi \, d\psi/ds}{dr/ds} = \frac{d\psi/ds}{dr/ds} \qquad (7.11.15)$$

so that equation (7.11.13) becomes

$$E_\alpha = 2p \tan \phi \frac{d\psi}{ds} \qquad (7.11.16)$$

and, hence, equation (7.11.7) takes the form

$$\frac{dp}{ds} - 4p \tan \phi \frac{d\psi}{ds} = \gamma \left(\frac{dz}{ds} - \tan \phi \frac{dr}{ds} \right) \qquad (7.11.17)$$

In the active case $\psi = 90°$ on $r = 0$ so that

$$\frac{\cos \psi}{r} = -\frac{\sin \psi \, d\psi/ds}{dr/ds} = -\frac{d\psi/ds}{dr/ds} \qquad (7.11.18)$$

and equation (7.11.7) again reduces to equation (7.11.17), which therefore applies equally to the passive and active states.

Similarly, we can evaluate E_β from equations (7.11.6) and (7.11.10) giving

$$E_\beta = -\frac{p \tan \phi}{r} \left[\frac{\kappa \sin(\psi - \varepsilon) + \sin(\psi + \varepsilon)}{\cos(\psi - \varepsilon)} \right] \frac{dr}{ds} \qquad (7.11.19)$$

For $\kappa = +1$, this becomes

$$E_\beta = -\frac{2p \tan \phi \, \sin \psi \cos \varepsilon}{r \cos(\psi + \varepsilon)} \frac{dr}{ds} \qquad (7.11.20)$$

and, when $\kappa = -1$,

$$E_\beta = -\frac{2p \tan \phi \cos \psi \sin \varepsilon}{r \cos(\psi + \varepsilon)} \frac{dr}{ds} \qquad (7.11.21)$$

On the centre-line, $r = 0$, equation (7.11.8) becomes

$$\frac{dp}{ds} + 4p \tan \phi \frac{d\psi}{ds} = \gamma \left(\frac{dz}{ds} + \tan \phi \frac{dr}{ds} \right) \qquad (7.11.22)$$

for both the active and passive cases.

Equations (7.11.5) to (7.11.8) can be put in finite difference form and used in exactly the same way as their Cartesian counterparts except on the centre-line where equations (7.11.17) and (7.11.22) replace equations (7.11.7) and (7.11.8). Clearly, the terms E_α and E_β will be small at large r and will increase in magnitude as the centre-line is approached. It is prudent to watch these terms when performing calculations at small r to detect round-off errors and to transfer to the form of equations (7.11.17) and (7.11.22) when either these errors become significant, or when the centre-line is actually reached.

Caution must also be exercised to avoid difficulties with the sign convention for shear stresses, always potentially present in axisymmetric cases. The wall condition $r = a$ refers to the left-hand wall since we must always pay attention to the positive quadrant. Here the shear stresses will be negative both in the active case and in the passive case with negative wall shear. Hence, in the active case

$$2\,\psi_w = 180 - \tfrac{1}{2}(\omega - \phi_w) \qquad (7.11.23)$$

and in the passive case

$$2\,\psi_w = \tfrac{1}{2}(\omega + \phi_w) \qquad (7.11.24)$$

Spherical co-ordinates

As with polar co-ordinates there are few, if any, circumstances in which it is advantageous to use the characteristic equations in spherical co-ordinates. It is best to perform the calculations using cylindrical co-ordinates and later convert the results to spherical co-ordinates. Moreea (1990) has, however, used the characteristic equation in spherical co-ordinates successfully in the analysis of a conical hopper.

7.12 The stress discontinuity

We saw in §5.9 that if a retaining wall is inclined to the vertical by more than some critical angle, the α-characteristics emanating from the wall are less steep than those coming from the top surface. Thus the characteristics can intersect as shown in figure 7.26(*a*). We can arbitrarily select a pair of converging α-characteristics, and find the β-characteristic that passes through the point of intersection. The stress parameters at this point can be evaluated by using the equation along the β-characteristic and the equation along *either* of the α-characteristics. If we do so we find that we obtain two different predictions of ψ and P depending on which of the two α-characteristics is used. This is a physically impossible situation and there is therefore a flaw in the

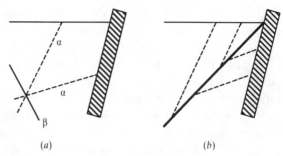

Figure 7.26 Intersecting α-characteristics and the formation of a discontinuity.

analysis of §7.6 which led to the derivation of the characteristic equations. The weak point in this analysis is the statement below equation (7.6.8) that the term in $\partial\psi/\partial y$ disappears when its coefficient is zero. This is only true provided $\partial\psi/\partial y$ is not infinite. Thus we conclude that $\partial\psi/\partial y$ is infinite whenever characteristics of the same family intersect and that the method of characteristics is only valid in regions in which there are no discontinuities. It therefore follows that characteristics terminate when they reach a stress discontinuity.

The existence of stress discontinuities can be seen qualitatively by considering the fan of characteristics that emerges from the top corner of a bunker. As these characteristics diverge, the initially discontinuous change in the principal stress direction gradually become less sudden. By the same token, if the characteristics of a particular family converge, the stress gradients steepen and eventually become discontinuous. Such discontinuities can be formed within the material but most commonly they are formed at a wall, particularly when there is a sudden change in the angle of wall friction or the wall slope. The discontinuity discussed in §7.9 is a typical example of this phenomenon.

Stress discontinuities can be analysed by considering the surface across which the stresses are discontinuous as shown in figure 7.27. We will take axes (n,s) normal and along the surface and denote the stress parameters on one side of the discontinuity by the superscript $^-$ and on the other side by $^+$. It is clear from the stability of a particle lying on the discontinuity that the normal and shear stresses σ_{nn} and τ_{ns} must be continuous, i.e.

$$\sigma_{nn}^- = \sigma_{nn}^+ \tag{7.12.1}$$

and

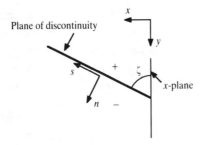

Figure 7.27 Definition of the discontinuity angle ζ.

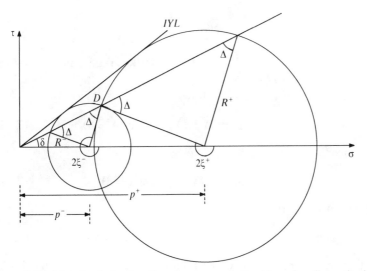

Figure 7.28 Mohr's circles for the stresses on both sides of a discontinuity.

$$\tau_{ns}^- = \tau_{ns}^+ \tag{7.12.2}$$

but there is no necessity that

$$\sigma_{ss}^- = \sigma_{ss}^+ \tag{7.12.3}$$

Indeed, if these stresses are equal, we have no discontinuity.

We can mark the plane of the discontinuity on the Mohr–Coulomb diagram of figure 7.28 by the point D, with co-ordinates (σ_{nn}, τ_{ns}). If the material on both sides of the discontinuity is in a state of incipient yield, the Mohr's circle passing through D must touch the Coulomb line, and clearly there are two such circles, one corresponding to each side of the discontinuity. Traditionally we use the superscript $^+$ to refer to the larger of the two circles and denote the centres of these circles by p^+ and p^-, and their radii by R^+ and R^-. Figure 7.28 and the analysis below considers the case of a cohesionless material. The corresponding results for a cohesive material can be obtained by replacing p by p^*.

The analysis of this situation is simplified by constructing the line joining D to the origin. Denoting the angle between this line and the σ axis by δ, we can perform an analysis analogous to that of §3.7 which shows that the four angles marked Δ are equal and given by

$$\sin \Delta = \frac{\sin \delta}{\sin \phi} \qquad (7.12.4)$$

It should be noted that D can lie below the σ-axis, in which case δ and Δ are taken to be negative quantities.

Examination of figure 7.28 shows that

$$\tau_{ns} = R^- \sin(\Delta + \delta) = R^+ \sin(\Delta - \delta) \qquad (7.12.5)$$

and since $R^- = p^- \sin \phi$ etc., we have

$$p^- \sin(\Delta + \delta) = p^+ \sin(\Delta - \delta) \qquad (7.12.6)$$

We can also see that

$$\sigma_{nn} = p^- + R^- \cos(\Delta + \delta) = p^+ - R^+ \cos(\Delta - \delta) \qquad (7.12.7)$$

but this equation gives us no extra information since equations (7.12.4), (7.12.6) and (7.12.7) form a redundant set.

The angle ξ^- measured anticlockwise from the plane of the discontinuity to the major principal plane is seen to be given by

$$2\xi^- = 360 - (\Delta + \delta) \qquad (7.12.8)$$

and, similarly,

$$2\xi^+ = 180 + (\Delta - \delta) \qquad (7.12.9)$$

Since δ lies in the range $- \phi < \delta < \phi$ and Δ lies in the range $- 90 < \Delta < 90$, it can be seen that ξ^- lies in the range

$$135 - \tfrac{1}{2} \phi < \xi^- < 225 + \tfrac{1}{2}\phi \qquad (7.12.10)$$

and that ξ^+ lies in the range

$$90 - \tfrac{1}{2}\phi < \xi^+ < 90 + \tfrac{1}{2}\phi \qquad (7.12.11)$$

In figure 7.27. we denote the angle between the x-plane and the plane of the discontinuity by ζ. This is clearly the difference between ψ, the angle between the x-plane and the major principal plane, and ξ, the angle between the discontinuity and the major principal plane. This applies on both sides of the discontinuity, so we can say that

$$\zeta = \psi^- - \xi^- \tag{7.12.12}$$

and

$$\zeta = \psi^+ - \xi^+ \tag{7.12.13}$$

and therefore, from equations (7.12.8) and (7.12.9),

$$\psi^- - \psi^+ = \xi^- - \xi^+ = 90 - \Delta \tag{7.12.14}$$

We have chosen to confine Δ to the range $-90 < \Delta < 90$ and therefore $\psi^- - \psi^+$ must lie in the range

$$0 < \psi^- - \psi^+ < 180 \tag{7.12.15}$$

If we know the value of ψ on one side of a discontinuity of known slope ζ, we can evaluate ξ from

$$\xi = \psi - \zeta \tag{7.12.16}$$

and comparing with the ranges (7.12.10) and (7.12.11), determine whether this is the value of ξ^+ or ξ^-. The values of δ and Δ can then be obtained from the simultaneous solution of equation (7.12.4) with either equation (7.12.8) or (7.12.9). All the other parameters follow directly from the remaining equations.

More commonly, we know the values of ψ on both sides of a discontinuity of unknown slope, and it is not immediately clear which side corresponds to the $^+$ side and which to the $^-$ side. It will be recalled that equation (7.12.15) specifies that $\psi^- > \psi^+$ but this does not resolve the problem since multiples of 180° can be added or subtracted from these angles at will. Thus if two values ψ_1 and ψ_2 are specified with $\psi_1 > \psi_2$, we can say

$$\left.\begin{array}{l} \textit{either } \psi^- = \psi_1 \text{ and } \psi^+ = \psi_2 \\[2ex] \textit{or } \psi^- = \psi_2 + 180 \text{ and } \psi^+ = \psi_1 \end{array}\right\} \tag{7.12.17}$$

Thus there are two solutions to this problem giving rise to discontinuities of different slope and we have to determine which solution is appropriate to the problem in hand.

We can illustrate the use of these equations by considering the active state in a material with $\phi = 30°$ in contact with a vertical wall which has a sudden change in ϕ_w from 25° to 10° at point A of figure 7.29. The value of ψ in the upper region is found from equation (7.7.6) to be 106.35° and, similarly, in contact with the lower part of the wall $\psi = 95.16°$. Thus, recalling that $\varepsilon = \frac{1}{2}(90-\phi) = 30°$, we see that the α-characteristics adjacent to the upper wall are at 76.35° to the horizontal and those adjacent to the lower wall are at 65.16°. Thus these characteristics converge and a discontinuity is formed.

Equation (7.12.15) shows that $\psi^- > \psi^+$ and we will therefore start by taking the upper region to be the $^-$ region giving $\psi^- = 106.35°$ and $\psi^+ = 95.16°$. Thus from equation (7.12.14) we find that $\Delta = 90+95.16-106.35 = 78.81°$ and from equation (7.12.4), we have $\delta = 29.37°$. We can evaluate ξ^- from equation (7.12.8) giving $\xi^- = 125.91°$ and equation (7.12.12) gives $\zeta = -19.56°$. Thus the plane of the discontinuity is 19.56° *clockwise* from the x-plane or 70.44° from the horizontal. The discontinuity is marked as the heavy line on figure 7.29 and it can be seen that the plane of the discontinuity lies between the two sets of α-characteristics which terminate when they reach the discontinuity.

From equation (7.12.6) we have

$$\frac{p^+}{p^-} = \frac{\sin(\Delta + \delta)}{\sin(\Delta - \delta)} = 1.251$$

and, since $\sigma_w = p(1 + \sin \phi \cos 2\psi)$,

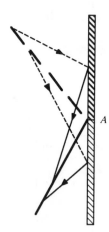

Figure 7.29 Formation of a discontinuity between converging characteristics.

$$\frac{\sigma_w^+}{\sigma_w^-} = 1.251 \times \frac{1 + \sin \phi \cos 2\psi^+}{1 + \sin \phi \cos 2\psi^-} = 1.097$$

Thus the sudden change in the angle of wall friction causes a 9.7% increase in the wall stress.

We must also consider what would have happened if we had taken the other alternative presented by equation (7.12.17), i.e. $\psi^+ = 106.35°$ and $\psi^- = 180 + 95.16 = 275.16°$. From equations (7.12.14) and (7.12.4) we find that $\Delta = -78.81°$ and $\delta = -29.37°$. These are simply the negative of the values obtained previously. Equation (7.12.8) gives $\xi^- = 234.09°$ and equation (7.12.12) gives $\zeta = 41.07°$. This lies between the upper and lower sets of β-characteristics as shown by the heavy dashed line in figure 7.29. Recalling that we must work downwards in situations in which the stresses are determined by gravitational effect, we see that this discontinuity bisects the *diverging* β-characteristics and so will not be formed spontaneously. The only situation in which this discontinuity is meaningful is when some sudden change in the surcharge forms a discontinuity which happens to meet the wall where there is a compatible change in the angle of wall friction angle. Thus this represents a mathematically permissible but physically improbable situation.

There seems to be no easy ritual for determining which of the two discontinuities is appropriate for a particular situation and the reader is advised to sketch the pattern of characteristics and possible discontinuities. Which of the two possibilities is correct can then be seen by inspection.

We can now re-consider the analysis of the inclined retaining wall of §7.9 for the case when the wall inclination η exceeds the critical value of $\frac{1}{2}(\omega - \phi_w)$, which for our usual example of $\phi = 30°$ and $\phi_w = 20°$, is 11.580°. We will consider a wall inclined at 25° to the vertical so that on the top surface $\psi = 90°$ and on the wall $\psi = 90 + 11.580 - 25 = 76.58°$. We will treat this problem in two stages, first with a very large surcharge Q_0 and then with zero surcharge.

In the former case we can put $\psi^- = 90°$ and $\psi^+ = 76.58°$, which by the method above gives $\Delta = 76.58°$, $\delta = 29.10°$, $\xi^- = 127.16°$ and $\zeta = -37.16°$. The discontinuity is inclined to the horizontal by $90 - 37.16 = 52.84°$ and therefore lies between the two sets of α-characteristics, which are at 60° and 46.58° to the horizontal, as it must. This is illustrated in figure 7.26(*b*). The other discontinuity is obtained by putting $\psi^- = 76.58 + 180$ and $\psi^+ = 90°$, from which we find that $\zeta = 23.73°$ and this discontinuity lies entirely outside the material and has no physical meaning. Above the discontinuity, p^- has the constant value of $Q_0/(1-\sin \phi \cos 2\psi^-) = 0.667Q_0$, since we can neglect the weight of the material, and from equation (7.12.6) we have

$$p^+ = 0.667 \, Q_0 \frac{\sin(76.58 + 29.10)}{\sin(76.58 - 29.10)} = 0.871 \, Q_0$$

Thus we have a zone of constant p ($= 0.667 \, Q_0$) above the discontinuity and a zone of constant p ($= 0.871 \, Q_0$) below the discontinuity. The wall stress is given by

$$\sigma_w = p^+[1 + \sin \phi \cos (2\psi^+ + 2\eta)]$$
$$= 0.871 \, Q_0[1 + \sin 30 \cos (2 \times 101.58)] = 0.471Q_0$$

For small or zero surcharges we must allow for the weight of the material and the regions above and below the discontinuity are no longer regions of constant stress. In the absence of a surcharge we can look for a solution based on the radial stress field. Above the discontinuity we have, as in §7.8 $\psi = 90°$ and $p = -\gamma r \cos \theta/(1 + \sin \phi)$. If we assume that a discontinuity occurs at an angle ζ, we can evaluate the values of ψ and p below the discontinuity from the equations derived earlier in this section, and then use these as the boundary conditions for the radial stress field equations below the discontinuity. Integrating these equations up to the wall, $\theta = 180 - \eta$, gives us ψ_w, from which we can find the value of ϕ_w corresponding to the chosen value of ζ. Thus, by trial and error, we can find the value of ζ appropriate to the specified values of ϕ_w and η.

As an example, let us assume that a discontinuity is formed at $\theta = 146°$, i.e. just within the Rankine zone. We have that $\zeta = 146 - 180 = -34°$ and $\psi = 90°$ from which we find, using equation (7.12.16) that, $\xi = \psi - \zeta = 124°$. This lies within the limits imposed by equation (7.12.10) and therefore represents the value of ξ^-. Thus from equation (7.12.8), $\Delta + \delta = 112°$ and hence we find, $\Delta = 82.298°$, $\delta = 29.702°$. Equation (7.12.14) then gives $\psi^+ = \Delta = 82.298°$.

Above the discontinuity,

$$p^- = -\gamma r \frac{\sin \theta}{1 + \sin \phi} = 0.5527 \, \gamma r$$

and hence from equation (7.12.6), $p^+ = 0.6451 \, \gamma r$.

Integrating the radial stress field equations with the boundary conditions $q = 0.6451$, $\psi^* = 82.298+90-146 = 26.298°$ on $\theta = 146°$, we find that ψ^* falls to the value of $11.580°$ on $\theta = 153.73°$ with the corresponding value of q being 0.797. Thus the stress discontinuity at $\theta = 146°$ is appropriate for a wall with $\phi_w = 20°$ inclined at $26.27°$ to the vertical.

Hence, by trial and error, we find that a wall inclined at $25°$ to the vertical gives a discontinuity at $\theta = 146.72°$. The corresponding values are $\zeta = -33.28°$, $\zeta^- = 123.23°$, $\Delta+\delta = 113.44°$, $\Delta = \psi^+ = 83.64°$, $\delta = 29.80°$, $q^- = 0.5573$, $q^+ = 0.6333$, $\psi^* = 26.92°$, $q_w = 0.786$ and $\psi_w^* = 11.58°$ (as it must).

The resulting wall stress is given by

$$\sigma_w = \gamma r q_w (1 - \sin\phi\cos 2\psi^*)$$
$$= 0.786\,\gamma(-y\sec 155)(1 - \sin 30\cos 23.16) = 0.385\,\gamma r$$

It will be noted that the angle $\zeta = 146.72-180 = -33.28°$ is not the same as that found above for a large surcharge, i.e. $-41.07°$. Thus we can deduce that for a modest surcharge the slope of the discontinuity will be $-41.07°$ at the top corner, where the weight of the material can be neglected, but the discontinuity will be curved, eventually reaching a slope of $-33.28°$ at great depth where the effect of the surcharge is negligible compared with the weight.

The prediction of the precise shape of a curved discontinuity is beyond the scope of this book and the interested reader is referred to Horne and Nedderman (1978a, b) who consider this matter in detail.

We can now give further consideration to the case of a vertical wall supporting a fill with a very large surcharge, which we first considered in §7.8.

Let us postulate that instead of the fan solution, a discontinuity is formed at the top corner. The boundary conditions in our previous solution were $\psi_0 = 90°$ and $\psi_w = 101.580°$. Using equations (7.12.4) to (7.12.13) we find, $\Delta = 78.420°$, $\delta = 29.328°$, $\xi^- = 126.125°$ and $\zeta = -24.54°$. Thus the discontinuity roughly bisects the fan which extends from $30°$ to $18.42°$ to the vertical. The ratios p^+/p^- is given by equation (7.12.6) as 1.2602.

In §7.8 we assumed that a fan was formed and from equation (7.8.12) we see that this assumption leads to the result $p^+/p^- = \exp[2\mu(\psi_w-\psi_0)] = 1.2629$ which barely differs from the value of 1.2602 predicted above. Clearly, if the only quantity of interest is the wall stress, either solution will give a result of adequate accuracy. It is, however, of more fundamental importance to discuss which of these two solutions is correct.

The pattern of characteristics for the discontinuous solution is shown in figure 7.30. It is seen that the discontinuity lies between the zones of influence of the top surface and the wall. If any attempt was made to obtain this solution by the method of characteristics, we would find that we were unable to continue the calculation beyond the base of the Rankine zone as the β-characteristics crossing the line *OB* have no α-characteristics with which to intersect. Thus we conclude that the discontinuous solution, though a perfect solution to the equations of static equilibrium, can only be achieved if information is propagated in inadmissible directions. For this reason a fan solution is to be preferred whenever possible and discontinuous solutions only used when no fan solution exists. Further consideration is given to this matter in §7.13.

7.13 Switch stresses

We showed in §5.9, that sudden stress changes can occur at the junction of a conical hopper with a cylindrical bunker or at the boundary between active and passive regions. Such stress changes are known as 'switch stresses', 'kick stresses' or, deplorably, 'kick pressures' and their magnitude can be determined approximately by the relationships developed in §5.9. The purpose of this section is to present a somewhat more rigorous analysis of this phenomenon.

Let us first consider a hopper, either wedge-shaped or conical, in which there is a sudden change in wall slope as shown in figure 7.31.

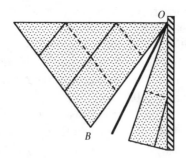

Figure 7.30 Non-overlapping zone of dependence. Under these circumstances there is insufficient information to complete the solution.

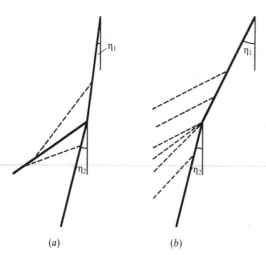

(a)   (b)

Figure 7.31 Characteristic meshes showing the formation of either a fan or a discontinuity depending on the nature of the change in wall slope.

Assuming that both sections have the same frictional properties, we can say that the angle between the major principal plane and the wall is constant, provided a passive state of stress occurs in both parts of the hopper. This seems probable, at least when discharge is occurring.

Thus if the walls become less steep as in figure 7.31(a), i.e. if $\eta_2 > \eta_1$, the β-characteristics leaving the wall converge and a discontinuity is formed. If the walls become more steep, figure 7.31(b), the β-characteristics diverge and we can have either a fan or a discontinuous solution. However, as discussed in the previous section, the discontinuous solution could not be obtained by the method of characteristics and the fan solution is to be preferred.

Further weight can be given to this choice by considering the analogous case of supersonic flow along a bent wall. If the wall is concave, as in figure 7.31(a), a shock wave is produced but if the wall is convex, figure 7.31(b), both fan and discontinuous solutions satisfy the equations of motion. The fan solution is known as the Prandtl–Meyer expansion. It can be shown that when both types of solution can be found, the discontinuous solution gives rise to a decrease in entropy and is therefore forbidden by the Second Law of Thermodynamics. Thus in compressible flow we must use the fan solution when this is possible.

There is no known equivalent to entropy in granular materials, but the arguments above lend weight to the conclusion we reached at the end of the previous section that we should use the fan solution whenever possible and confine discontinuous solution to circumstances in which there are no fan solutions. We did, however, show in §7.12 that little error occurs when the wrong type of solution is used.

We can now consider the junction of a bunker and a hopper as shown in figure 7.32. In a discharging hopper the stress state is almost certainly passive so that on the right-hand wall we have, from equation (7.9.11),

$$(\psi_w)_h = 180 - \tfrac{1}{2}(\omega + \phi_w) - \eta \tag{7.13.1}$$

Within a truly parallel-sided bunker, the stress state can be anywhere between the active and passive states but commonly it approximates to the active state so that, from equation (7.7.8),

$$(\psi_w)_b = 90 + \tfrac{1}{2}(\omega - \phi_w) \tag{7.13.2}$$

In these expressions the subscripts h and b refer to the walls of the hopper and bunker.

Clearly, the form of the solution will depend on whether $(\psi_w)_h$ is greater or less than $(\psi_w)_b$ and we can therefore define a critical value of η at which $(\psi_w)_b = (\psi_w)_h$. From equations (7.13.1) and (7.13.2) this is given by

$$\eta_{crit} = 90 - \omega \tag{7.13.3}$$

Hoppers for which $\eta > \eta_{crit}$ are called 'shallow-walled hoppers' and

Figure 7.32 Cylindrical plus conical bin.

have $(\psi_w)_h < (\psi_w)_b$ and hoppers for which $\eta < \eta_{crit}$ are called 'steep-walled' hoppers and have $(\psi_w)_h > (\psi_w)_b$.

For the case of $\phi = 30°$ and $\phi_w = 20°$, $\eta_{crit} = 46.84°$, a value which is somewhat larger than the normal range of hopper slopes. None-the-less, we will first consider the case of a shallow-walled hopper in which $(\psi_w)_h < (\psi_w)_b$ and a fan solution cannot be obtained.

As an example we will consider the case of $\eta = 50°$ which gives from equation (7.13.1), $(\psi_w)_h = 98.42°$ and from equation (7.13.2), $(\psi_w)_b = 101.58°$. Using the method of the previous section we can put $\psi^+ = 98.42°$ and $\psi^- = 101.58°$ and find that $\zeta = -20.03°$ which represents a discontinuity that lies entirely outside the material and therefore has no physical significance. Putting $\psi^+ = 101.58°$ and $\psi^- = 98.42+180$ gives $\zeta = 40.03°$ as shown in figure 7.33. However, this is just the sort of discontinuity we dismissed in the previous section as unrealistic since on working downwards we find that the β-characteristics are diverging from the discontinuity as can be seen from the figure. Both Horne and Nedderman (1978a, b), and Bransby and Blair-Fish (1974) conclude that there is no permissible solution for the shallow-walled hopper and that the material in this corner cannot be in a state of critical equilibrium. An elastic–plastic analysis is therefore required to analyse the stresses in such situations. Bransby and Blair-Fish further propose that η_{crit} is the critical wall angle for the transition from core to mass flow. The argument that mass flow cannot occur if $\eta > \eta_{crit}$ seems convincing but it does not follow that mass flow will occur when $\eta < \eta_{crit}$. As we will see in the next chapter mass flow seems to be confined to narrower angles than η_{crit}.

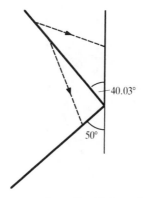

Figure 7.33 Formation of a discontinuity in a shallow-walled hopper.

Since flow is not occurring along the hopper wall we have that

$$\tau_w < \sigma_w \tan \phi_w \qquad (7.13.4)$$

and therefore,

$$(\psi_w)_h > 180 - \tfrac{1}{2}(\omega + \phi_w) - \eta \qquad (7.13.5)$$

Bransby and Blair-Fish argue that an upper limit can be placed on the value of σ_w immediately below the corner by assuming that $(\psi_w)_h = (\psi_w)_b$ and therefore that $p_h = p_b$. Hence,

$$\frac{(\sigma_w)_h}{(\sigma_w)_b} = \frac{1 + \sin\phi \cos[2(\psi_w)_b - 2\eta]}{1 + \sin\phi \cos 2(\psi_w)_b} \qquad (7.13.6)$$

but this result is not convincing. The material is not slipping along either wall and we cannot therefore be sure that the material is in a state of incipient failure and that the Mohr's circles touch the internal yield locus.

For a steep-walled hopper we have $(\psi_w)_h > (\psi_w)_b$ and a fan of β-characteristics can be drawn as shown in figure 7.34. Thus, considering the degenerate α-characteristic passing through the corner, we have from equation (7.6.22), with $dx = dy = 0$,

$$p = A \exp(2\mu\,\triangle\psi) \qquad (7.13.7)$$

or

$$p_h = p_b \exp 2\mu[(\psi_w)_h - (\psi_w)_b] \qquad (7.13.8)$$

Figure 7.34 Formation of a fan in a steep-walled hopper.

In these equations ψ must be measured in radians and appropriate conversion factors are given in the following equations.

Substituting from equation (7.13.1) and (7.13.2), we have

$$p_h = p_b \exp \left[\frac{2\mu\pi(90 - \omega - \eta)}{180} \right] \qquad (7.13.9)$$

and hence,

$$\frac{(\sigma_w)_h}{(\sigma_w)_b} = \exp \left[\frac{2\mu\pi(90-\omega-\eta)}{180} \right] \frac{1 + \sin\phi \cos 2[(\psi_w)_h - \eta]}{1 + \sin\phi \cos 2(\psi_w)_b} \qquad (7.13.10)$$

For the case of $\phi = 30°$, $\phi_w = 20°$ and $\eta = 15°$ we find from equations (7.13.1) and (7.13.2) that $(\psi_w)_b = 101.58°$ and $(\psi_w)_h = 133.42°$. Hence, the β-characteristics range from $131.58°$ to $163.42°$ to the horizontal as shown in figure 7.34. From equation (7.3.10) we have that

$$\frac{(\sigma_w)_h}{(\sigma_w)_b} = 2.56$$

showing that the corner causes a considerable increase in the stress.

Philosophical problems arise in this solution since the β-characteristics in the fan are directed upwards and may therefore interact in some awkward way with the boundary conditions on the top surface or with the corresponding fan of α-characteristics coming from the opposite corner. The latter case has been considered by Horne and Nedderman (1978a, b) and interested readers are referred to that work. Alternatively one can argue that since there is no satisfactory solution even for the steep-walled hopper, the material within a sharp corner can never be in a state of critical equilibrium and elastic effects must always be considered.

None-the-less, people have postulated that a discontinuity is formed at the corner. Its slope at the wall can be found by the method of the previous section but it is not possible to predict the rest of its shape since the characteristics are not appropriately placed. It is therefore necessary to assume a shape for the discontinuity. One possibility is to take inspiration from the Enstad element and assume that the discontinuity is part of a spherical surface. Prakesh and Rao (1991), on the other hand, assume that the discontinuity is parabolic.

The stresses on the upper surface of the discontinuity can be found by the method of characteristics and those immediately below the discontinuity can then be found from the equations of §7.12. The calculations can then be continued downwards by the method of characteristics using these results as boundary conditions. Though far from ideal, this seems an adequate working hypothesis in our present state of knowledge. Sometimes, however, the characteristics in the lower part intersect awkwardly, and this confirms that the assumed shape of the discontinuity is unsatisfactory.

The methods outlined above can be used to re-work Walters' analysis simply by putting $\eta = 0$. As in the previous case, we find that we do not have a discontinuity but a fan is formed across which the values of p are in the ratio given by equation (7.13.9), i.e.

$$p_1 = p_u \exp\left[\frac{2\mu\pi(90 - \omega)}{180}\right] \qquad (7.13.11)$$

Here we have used the subscripts l and u to denote the lower and upper parts of the bunker. For our typical material, with $\phi = 30°$ and $\phi_w = 20°$,

$$p_1 = 2.57 \, p_u$$

and the wall stress ratio is given by

$$\frac{(\sigma_w)_l}{(\sigma_w)_u} = \frac{p_1}{p_u} \frac{[1 + \sin\phi \cos 2(\psi_w)_l]}{[1 + \sin\phi \cos 2(\psi_w)_u]} \qquad (7.13.12)$$

From equations (7.13.1) and (7.13.2), $(\psi_w)_u = 101.58°$ and $(\psi_w)_l = 148.42°$ giving

$$\frac{(\sigma_w)_l}{(\sigma_w)_u} = 5.83$$

If we were to assume a discontinuity, we would find that the upper region was the $^-$ region, that $\zeta = 80.06°$ (instead of 90° as assumed by Walters), $p_1 = 2.27 \, p_u$ and

$$\frac{(\sigma_w)_l}{(\sigma_w)_u} = 5.15$$

As usual the fan and discontinuity solutions predict similar stress ratios. However, both are considerably less than the ratio of 9 given by the simplest form of Walters' analysis.

8

Velocity distributions

8.1 Introduction

The prediction of the velocity profile in a discharging silo is an essential preliminary to several topics of industrial importance. A knowledge of the velocities is required for the determination of the residence time distribution, the mixing properties of the silo and the rate of wall wear. However, the most important aspect of velocity analysis lies in the determination of whether the silo discharges in mass or core flow. As explained in chapter 1, each of these flow modes has its own advantages and it is therefore important to be able to predict in advance which will occur and how wide the flowing core will be in the case of core flow.

It must, however, be admitted that these objectives have not yet been achieved. There has been some success at predicting the velocity distribution in mass flow but as yet there is no satisfactory determination of the core/mass flow transition.

The main problem lies in the formulation of the mechanisms controlling the flow. In the stress analysis of a Coulomb material, we are confident that we understand the governing equations, the remaining problems lie solely in the solution of these equations. In velocity prediction there is less general agreement about the form of the governing equations. In particular, there are two radically different starting points. In one, known as kinematic modelling, it is assumed that the particles flow by falling into the spaces vacated by the departing particles in the layer beneath. In this case the velocity profiles will depend on geometric factors alone and will be independent of the stress distribution.

Alternatively, one can take inspiration from plasticity theory and assume that the stress and velocity profiles are linked by some flow rule. Current opinion seems to favour this latter approach, but, none-the-less, the kinematic models will be discussed briefly in §8.2 before moving on to an account of plasticity theory. The concepts of a plastic potential and a flow rule will be discussed in §8.3 and we will pay particular attention to the flow rule known as the '*Principle of Co-axiality*'. In §8.4 we will use this flow rule to evaluate the radial velocity field, which has close analogies with the radial stress field of §7.6 and we will use this solution in §8.5 to give a preliminary prediction of the core/mass flow transition. More general analyses are given in §8.6 which uses the method of characteristics and finally in §8.7 we consider discontinuous velocity fields. First, however, we will give a qualitative account of the flow patterns that are observed in bins.

In a mass flow hopper discharging a relatively coarse free-flowing material, the velocities seem to be steady and reproducible. There is evidence of small local fluctuations as particles jostle past each other but there is little difficulty in defining a mean velocity at a point just as in the turbulent flow of a liquid. In a mass flow the highest velocities occur on the centre-line with the velocity decreasing towards the wall. As a result the centre of the top surface descends more rapidly than the outer portions and a central depression is formed. As this depression deepens, its sides become more steep and when the slope reaches the angle of repose, material cascades down the surface so that the faster central core is fed from the slow moving outer parts of the top surface.

In a bin consisting of a bunker surmounting a hopper, unsteady velocities are sometimes found near the transition plane. Bransby and Blair-Fish (1974) studied the discharge of a compacted angular sand through a bunker of square cross-section feeding a wedge-shaped hopper. Within the bunker the material descended as a solid plug and rupture zones were formed alternately at the two corners, with material breaking off the plug like icebergs calving off a glacier. As a result gross variation in the velocity and stress distributions occurred within the hopper. In such situations theoretical prediction of the velocity distribution is beyond our present knowledge. However, in axial symmetry, the alternate formation of rupture zones is improbable and more regular flow is observed.

The flow in a mass flow bin consisting of a cylindrical bunker and a conical hopper is sometimes modelled as shown in figure 8.1. The material in the bunker is assumed to descend as a compacted mass in plug flow. A rupture surface is postulated which spans the bin at the level of the bunker/hopper transition and as the material flows through

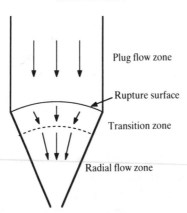

Figure 8.1 Flow zones in a cylindrical plus conical bin.

this surface it dilates and its velocity changes suddenly to a near-radical direction. Below this there is assumed to be a transition zone in which the velocity adjusts to the truly radial flow in the lower part of the hopper. The experimental work of Cousens (1980) gives some support to this model.

The flow of coarse rounded materials tends to be regular whereas increasing the angularity and decreasing the particle size favours the periodic formation of rupture zones. It may be that in the case of fine particles the rupture zones are stabilized by the percolation of the interstitial air. Dilation of the material takes place within a rupture zone and this will give rise to a reduction in the interstitial pressure which will cause the particles to lock together. Air percolation is slow within a compacted fine material inhibiting the formation of a new rupture zone. However, the dilated material within a rupture zone provides an easy path for the air enabling an existing rupture zone to grow. Thus within fine materials there is a tendency to form a few large rupture zones. By contrast, within coarse materials where air percolation is easy, there tend to be many small rupture surfaces giving an effectively continuous velocity field.

There are no sharp corners in the flowing zone of a core flow bunker and therefore a lesser tendency for the formation of rupture zones. The flow is more regular, and if the flowing core does not reach the walls it will be fed by material cascading off the top surface of the stagnant zone. Thus the flowing core is fed by dilated material which also promotes regular behaviour.

Brown and Hawksley (1947) described the flow in a typical core flow bunker in terms of four zones as illustrated in figure 8.2. They distinguish between a fast flowing core, a creeping zone and a stagnant zone in the lower part of the bunker and note the existence of a cascading zone along the top surface. The reader may find it helpful to keep this model in mind while reading the later parts of this chapter.

8.2 Kinematic models

The kinematic model for predicting the velocity distribution received some support in the period up to 1982. Since then is has fallen out of fashion and this account of the model will therefore be brief.

The model can be formulated by starting from any of three hypotheses. Though quite different in concept, they all give rise to identical equations and therefore measurements of the velocity profiles can give no information about which hypothesis is correct.

Litwiniszyn (1971) postulated that the particles are confined to a set of hypothetical cages such as those shown in figure 8.3. When the particle leaves cage 1 there is a probability **p** that it is replaced by the particle in cage 2 and probability $1 - $ **p** that it is replaced from cage 3. Commonly **p** $= 0.5$ but **p** can be ± 1 adjacent to a wall.

Mullins (1972) modelled the flow of particles in terms of a counter-diffusion of voids, which propagate through the material according to

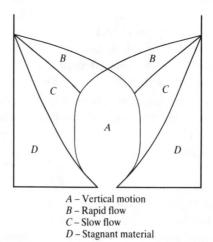

A – Vertical motion
B – Rapid flow
C – Slow flow
D – Stagnant material

Figure 8.2 Flow zones proposed by Brown and Hawksley.

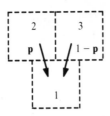

Figure 8.3 Litwiniszyn's hypothetical cages.

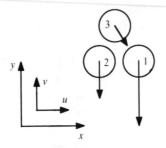

Figure 8.4 The kinematic model of Nedderman and Tüzün.

laws analogous with those for the diffusion of heat or mass in flowing systems.

Nedderman and Tüzün (1979*a*,*b*) formulated the model by considering three particles as in figure 8.4. If the downward velocity of particle 1 is greater than that of particle 2 there will be a tendency for particle 3 to move to the right, i.e. there is a relationship between the horizontal velocity u and the gradient of the vertical velocity $\partial v/\partial x$. The simplest possible relationship is

$$u = -B\frac{\partial v}{\partial x}$$ (8.2.1)

where B is a constant dependent on the nature of the material. For an incompressible material this can be coupled with the two-dimensional continuity equation

$$\frac{\partial u}{\partial x} + \frac{\partial v}{\partial y} = 0$$ (8.2.2)

to give

$$\frac{\partial v}{\partial y} = B\frac{\partial^2 v}{\partial x^2}$$ 8.2.3)

This equation is identical in form to the result of Mullins' analysis and also of Litwiniszyn's model for the case of $p = 0.5$. From this point on the models are identical.

In axi-symmetric situations equation (8.2.3) takes the form

$$\frac{\partial v}{\partial z} = \frac{B}{r} \frac{\partial}{\partial r} \left(r \frac{\partial v}{\partial r} \right) \tag{8.2.4}$$

The fact that equations (8.2.3) and (8.2.4) are not invariant with respect to rotation of co-ordinates need not bother us, since the nature of the model requires that the y- (or z-) axis is directed vertically upwards.

Unfortunately, there is no solution to these equations giving radial flow in a wedge-shaped or conical hopper, contrary to common observation. This can be seen by expressing equation (8.2.3) in terms of the stream function ψ, defined by

$$u = \frac{\partial \psi}{\partial y} , v = -\frac{\partial \psi}{\partial x} \tag{8.2.5}$$

Substituting into equation (8.2.3), and integrating with respect to x, gives

$$\frac{\partial \psi}{\partial y} = B \frac{\partial^2 \psi}{\partial x^2} \tag{8.2.6}$$

If the flow is radial the stream function must be a function of the ratio x/y only. Denoting this ratio by α, we can relate the derivatives with respect to x and y to the derivative with respect to α as follows

$$\frac{\partial \psi}{\partial y} = \frac{d\psi}{d\alpha} \frac{\partial \alpha}{\partial y} = -\frac{x}{y^2} \frac{d\psi}{d\alpha} \tag{8.2.7}$$

$$\frac{\partial^2 \psi}{\partial x^2} = \frac{d^2 \psi}{d\alpha^2} \left(\frac{\partial \alpha}{\partial x} \right)^2 = \frac{1}{y^2} \frac{d^2 \psi}{d\alpha^2} \tag{8.2.8}$$

Hence, from equation (8.2.6),

$$-x \frac{d\psi}{d\alpha} = B \frac{d^2 \psi}{d\alpha^2} \tag{8.2.9}$$

from which it is apparent that ψ is *not* a function of α only and therefore that radial flow is not compatible with these models.

Equation (8.2.3) is identical in form to the equations of unsteady diffusion or heat conduction and many solutions can be found in standard texts such as Crank (1956). The solution appropriate for a narrow slot orifice in an infinitely wide two-dimensional bed is

$$v = -\frac{Q^*}{\sqrt{(4\pi By)}} \exp\left(-\frac{x^2}{4By}\right)$$

(8.2.10)

and that for a small circular orifice in an infinite three-dimensional bed is

$$v = -\frac{Q}{4Bz} \exp\left(-\frac{r^2}{4Bz}\right)$$

(8.2.11)

where Q is the volumetric flow rate through the circular orifice and Q^* is the volumetric flow rate per unit length through the slot orifice.

There is also a solution for the flow in a parallel sided bunker of width $2a$ discharging through a slot orifice of width $2b$ as shown in figure 8.5. At great heights the velocity will be uniform and can be represented by V_p and by continuity the velocity through the orifice is given by aV_p/b. Hence we must solve equation (8.2.3) subject to the following boundary conditions:

(i) At the base the vertical velocity is zero except through the orifice where it is assumed constant, i.e.

$$\text{on } y = 0, v = \frac{aV_p}{b} \text{ for } 0 < x < b; v = 0 \text{ for } b < x < a$$

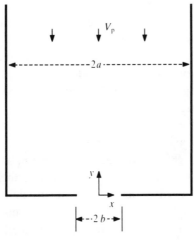

Figure 8.5 Two-dimensional bunker; definition of symbols.

(ii) On both the wall, $x = a$, and on the centre-line, $x = 0$, the horizontal velocity u must be zero and hence from equation (8.3.1) we have

$$\frac{\partial v}{\partial x} = 0 \text{ on both } x = 0 \text{ and } x = a$$

The following solution fits these boundary conditions

$$v = V_\text{p} + \sum_n \frac{2aV_\text{p}}{\pi nb} \exp\left(-n^2\pi^2 By/a^2\right) \sin\frac{n\pi b}{a} \cos\frac{n\pi x}{a} \quad (8.2.12)$$

The corresponding result for a circular orifice of radius b in the base of a cylindrical bunker of radius a is

$$v = V_\text{p} + \sum_n \frac{2V_\text{p}}{(\beta_n b)J_0^2(\beta_n a)} \exp\left(-\beta_n^2 Bz\right) J_1(\beta_n b) J_0(\beta_n r) \quad (8.2.13)$$

where J_0 and J_1 are the Bessel functions of the first kind of the zeroth and first orders and $\beta_n a$ are the roots of $J_1 = 0$.

It will be noted that the parameter B has the dimensions of length and must therefore be related to some length scale of the material or the container. All three starting hypotheses are based on arguments with a length scale of the order of the particle dimension and it is therefore reasonable to assume that B is proportional to the particle diameter d.

Tüzün (1979) conducted experiments in a two-dimensional bin with faces consisting of smooth glass sheets. Close to the orifice, equation (8.2.10) applies and a best fit value of B was obtained from the measured velocities. He found that $B \approx 2.3\,d$ confirming the ideas presented in the previous paragraph. However, larger values of B were required when using equation (8.2.12) to model the flow in the upper part of the bin. Later work showed that B was also a function of the spacing between the glass sheets and even larger values of B are required to model axi-symmetric flows. Thus there seems to be no way of estimating B a priori and the kinematic approach seems therefore to have little predictive value. Consequently, it has fallen out of favour. On the other hand, realistic streamline patterns can be generated by a suitable choice of B so that this model can be used to provide a convenient one-parameter description of an observed flow pattern.

8.3 The yield function, the plastic potential and flow rules

The use of plasticity theory to predict the velocity distribution in a granular material involves the concepts of a *Yield Function*, a *Plastic Potential* and a *Flow Rule* and we will start this section by defining these quantities.

The yield function is no more than a formalised way of expressing the failure criterion, which for a cohesionless Coulomb material is usually expressed in the form

$$|\tau_{ns}| = \sigma_{nn} \tan \phi \qquad (8.3.1)$$

where n and s are co-ordinates normal and along the slip plane. Alternatively, this criterion can be expressed in terms of the principal stresses as in §3.4, giving

$$\sigma_1(1 - \sin \phi) = \sigma_3(1 + \sin \phi) \qquad (8.3.2)$$

These formulations, and indeed any other formulation for Coulomb or non-Coulomb materials, can be expressed in the form of a *Yield Function Y* which, by definition, satisfies the relationship

$$Y = 0 \qquad (8.3.3)$$

Thus the Coulomb failure criterion can be expressed either as

$$Y = |\tau_{ns}| - \sigma_{nn} \tan \phi \qquad (8.3.4)$$

or

$$Y = \sigma_1(1 - \sin \phi) - \sigma_3(1 + \sin \phi) \qquad (8.3.5)$$

The Warren Spring relationship, equation (6.5.8), can similarly be written as

$$Y = \left(\frac{|\tau_{ns}|}{C}\right)^n - \frac{\sigma_{nn}}{T} - 1 \qquad (8.3.6)$$

It is always possible to define a *Plastic Potential G* so that the strain rates $\dot{\varepsilon}_{ij}$ in any arbitrary direction are proportional to the derivatives of the plastic potential with respect to the corresponding stress, i.e.

$$\dot{\varepsilon}_{ij} = \lambda \frac{\partial G}{\partial \sigma_{ij}} \qquad (8.3.7)$$

where λ is a scalar constant. In this expression i and j can take any combination of the co-ordinate variables, (x,y,z), (r,z,θ) etc. Thus in Cartesian co-ordinates

$$\dot{\varepsilon}_{xx} = \lambda \frac{\partial G}{\partial \sigma_{xx}} \qquad (8.3.8)$$

$$\dot{\varepsilon}_{yy} = \lambda \frac{\partial G}{\partial \sigma_{yy}} \qquad (8.3.9)$$

and

$$\dot{\gamma}_{xy} = \lambda \frac{\partial G}{\partial \tau_{xy}} \qquad (8.3.10)$$

The value and meaning of the constant λ need not concern us since it will subsequently cancel from our expressions and therefore need not be evaluated. It must, however, be appreciated that λ is not a property of the material but depends on the flow rate.

The relationship between the plastic potential and the yield function is known as the *Flow Rule* and the particular flow rule known as the *Associated Flow Rule* has considerable theoretical and historical importance. This flow rule postulates that the plastic potential is identical to the yield function, i.e.

$$Y = G \qquad (8.3.11)$$

Hence, for a Coulomb material we can say, from equations (8.3.4) and (8.3.7),

$$\dot{\varepsilon}_{nn} = - \lambda \tan \phi \qquad (8.3.12)$$

$$|\dot{\gamma}_{ns}| = \lambda \qquad (8.3.13)$$

and

$$\dot{\varepsilon}_{ss} = 0 \qquad (8.3.14)$$

Thus if slip occurs along a narrow failure zone in the s-direction, the normal strain rate is related to the shear strain rate by

$$\dot{\varepsilon}_{nn} = - |\dot{\gamma}_{ns}| \tan \phi \qquad (8.3.15)$$

Since the volumetric strain rate is the sum of the direct strain rates we can also say that

$$\dot{\varepsilon}_V = \dot{\varepsilon}_{nn} + \dot{\varepsilon}_{ss} = - |\dot{\gamma}_{ns}| \tan \phi \qquad (8.3.16)$$

and we therefore see that this flow rule predicts that shear causes dilation of the material, an observation which is sometimes known as *Reynolds' Principle of Dilatancy* and was discussed in §6.5. The associated flow rule has its attractions, first because the bound theorems mentioned in §7.8 can be proved if this rule is obeyed, and secondly because it predicts that dilation will occur during shear.

However, the observations on the shear of a consolidated material, presented in §6.5, indicate that the extent of the dilation is not usually as great as that predicted by equation (8.3.16). It is therefore usual to define an *Angle of Dilation v* by

$$\dot{\varepsilon}_V = - |\dot{\gamma}_{ns}| \tan v \qquad (8.3.17)$$

Thus for materials that obey the associated flow rule we have

$$v = \phi \qquad (8.3.18)$$

but for most real materials we must use a *Non-associated Flow Rule* in which $v < \phi$.

We also noted in §6.5 that the dilation occurring during the initial stages of the shear of a consolidated material is not maintained for shear strains greater than about 1 or 2%. Beyond this the rate of dilation decreases and eventually reaches zero as the material approaches the critical state. Thus for the *initial displacements* of a compacted material we must use a non-associated flow rule with some experimentally determined value of v, but *during steady flow* we can take v to be zero and consider the material to be incompressible.

The associated flow rule is sometimes expressed in terms of the *Principle of Normality*. This can be illustrated by superimposing a plot of the rate of strain vector $(\dot{\gamma}_{ns}, \dot{\varepsilon}_{nn})$ on a plot of the yield locus as shown in figure 8.6. If the principle of normality applies, the rate of displacement vector will be normal to the yield locus, a result expressed algebraically by equations (8.3.7) and (8.3.11). However, as we saw in §6.6, the stresses in steady flow are given by the end point of the incipient yield locus and people have argued that the normal to a line is undefined at its end. Thus there is a range of permissible directions for the rate of displacement vector in steady flow as illustrated in figure 8.6. Other people have argued that the principle of normality constrains the yield locus to become horizontal as it approaches its end point and therefore that the Coulomb yield locus cannot apply over the whole range of stresses. Arguments of this sort are used to justify the modification to the Coulomb yield locus proposed by Prakesh and Rao (1991) as presented in §6.6.

We see therefore that we cannot use the combination of the Coulomb yield criterion and the associated flow rule in steady flow and some

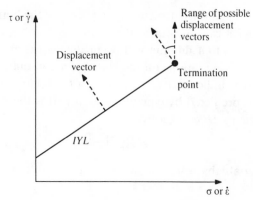

Figure 8.6 Indeterminacy of the principle of normality at the end point of the yield locus.

other flow rule must be invoked. Jenike advocates the *Principle of Co-axiality*, which states that the principal axes of stress and strain rate are co-incident. This is based on arguments of symmetry and is equivalent to saying that there can be no shear strain on planes on which there is no shear stress. It is clearly analogous to *St. Venant's Principle* which is commonly used in the theory of elasticity. Jenike calls this principle, the *Principle of Isotropy*. However, the word co-axiality is to be preferred in this context since the principle of co-axiality is an exact description of the assumption in question whereas there is no guarantee that an isotropic material will obey the principle of isotropy.

In the remaining sections of this chapter, we will present a series of analyses appropriate to incompressible materials that obey the principle of co-axiality. Though all materials are compressible to some extent, the actual volume changes are small and have little effect on the velocities, though there may be a considerable effect on the frictional properties of the material as explained in §6.5.

Assuming the material is incompressible, we can write the two-dimensional continuity equation as

$$\frac{\partial u}{\partial x} + \frac{\partial v}{\partial y} = 0 \qquad (8.3.19)$$

The principle of co-axiality can be put into algebraic form by considering the Mohr's circles for stress and strain rate as in figure 8.7. Since the angles between the x-direction and the principal directions

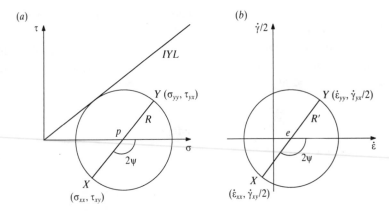

Figure 8.7 Mohr's circles for (a) stress and (b) strain rate.

of stress and strain rate are the same, the two angles marked 2ψ must be equal.

The stress components are given by equations (3.3.8) to (3.3.10) as

$$\sigma_{xx} = p + R \cos 2\psi \qquad (8.3.20)$$

$$\sigma_{yy} = P - R \cos 2\psi \qquad (8.3.21)$$

$$\tau_{xy} = - R \sin 2\psi \qquad (8.3.22)$$

and similarly the strain rate components are given by

$$\dot{\varepsilon}_{xx} = e + R' \cos 2\psi \qquad (8.3.23)$$

$$\dot{\varepsilon}_{yy} = e - R' \cos 2\psi \qquad (8.3.24)$$

$$\tfrac{1}{2}\dot{\gamma}_{xy} = - R' \sin 2\psi \qquad (8.3.25)$$

Thus we see that

$$\frac{- \tau_{xy}}{\tfrac{1}{2}(\sigma_{xx} - \sigma_{yy})} = \tan 2\psi = \frac{- \tfrac{1}{2}\dot{\gamma}_{xy}}{\tfrac{1}{2}(\dot{\varepsilon}_{xx} - \dot{\varepsilon}_{yy})} \qquad (8.3.26)$$

Substituting for the strain rates from equations (2.5.2) to (2.5.4) gives

$$\left(\frac{\partial u}{\partial x} - \frac{\partial v}{\partial y}\right) \tan 2\psi = \frac{\partial u}{\partial y} + \frac{\partial v}{\partial x} \qquad (8.3.27)$$

and similar expressions in other co-ordinate systems are listed below.

In all systems of plain strain, i.e. when the strain rate normal to the plane of interest is zero, the volumetric strain rate is equal to the sum of the two remaining strain rates, which for an incompressible material must be zero. Thus the centre of the Mohr's strain rate circle will be at the origin. However, for axi-symmetric systems in which all three principal strain rates can be non-zero, the centre of Mohr's strain rate circle need not be at the origin and e can be non-zero.

Polar co-ordinates

The strain rates are

$$\dot{\varepsilon}_{rr} = -\frac{\partial v_r}{\partial r} , \dot{\varepsilon}_{\theta\theta} = -\frac{v_r}{r} - \frac{1}{r}\frac{\partial v_\theta}{\partial \theta} ,$$

$$\dot{\gamma}_{r\theta} = -\dot{\gamma}_{\theta r} = r\frac{\partial}{\partial r}\left(\frac{v_\theta}{r}\right) + \frac{1}{r}\frac{\partial v_r}{\partial \theta} \tag{8.3.28}$$

the continuity equation takes the form

$$\frac{\partial v_r}{\partial r} + \frac{v_r}{r} + \frac{1}{r}\frac{\partial v_\theta}{\partial \theta} = 0 \tag{8.3.29}$$

and the principle of co-axiality can be expressed as

$$\left(\frac{\partial v_r}{\partial r} - \frac{v_r}{r} - \frac{1}{r}\frac{\partial v_\theta}{\partial \theta}\right)\tan 2\psi^* = r\frac{\partial}{\partial r}\left(\frac{v_\theta}{r}\right) + \frac{1}{r}\frac{\partial v_r}{\partial \theta} \tag{8.3.30}$$

Cylindrical co-ordinates

The strain rates are

$$\dot{\varepsilon}_{rr} = -\frac{\partial v_r}{\partial r} , \dot{\varepsilon}_{zz} = -\frac{\partial v_z}{\partial z} , \dot{\varepsilon}_{\theta\theta} = -\frac{v_r}{r} ,$$

$$\dot{\gamma}_{rz} = -\dot{\gamma}_{zr} = \frac{\partial v_r}{\partial z} + \frac{\partial v_z}{\partial r} \tag{8.3.31}$$

the continuity equation takes the form

$$\frac{\partial v_r}{\partial r} + \frac{v_r}{r} + \frac{\partial v_z}{\partial z} = 0 \tag{8.3.32}$$

and the principle of co-axiality can be expressed as

$$\left(\frac{\partial v_r}{\partial r} - \frac{\partial v_z}{\partial z}\right)\tan 2\psi = \frac{\partial v_r}{\partial z} + \frac{\partial v_z}{\partial r} \tag{8.3.33}$$

Spherical co-ordinates

The strain rates are

$$\dot{\varepsilon}_{rr} = -\frac{\partial v_r}{\partial r} \ , \ \dot{\varepsilon}_{\theta\theta} = -\frac{v_r}{r} - \frac{1}{r}\frac{\partial v_\theta}{\partial \theta} \ , \ \dot{\varepsilon}_{xx} = -\frac{v_r}{r} - \frac{v_\theta \cot \theta}{r} \ ,$$

$$\dot{\gamma}_{r\theta} = -\dot{\gamma}_{\theta r} = r\frac{\partial}{\partial r}\left(\frac{v_\theta}{r}\right) + \frac{1}{r}\frac{\partial v_r}{\partial \theta} \tag{8.3.34}$$

the continuity equation takes the form

$$\frac{\partial v_r}{\partial r} + 2\frac{v_r}{r} + \frac{1}{r}\frac{\partial v_\theta}{\partial \theta} + \frac{v_\theta \cot \theta}{r} = 0 \tag{8.3.35}$$

and the principle of co-axiality can be expressed as

$$\left(\frac{\partial v_r}{\partial r} - \frac{v_r}{r} - \frac{1}{r}\frac{\partial v_\theta}{\partial \theta}\right)\tan 2\psi^* = r\frac{\partial}{\partial r}\left(\frac{v_\theta}{r}\right) + \frac{1}{r}\frac{\partial v_r}{\partial \theta} \tag{8.3.36}$$

8.4 Particular solutions, the radial velocity field

In this section we will consider special cases of the flow patterns in bunkers of constant cross-section and in conical or wedge-shaped hoppers. In the former case the geometry permits plug flow and consideration of equations (8.6.14) and (8.6.16) below shows that these are always satisfied by $du = 0$ and $dv = 0$ whatever the stress distribution. Thus plug flow is possible for all combinations of wall and material properties. It does not follow that plug flow is the only permissible velocity distribution and the methods for determining the rate of convergence to plug flow are discussed in §8.6. In practice it is commonly found that plug flow exists in the upper part of a bin and continues down to the section at which the influence of the hopper walls is first felt.

In a conical hopper, radial flow is clearly a possibility and we will investigate whether such a flow can satisfy the principle of co-axiality. If we assume that the material is incompressible, the radial velocity will be inversely proportional to r^2 and all other velocity components will be zero, i.e.

$$v_r = -\frac{f(\theta)}{r^2} \ , \ v_\theta = 0 \ , \ v_\chi = 0 \tag{8.4.1}$$

where $f(\theta)$ is some unknown function of θ. The same result could have been obtained from equation (8.3.35) by putting $v_\theta = 0$ and integrating. The minus sign in the expression for v_r appears since, in a discharging hopper, the flow is in the direction of r decreasing and it is convenient to have $f(\theta)$ positive. This flow pattern is known as the *Radial Velocity Field* and we will see later that it is closely related to the radial stress field discussed in §7.4

The rate of strain components in spherical co-ordinates are given by equations (8.3.34) and hence from equation (8.4.1)

$$\dot{\varepsilon}_{rr} = -\frac{\partial v_r}{\partial r} = -\frac{2f(\theta)}{r^3} \tag{8.4.2}$$

$$\dot{\varepsilon}_{\theta\theta} = \dot{\varepsilon}_{xx} = -\frac{v_r}{r} = \frac{f(\theta)}{r^3} \tag{8.4.3}$$

$$\dot{\gamma}_{r\theta} = -\dot{\gamma}_{\theta r} = \frac{1}{r}\frac{\partial v_r}{\partial \theta} = -\frac{f'(\theta)}{r^3} \tag{8.4.4}$$

where

$$f'(\theta) \equiv \frac{df(\theta)}{d\theta} \tag{8.4.5}$$

These strain rates can be represented on the Mohr's circle of figure 8.8. The point O, representing the centre of the circle, has co-ordinates $(e,0)$ where

$$e = \tfrac{1}{2}\left(\dot{\varepsilon}_{rr} + \dot{\varepsilon}_{\theta\theta}\right) = -\frac{f(\theta)}{2r^3} \tag{8.4.6}$$

Hence, the angle ROA is given by

$$\tan^{-1}\frac{\dot{\gamma}_{r\theta}}{2(e - \dot{\varepsilon}_{rr})} = 2\psi^* - 180 \tag{8.4.7}$$

or

$$-\frac{f'(\theta)}{r^3} = 2\left[-\frac{f(\theta)}{2r^3} + \frac{2f(\theta)}{r^3}\right]\tan 2\psi^* \tag{8.4.8}$$

i.e.

$$\frac{f'(\theta)}{f(\theta)} = -3\tan 2\psi^* \tag{8.4.9}$$

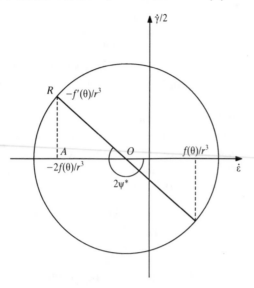

Figure 8.8 Mohr's circle for the strain rates in radial flow.

It can readily be checked that the same result can be obtained by substituting equations (8.4.2) to (8.4.4) into equation (8.3.36).

Thus we see that the assumptions of incompressibility and co-axiality show that within the radial velocity field the principal stress direction is a function of angular position only. This was one of the two assumptions we made when deriving the radial stress field in §7.4 and we therefore see that the radial stress field solutions are compatible with the assumption of radial flow. However, the converse is not true, and the flow need not be radial throughout the radial stress field. There is every reason to believe that the radial velocity field represents the asymptote within the radial stress field to which the velocities tend near the apex. Recalling that the radial stress field is itself the asymptote to which the stresses tend near the apex we see that we have an asymptote within an asymptote. None-the-less, Cleaver's (1991) experiments seem to show that in a steep-walled conical hopper, the flow is effectively radial throughout at least the lower half of the hopper. This rather surprising observation is discussed further in §8.6 and §8.7.

We can use the values of ψ^* determined by the methods of §7.4 and obtain the velocity function $f(\theta)$ from equation (8.4.9) which can be integrated to give

$$f(\theta) = A \exp\left(-3 \int_0^\theta \tan 2\psi^* \, d\theta\right) \qquad (8.4.10)$$

The integral in this relationship is easily evaluated using either the trapezium rule or Simpson's rule and A is an arbitrary constant of integration.

We cannot evaluate this arbitrary constant from consideration of plasticity alone, since, as can be seen from equation (8.3.26), this gives us information only about the ratio of the shear and direct strain rates. The results can therefore be scaled arbitrarily. It is therefore convenient to rewrite equation (8.4.10) in the form

$$f(\theta) = AF(\theta) \qquad (8.4.11)$$

where

$$F(\theta) = \exp\left(-3 \int_0^\theta \tan 2\psi^* \, d\theta\right) \qquad (8.4.12)$$

It is seen that $F(\theta)$ is the velocity profile normalised with respect to the centre-line velocity, i.e.

$$F(\theta) = \frac{f(\theta)}{f(0)} = \frac{v_r(\theta)}{v_r(0)} \qquad (8.4.13)$$

The value of A appropriate for a particular situation can be obtained by scaling the predictions to match the volumetric flow rate determined independently, either experimentally or by the methods presented in chapter 10. Since the volumetric flow rate is the integral of the velocity with respect to area, we have

$$Q = -\int_0^\alpha 2\pi r \sin \theta \, v_r \, r \, d\theta = 2\pi A \int_0^\alpha F(\theta) \sin \theta \, d\theta \qquad (8.4.14)$$

and A can be found from the ratio of the independently determined flow rate and the integral of $F(\theta) \sin \theta$ with respect to θ.

We see therefore that plasticity theory can only predict the shape of the velocity profiles but it follows from this that the shape must be independent of the flow rate. Thus if measured profiles are scaled in proportion to the mass flow rates, they should become identical. Cleaver (1991) has found that this is so, within the limits of experimental error, and his results give strong support to the use of plasticity theory for the prediction of velocity distributions.

In §7.4 we used the radial stress field analysis to determine the variation of ψ^* with θ for particular values of the stress parameter q_0 in a material with an angle of internal friction of 30°. The results were presented in figure 7.5 and have been used to evaluate the normalised velocity distributions $F(\theta)$ shown in figure 8.9. It is seen that for a cohesionless material $F(\theta)$ depends on ϕ and q_0 only. However, as explained in §7.4, q_0 is itself a function of ϕ, ϕ_w and the hopper angle α. Thus for example, if we consider a hopper with half-angle 25°, we can determine ψ_w^* from figure 7.5 and calculate the corresponding value of ϕ_w from equation (7.4.27). Table 8.1 gives the values of ϕ_w corresponding to the values of q_0 presented in figure 8.9.

It can be seen that the velocity profile is predicted to be a strong function of the angle of wall friction. This is in contrast to the kinematic model, presented in §8.2, which predicts that the velocity profile is independent of the angle of wall friction. Cleaver's experiments show that in practice the velocity profile is indeed dependent on wall

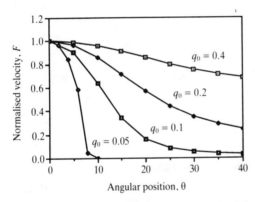

Figure 8.9 Dependence of the dimensionless velocity F with angular position θ.

Table 8.1

	$\phi = 30°$		$\alpha = 25°$	
q_0	0.4	0.2	0.1	0.05
ψ_w^* (°)	97.08	111.23	122.54	126.11
ϕ_w (°)	4.71	13.85	20.53	22.44

roughness and he argues that this proves that kinematic modelling is inappropriate. His results are, however, not in perfect numerical agreement with the predictions based on the Coulomb yield criterion and the principle of co-axiality. He finds that plasticity calculations starting from the so-called conical yield function are in closer agreement with experiment and this will be considered in greater detail in chapter 9.

The procedure for wedge-shaped hoppers is very similar. Here the radial velocity is inversely proportional to r, so that equation (8.4.1) becomes

$$v_r = -\frac{f(\theta)}{r}, \quad v_\theta = 0 \qquad (8.4.15)$$

Hence, from equation (8.3.30),

$$\frac{f'(\theta)}{f(\theta)} = -2 \tan 2\psi^* \qquad (8.4.16)$$

and therefore,

$$f(\theta) = A \exp\left(-2 \int_0^\theta \tan 2\psi^* \, d\theta\right) \qquad (8.4.17)$$

The volumetric flow rate per unit width, Q^*, is given by

$$Q^* = 2 \int_0^\alpha f(\theta) \, d\theta \qquad (8.4.18)$$

8.5 The mass/core flow criterion

It can be seen from equation (8.4.12) that the velocity function $F(\theta)$ falls to zero when ψ^* reaches $135°$; since under these circumstances $\tan 2\psi^*$ is infinite. Figure 7.5 shows typical plots of ψ^* as a function of angular position and it can be seen that these curves can, but do not necessarily, pass through maxima. Thus the critical value of ψ^* can be reached either at the wall or part-way across the hopper.

The critical value of $\psi^* = 135°$ occurs at the wall when

$$\tan \phi_w = \sin \phi \qquad (8.5.1)$$

a result which can be obtained from equation (7.4.27), or more conveniently by inspection from figure 8.10. Since no greater value of ψ^* can be obtained next to a wall during flow, equation (8.5.1) can

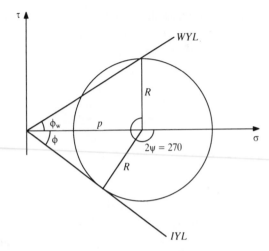

Figure 8.10 Mohr's circle for the stresses adjacent to a 'fully rough' wall.

be taken to give the maximum realistic value of ϕ_w. It is seen that this is somewhat less than the value given by the concept of the fully rough wall as presented in §3.7 and justifies, to some extent, Jenike's assertion that 'experiments indicate' that the maximum value of ϕ_w is given by equation (8.5.1). Certainly, no greater value of ϕ_w can be detected by measurement devices in which motion actually occurs, but the arguments above do not prove that greater values of ϕ_w cannot occur in incipient failure. However, the absence of peak stresses in typical wall friction tests suggests that equation (8.5.1) may be a better definition of a fully rough wall than that traditionally used. This has great practical convenience since, as we have seen in previous chapters, the results of many analyses either become unworkable or very sensitive to the precise value of ϕ_w as $\phi_w \rightarrow \phi$.

If ψ^* reaches 135° part-way across the hopper the velocity will fall to zero and the material between this point and the wall will be stationary. The criterion for the difference between mass and core flow is therefore whether ψ^* reaches 135° at any point within the hopper. For the case of a material with $\phi = 30°$, the value of q_0 at which ψ^* first reaches 135° is 0.0565 and this occurs at $\theta = 12°$. If the hopper half-angle α is less than this value, mass flow will occur provided ϕ_w is less than the critical value given by equation (8.5.1). For greater values of α, the transition between mass and core flow is given by the results for $q_0 = 0.0565$. The dependence of ψ^* on θ can be found from the radial stress field and the corresponding values of ϕ_w from equation (7.4.27). The results of this analysis are given in figure 8.11 for the case of $\phi = 30°$. This analysis has greater rigour than that originally presented by Jenike (1961), who gave the existence of a radial stress field as the criterion for mass flow. The two predictions, however, do not differ significantly.

If, for example, we consider a hopper of half-angle 25°, we see from figure 8.11 that core flow occurs for angles of wall friction greater than 22.39°. The transition is however very sudden. For the slightly smaller value of ϕ_w of 21.1° the value of $F(\alpha)$ is 0.05, i.e. the velocity at the wall is 5% of the centre-line velocity. This rather rapid change in the wall velocity with angle of wall friction causes problems during operation. It is most unlikely that the wall will be perfectly uniform round its circumference and minor changes of wall roughness with angular position can therefore result in markedly asymmetric flow patterns. Furthermore, any difference between static and dynamic frictional properties will cause irreproducible behaviour in a hopper operated close to the mass/core flow boundary. Thus Jenike recommends that such hoppers should be avoided if at all possible.

The critical value of θ at which ψ^* first reaches 135° (12° for the case of $\phi = 30°$) is a prediction of the widest possible flowing core in core flow. In practice, flowing cores are found to be considerably wider than this and the prediction of the mass/core flow transition and the associated core flow angle is generally regarded as the poorest of the results of the combination of the Coulomb yield criterion and the

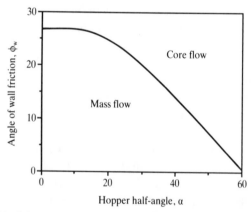

Figure 8.11 Criterion for the transition between core and mass flow.

principle of co-axiality. For this reason Jenike (1987) has advocated the use of the *Conical*, or *Extended von Mises*, *Criterion* and this forms the subject matter of chapter 9.

8.6 The method of characteristics

In this section we will develop a method for predicting velocity distributions in situations other than the radial velocity field. As with the analysis we assume that the material is incompressible and obeys the principle of co-axiality.

The principle of co-axiality can be expressed by equation (8.3.26), which in Cartesian axes takes the form

$$\frac{\partial u}{\partial y} + \frac{\partial v}{\partial x} = \left(\frac{\partial u}{\partial x} - \frac{\partial v}{\partial y}\right) \tan 2\psi \qquad (8.6.1)$$

and the incompressibility condition is given by equation (8.3.19),

$$\frac{\partial u}{\partial x} + \frac{\partial v}{\partial y} = 0 \qquad (8.6.2)$$

Like the stress equations of §7.6, these equations can be solved by the method of characteristics. The characteristic directions and the equations along the characteristics are derived below after which we will illustrate the use of these equations. The method follows closely that of §7.6 and therefore will be considered only briefly.

From equation (8.6.2) we have

$$\frac{\partial u}{\partial x} = -\frac{\partial v}{\partial y} \qquad (8.6.3)$$

and eliminating $\dfrac{\partial u}{\partial x}$ from equations (8.6.1) and (8.6.3) gives

$$\frac{\partial u}{\partial y} = -\frac{\partial v}{\partial x} - 2\frac{\partial v}{\partial y} \tan 2\psi \qquad (8.6.4)$$

Defining ζ to be the angle measured anticlockwise from the x-axis to the characteristic direction, we have that the total derivative of u along the characteristic is given by

$$\frac{du}{ds} = \frac{\partial u}{\partial x} \cos \zeta + \frac{\partial u}{\partial y} \sin \zeta = -\frac{\partial v}{\partial x} \sin \zeta - \frac{\partial v}{\partial y} (\cos \zeta$$

$$+ 2 \tan 2\psi \sin \zeta) \qquad (8.6.5)$$

We can now eliminate $\partial v / \partial x$ since

$$\frac{dv}{ds} = \frac{\partial v}{\partial x} \cos \zeta + \frac{\partial v}{\partial y} \sin \zeta \qquad (8.6.6)$$

giving

$$\frac{du}{ds} = -\frac{dv}{ds} \tan \zeta - \frac{\partial v}{\partial y} (\cos \zeta + 2 \tan 2\psi \sin \zeta$$

$$- \sin \zeta \tan \zeta) \qquad (8.6.7)$$

This equation becomes an ordinary differential equation when the coefficient of the $\partial v / \partial y$ term is zero i.e. when

$$\cos^2\zeta + 2 \tan 2\psi \sin \zeta \cos \zeta - \sin^2\zeta = 0 \qquad (8.6.8)$$

or

$$\cos 2\zeta + \sin 2\zeta \tan 2\psi = 0 \qquad (8.6.9)$$

Thus

$$\tan 2\psi = - \cot 2\zeta \qquad (8.6.10)$$

or

$$\zeta = \psi \pm 45° \qquad (8.6.11)$$

We therefore have two characteristic directions and we will call the one for which $\zeta = \psi - 45°$, the α-direction and the one for which $\zeta = \psi + 45°$, the β-direction. The stress and velocity characteristics cannot be coincident since the stress characteristics are inclined to the x-axis by the angles $\psi \pm \varepsilon$, where $\varepsilon = 45 - \phi/2$, and is therefore always less than 45°.

Putting $\zeta = \psi \pm 45°$ into equation (8.6.7) gives

$$\frac{du}{ds} + \frac{dv}{ds} \tan (\psi \pm 45) = 0 \qquad (8.6.12)$$

Thus we have:

along the α-characteristics

$$\frac{dy}{dx} = \tan (\psi - 45) \qquad (8.6.13)$$

$$\frac{du}{ds} + \frac{dv}{ds} \tan (\psi - 45) = 0 \qquad (8.6.14)$$

and along the β-characteristics

$$\frac{dy}{dx} = \tan(\psi + 45) \tag{8.6.15}$$

$$\frac{du}{ds} + \frac{dv}{ds} \tan(\psi + 45) = 0 \tag{8.6.16}$$

As before, these can be put in finite difference form:

along the α-characteristic

$$y - y_1 = (x - x_1) \tan(\overline{\psi}_1 - 45) \tag{8.6.17}$$

$$u - u_1 + (v - v_1) \tan(\overline{\psi}_1 - 45) = 0 \tag{8.6.18}$$

and along the β-characteristic

$$y - y_2 = (x - x_2) \tan(\overline{\psi}_2 + 45) \tag{8.6.19}$$

$$u - u_2 + (v - v_2) \tan(\overline{\psi}_2 + 45) = 0 \tag{8.6.20}$$

These equations can be used in exactly the same way as the stress equations in finite difference form as presented in §7.6. To solve these equations ψ must already be known as a function of position and since $\overline{\psi}_1$ is the average value of ψ along the interval (x_1,y_1) to (x,y), iteration will be required to obtain the point (x,y). However, no iteration will be required when solving equations (8.6.18) and (8.6.20) for u and v as these equations are linear and $\overline{\psi}_1$ and $\overline{\psi}_2$ are already known.

Besides the basic calculation, outlined above, we must also consider the wall calculation. For the case of an α-characteristic meeting a wall we have equations (8.6.17) and (8.6.18), the equation of the wall which can be locally linearised to

$$y = mx + c \tag{8.6.21}$$

and the criterion that the component of the velocity normal to the wall is zero. This last result can be put in the more convenient form

$$v = mu \tag{8.6.22}$$

These four equations are sufficient for solution.

It can be seen that difficulties arise in the solution of these equations when ψ approaches $45 + 90n$ when n is any integer, as here the tan terms tend to either zero or infinity and it may be necessary to rearrange the equations to avoid excessive round-off errors. No fundamental difficulty arises since it can be seen that in the particular case when $\overline{\psi}_1 = \overline{\psi}_2 = 45°$, $y = y_1$, $u = u_1$, $x = x_2$ and $v = v_2$.

We can illustrate the use of the equations derived above by considering the flow pattern between parallel planes a distance $2a$ apart. The variation of ψ with x can be found by eliminating p^* from equations (7.3.6) and (7.3.7) giving, for the case of $\phi = 30°$, $\phi_w = 20°$ and $c = 0$,

$$\frac{\sin 2\psi}{2 + \cos 2\psi} = 0.3640 \, X \qquad (8.6.23)$$

where $X = x/a$.

Let us assume that at some point A, $X_A = -0.8$, $Y_A = 0$, $u_A = 0.1 \, V$ and $v_A = V$ and that at point B, $X_B = -0.9$, $Y_B = 0$, $u_B = 0.05 \, V$ and $v_B = 0.9 \, V$ as shown in figure 8.12, where V is an arbitrary velocity. The α-characteristic through B cuts the β-characteristic through A at point C with co-ordinates and velocity components (X_C, Y_C, u_C, v_C).

From equation (8.6.23) we have that $\psi_A = 98.88°$ and $\psi_B = 100.18°$. Taking as a first approximation that ψ_C is the mean of these two values i.e. 99.53° we have

$$\overline{\psi}_A = \tfrac{1}{2}(\psi_A + \psi_C) = 99.21° \quad \text{and} \quad \overline{\psi}_B = \tfrac{1}{2}(\psi_B + \psi_C) = 99.86°$$

From equations (8.6.17) and (8.6.19) we have

$$Y_C = (X_C + 0.9)\tan 54.86$$

$$Y_C = (X_C + 0.8)\tan 144.21$$

which have solution $X_C = -0.8663$, $Y_C = 0.0479$. Thus from equation (8.6.23) we have $\psi_C = 99.74°$. A second iteration gives

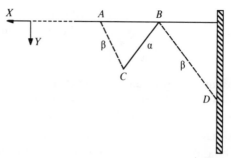

Figure 8.12 Characteristic mesh for velocity determination.

$\overline{\psi}_A = 9.31°, \overline{\psi}_B = 99.96°$, $X_C = -0.8665$ and $Y_C = 0.0478$. Further iterations seem unnecessary but can be performed if desired.

From equations (8.6.18) and (8.6.20) we have

$$u_C - 0.05\,V + (v_C - 0.9\,V)\tan 54.96 = 0$$

$$u_C - 0.10\,V + (v_C - V)\tan 144.31 = 0$$

giving $u_C = 0.035\,V$ and $v_C = 0.910\,V$. No iteration is required in this second part of the calculation.

The β-characteristic through point B cuts the wall at point D with co-ordinates and velocity components X_D, Y_D, u_D and v_D, where $X_D = -1.0$ and $u_D = 0$.

From equation (8.6.23) $\psi_D = 101.68°$ and hence $\overline{\psi}_B = \frac{1}{2}(101.68 + 100.18) = 100.93°$.

Thus from equation (8.6.19),

$$Y_D = (-1 + 0.9)\tan 145.93 \text{ giving } Y_D = 0.0676$$

and from (8.6.20)

$$0 - 0.05\,V + (V_D - 0.9\,V)\tan 145.93 = 0 \text{ giving } v = 0.826\,V$$

It is seen that iteration is not required in a wall calculation, just as in the corresponding calculation for stresses, as presented in §7.6.

It can be noted that the results are independent of the value of V and therefore can be scaled by any arbitrary factor, just as in the prediction of the radial velocity field presented in §8.4.

In cylindrical co-ordinates, we have from equation (8.3.32) and (8.3.33)

$$\frac{\partial v_r}{\partial r} + \frac{v_r}{r} + \frac{\partial v_z}{\partial z} \tag{8.6.24}$$

and

$$\left(\frac{\partial v_r}{\partial r} - \frac{\partial v_z}{\partial z}\right)\tan 2\psi = \frac{\partial v_r}{\partial z} + \frac{\partial v_z}{\partial r} \tag{8.6.25}$$

It can be seen that these equations have the same form as equations (8.6.1) and (8.6.2), except for the extra term v_r/r in the continuity equation, and can be manipulated in the same manner. Following the procedure of the first part of this section, we find that the equations of the characteristics are given by

$$\frac{dz}{dr} = \tan \zeta \qquad (8.6.26)$$

where again $\zeta = \psi \pm 45°$, and the equations along the characteristics are

$$\frac{dv_r}{ds} + \frac{dv_z}{ds} \tan \zeta + \frac{v_r}{r} (\tan 2\psi \sin \zeta + \cos \zeta) = 0 \qquad (8.6.27)$$

From these we can show that:

along the α-characteristic

$$dz = \tan (\psi - 45) \, dr \qquad (8.6.28)$$

$$dv_r + \tan (\psi - 45) \, dv_z - \frac{v_r}{r \cos 2\psi} \, dz = 0 \qquad (8.6.29)$$

and along the β-characteristic

$$dz = \tan (\psi + 45) \, dr \qquad (8.6.30)$$

$$dv_r + \tan (\psi + 45) \, dv_z + \frac{v_r}{r \cos 2\psi} \, dz = 0 \qquad (8.6.31)$$

It can be seen that these are identical to their Cartesian equivalents except for the extra term in v_r/r. This term gives rise to arithmetical difficulties on the centre-line where both v_r and r are zero. However, applying l'Hopital's rule to this fraction, and recalling that ψ is equal to either 0 or 90° on the centre-line, we find that equation (8.6.29) becomes

$$2 \, dv_r + \tan (\psi + 45) \, dv_z = 0 \qquad (8.6.32)$$

and equation (8.6.31) becomes

$$2 \, dv_r + \tan (\psi + 45) \, dv_z = 0 \qquad (8.6.33)$$

Equations (8.6.29) to (8.6.33) can be put in finite difference form as with the Cartesian form, remembering to take the average values of ψ *and* r over the interval.

The characteristic equations in polar and spherical co-ordinates can be derived from the equations presented in §8.3 using the same techniques. The results are not given here since the use of Cartesian or cylindrical co-ordinates is usually sufficiently convenient.

It should be noted that the characteristic directions are given by $\zeta = \psi \pm 45$ and therefore velocity characteristics only converge when the stress characteristics are also converging. Thus a velocity discontinuity can only be formed *within the material* when a stress discontinuity is also being formed. However, a velocity discontinuity can originate at a boundary and can propagate through the material in the absence of a stress discontinuity. Under these circumstances the characteristics are not converging and hence the velocity discontinuity must lie along a velocity characteristic and hence at $\pm$ 45° to the major principal stress direction. The equations that must be satisfied across a velocity discontinuity are considered in §8.7.

Baldwin and Hampson (1988) studied the convergence to the radial velocity field in a wedge-shaped hopper using these techniques. The dependence of ψ on position was obtained from the radial stress field and they used as the upper boundary condition that the velocity was uniform and vertical. Their results are shown qualitatively in figure 8.13.

Within the region ABA' where ABC' and $A'BC$ are the velocity characteristics from A and A', the velocity remained constant. Across the boundaries AB and $A'B$, the velocity changed suddenly in magnitude and direction and within the regions ABC and $A'BC'$ were found to be almost but not quite parallel to the wall. Within the region $BCDC'$, the velocities were nearly vertical and within the regions CDE and $C'DE'$ the velocities were again nearly parallel to the walls. Though the velocities did tend towards the values predicted from the radial velocity field, the convergence was neither rapid nor monotonic.

Moreea (1990) extended this analysis to the case of a conical hopper and found similar results, though convergence towards the radial

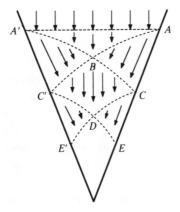

Figure 8.13 Pattern of velocity discontinuities in a wedge-shaped hopper.

velocity field was more rapid but not as rapid as suggested by the experimental results of Cleaver (1991). The reason for this may lie in the choice of the boundary conditions. The calculations of both Baldwin and Hampson (1988) and Moreea (1990) started with an assumed velocity distribution at the top of the hopper and worked downwards. We saw in §7.10 that the non-linearity of the stress equations requires that the stress calculations proceed downwards, a result which we justified physically on the grounds that the stresses depend mainly on the weight of the material above. However, the velocity equations are linear and equally stable for downward and upward calculation. Thus the calculations could start at the base of the system and work upwards. This can be justified physically by postulating that the direction of motion is determined by whether there is an appropriate space into which the material can flow. This argument is supported by the observations of Davies *et al.* (1980) who studied the flow round obstacles. They found that inserting an obstacle into a plug flow region creates more disturbance above the obstacle than below. Furthermore, changes to the shape of the underside of the obstacle affected the flow pattern above it.

Unfortunately, the velocity distribution through an orifice is not known and there is therefore no certain starting point for upward calculations in a discharging hopper. It is, however, possible to use the velocity distribution found from the radial velocity field as the boundary condition at the base of the cylindrical section of a cylindrical plus conical bin. However, in this case complications arise since there is some evidence that a velocity discontinuity is formed as mentioned in §8.1. The velocity discontinuity is discussed further in §8.7.

8.7 The velocity discontinuity

Since the velocity distribution in a flowing granular material can be found by the method of characteristics, velocity discontinuities can occur similar to the stress discontinuities discussed in §7.12. A velocity discontinuity can be coincident with a stress discontinuity or can occur in a continuous stress field. In the latter case the characteristics will not be converging and therefore such a velocity discontinuity must start at one of the boundaries of the system and propagate along a velocity characteristic.

It has been argued with some conviction that velocity discontinuities cannot exist since this would involve infinite accelerations. Though strictly true, the net result of this argument is to show that the 'discontinuity' must be spread over some small distance, perhaps equal to half-a-dozen particle diameters, giving a region of rapid velocity change. The velocity discontinuity relationship presented below gives the change across such a zone but gives no information about the velocity distribution within the zone. Once again we must bear in mind that the use of continuum analyses can only give meaningful results over distances considerably greater than a particle diameter. The situation in supersonic gas flow is no different. The shock wave is not a true discontinuity but a zone of rapid change, a few mean free paths in width.

As with the stress discontinuity it is convenient to denote the conditions on one side of the discontinuity by the superscript $^-$ and on the other side by $^+$. The velocity change across the discontinuity is denoted by the superscript $*$. Let us consider a velocity discontinuity as shown in figure 8.14 and take axes n and s normal to and along the discontinuity. We will denote by u and v the velocities in the n and s directions so that on the $^-$ side we have velocity components (u^-, v^-) which can be denoted by the vectorial velocity $\mathbf{V}^-$.

The normal strain rate across the discontinuity $\dot{\varepsilon}_{nn}$ is given by

$$\dot{\varepsilon}_{nn} = \frac{u^+ - u^-}{\delta n} \tag{8.7.1}$$

where δn is the width of the discontinuity, and the shear strain rate $\dot{\gamma}_{ns}$ is given by,

$$\dot{\gamma}_{ns} = \frac{v^+ - v^-}{\delta n} \tag{8.7.2}$$

There may also be a strain rate $\dot{\varepsilon}_{ss}$ along the discontinuity due to the velocity distribution in the $^-$ region, but this is likely to be very much

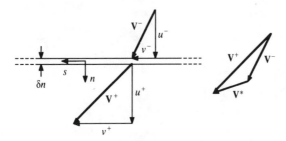

Figure 8.14 Vector diagrams for the velocities on either side of a discontinuity.

less than $\dot{\varepsilon}_{nn}$ or $\dot{\gamma}_{ns}$ since δn is very small. Thus the rate of volumetric strain $\dot{\varepsilon}_v \approx \dot{\varepsilon}_{nn}$.

By definition the ratio $\dot{\varepsilon}_v/\dot{\gamma}_{ns}$ is the tangent of the angle of dilation v, so that

$$\tan v = \frac{\dot{\varepsilon}_v}{\dot{\gamma}_{ns}} = \frac{\dot{\varepsilon}_{nn}}{\dot{\gamma}_{ns}} = \frac{u^+ - u^-}{v^+ - v^-} = \frac{u^*}{v^*} \tag{8.7.3}$$

Thus the jump velocity $\mathbf{V}^*$ (defined as $\mathbf{V}^+ - \mathbf{V}^-$) is inclined at angle v to the discontinuity. Hence, for an incompressible material for which $v = 0$, there can only be a discontinuity in the tangential component of the velocity, as occurs for example in a shear cell.

If we have a curved discontinuity as shown in figure 8.15, we can evaluate the rates of direct strain along the discontinuity by considering a small section of the discontinuity as the arc of a circle of radius R subtending the angle $\delta\theta$. The rate of direct strain in the s-direction is given by

$$\dot{\varepsilon}_{ss} = \frac{v^- - (v^- + \delta v^-)\cos\delta\theta - (u^- + \delta u^-)\sin\delta\theta}{R\delta\theta} \tag{8.7.4}$$

or, neglecting second order quantities

$$\dot{\varepsilon}_{ss} = -\frac{\delta v^- + u^-\delta\theta}{r\delta\theta} \tag{8.7.5}$$

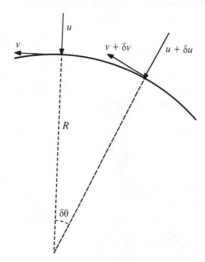

Figure 8.15 Curved velocity discontinuity.

Similarly, on the $^+$ side

$$\dot{\varepsilon}_{ss} = -\frac{\delta v^+ + u^+ \delta\theta}{r\delta\theta} \qquad (8.7.6)$$

These two quantities must be equal, since one side of the discontinuity cannot grow in length more rapidly than the other. Thus

$$\delta v^+ + u^+ \delta\theta = \delta v^- + u^- \delta\theta \qquad (8.7.7)$$

or

$$\delta v^* + u^* \delta\theta = 0 \qquad (8.7.8)$$

Substituting from equation (8.7.3) gives

$$\delta v^* = - v^* \tan v \, \delta\theta \qquad (8.7.9)$$

which on integration yields

$$\ln (v^*/v_0^*) = (\theta_0 - \theta) \tan v \qquad (8.7.10)$$

where v_0^* is the value of v^* at $\theta = \theta_0$. Similarly, from equation (8.7.3)

$$\ln (u^*/u_0^*) = (\theta_0 - \theta) \tan v \qquad (8.7.11)$$

Drescher, Cousens and Bransby (1978) have used these equations to predict the velocity distribution in a wedge-shaped hopper using the model shown in figure 8.16. In the regions ABC and $A'BC'$ the material was assumed to flow with a constant velocity parallel to the wall. Below the velocity discontinuity $C'BC$, the material was assumed to flow radially. These assumptions, plus the equations presented above are sufficient to predict both the shape of the discontinuity and the

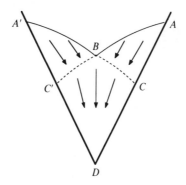

Figure 8.16 Drescher's model for the flow in a wedge-shaped hopper.

velocity in the radial flow region $BCDC'$. By suitable selection of the value of the angle of dilation, they were able to predict both a shape of discontinuity in accord with that revealed by X-radiography and a realistic velocity profile in the radial flow zone. The analysis is, however, not wholly satisfactory since it is not certain that the flow is radial throughout the zone $BCDC'$. It would be more realistic to predict the shape of the discontinuity from the velocity characteristics and find the resulting velocity distribution below it.

None-the-less, this analysis shows that a discontinuity of roughly the observed shape can produce a velocity distribution close to that predicted by the radial stress field. Thus while adjustments to the final asymptote may be slow, as found by Baldwin and Hampson (1988), the velocities start much closer to their final values than they assumed. The velocities therefore approximate to their final value for a greater proportion of the hopper than Baldwin and Hampson results seem to predict. This may account for the observation by Cleaver (1991) that the asymptotic velocity field extends throughout the lower half of the hopper.

Velocity discontinuities are also found near the transition plane of a cylindrical plus conical bin. In the cylindrical section the material descends as a compacted plug. On passing through the discontinuity some dilation occurs and the velocities change to values close to their final value, with the remaining adjustments taking place relatively slowly as illustrated in figure 8.1. Prakesh and Rao (1991) have analysed the two-dimensional analogue of this situation by assuming that the velocity discontinuity coincides with a stress discontinuity and takes the form of a parabolic arc.

9

The Conical yield function

9.1 Introduction

All the analyses of stress and velocity distributions so far presented in this book have been based on the Coulomb yield criterion. It must be borne in mind that this is a mathematical model and not a universal truth. Therefore we must always check the predictions of this model by comparing them with experimental observations. The Coulomb yield criterion does seem to give an excellent prediction of the wall stresses for many materials but it does not follow from this that it is valid for all granular materials or for the prediction of other phenomena in the materials for which it gives reliable stress distributions. In particular we saw in §8.5 that the Coulomb model in conjunction with the assumption of radial flow, does not give a good prediction of the position of the stagnant zone boundary in core flow. It is not, however, clear whether the fault lies in the Coulomb criterion or in the assumption of radial flow.

Inspired by the failure of the analysis mentioned above, Jenike (1987) has recently revived interest in the *Conical* or *Extended von Mises Failure Criterion*. The purpose of this chapter is to present the background to this yield criterion and to establish the stress and velocity equations for radial flow of an incompressible material. The equations for a compressible material are given by Jenike. We will show that this criterion predicts markedly different velocity distributions from those resulting from the Coulomb criterion, but that the predicted stresses do not differ much. Thus experimental stress determinations are not helpful in deciding which is the better of the two models. On the other hand, if the interest is solely in the prediction of stress distributions, one does not need to know which is better and the classic results based on the Coulomb model can be used with confidence.

9.2 The Conical yield function

Coulomb's yield criterion is based on the assumption that along the slip planes the shear stress is a linear function of the normal stress, and for a cohesionless material this can be written in the form,

$$\tau_{ns} = \sigma_{nn} \tan \phi \qquad (9.2.1)$$

where (n,s) is a set of Cartesian axes normal and along the slip plane. As shown in §3.3 this can be written as

$$\frac{\sigma_1}{\sigma_3} = \frac{1 + \sin \phi}{1 - \sin \phi} \qquad (9.2.2)$$

where σ_1 and σ_3 are the major and minor principal stresses. It is seen that this failure criterion is independent of the intermediate principal stress σ_2.

As an alternative, we could use some other measure of the shear and normal stresses obtaining at a point within the material. One such possibility is to work in terms of the octahedral stresses defined by,

$$\sigma_{\text{oct}} = \tfrac{1}{3} (\sigma_1 + \sigma_2 + \sigma_3) \qquad (9.2.3)$$

and

$$\tau_{\text{oct}} = \tfrac{1}{3} \sqrt{[(\sigma_1 - \sigma_2)^2 + (\sigma_2 - \sigma_3)^2 + (\sigma_3 - \sigma_1)^2]} \qquad (9.2.4)$$

It is seen that these are closely related to the stress invariants mentioned in §6.2. The octahedral normal stress is the mean of the principal stresses and, by definition, is equal to the isotropic pressure. Recalling that each principal stress difference is equal to the diameter of the appropriate Mohr's circle, which in turn is equal to twice the greatest shear stress in that set of planes, we see that the octahedral shear stress is proportional to the root mean square of the three maximum shear stresses.

We can postulate a yield function in which τ_{oct} is proportional to σ_{oct} in contrast to the Coulomb criterion which postulates that τ_{ns} is proportional to σ_{nn} as expressed by equation (9.2.1). For subsequent algebraic convenience the constant of proportionality is usually written as $\sqrt{(2M/3)}$, so that

$$\tau_{\text{oct}} = \sqrt{\left(\frac{2M}{3}\right)} \sigma_{\text{oct}} \qquad (9.2.5)$$

or from equation (9.2.4)

$$(\sigma_1 - \sigma_2)^2 + (\sigma_2 - \sigma_3)^2 + (\sigma_3 - \sigma_1)^2 = 6\,M^2\,\sigma^2 \qquad (9.2.6)$$

where, for simplicity, we have written σ_{oct} as σ.

It is convenient to write equation (9.2.6) in terms of the deviatoric stresses s, defined by

$$s_{ij} = \sigma_{ij} - \sigma\delta_{ij} \qquad (9.2.7)$$

where δ_{ij} is the Kronecke delta function, having the value 1 when $i = j$ and the value 0 when $i \neq j$. Thus for normal stresses the deviatoric stress is the difference between the actual stress and the mean stress, i.e.

$$s_{xx} = \sigma_{xx} - \sigma \qquad (9.2.8)$$

$$s_1 = \sigma_1 - \sigma \quad \text{etc.} \qquad (9.2.9)$$

but for shear stresses the actual and deviatoric stress are the same, i.e.

$$s_{xy} = \tau_{xy} \quad \text{etc.} \qquad (9.2.10)$$

With this notation, we can rewrite equation (9.2.3) as

$$s_1 + s_2 + s_3 = 0 \qquad (9.2.11)$$

and equation (9.2.6) as

$$(s_1 - s_2)^2 + (s_2 - s_3)^2 + (s_3 - s_1)^2 = 6\,M^2\,\sigma^2 \qquad (9.2.12)$$

Multiplying out this equation and eliminating the product terms, $s_1 s_2$, $s_2 s_3$, etc. by adding the square of equation (9.2.11), we obtain

$$s_1^2 + s_2^2 + s_3^2 = 2\,M^2\,\sigma^2 \qquad (9.2.13)$$

This result is known as the *Conical Yield Function*, since equation (9.2.13) describes a cone in principal stress space.

It is convenient to express the Coulomb yield criterion for a cohesionless material in similar form. From equation (9.2.2) we have

$$\sigma_1 - \sigma_3 = (\sigma_1 + \sigma_3) \sin \phi \qquad (9.2.14)$$

and recalling that the Coulomb criterion is essentially a two-dimensional concept we can define a mean stress p by

$$p = \tfrac{1}{2}(\sigma_1 + \sigma_3) \qquad (9.2.15)$$

and deviatoric stresses by

$$s_{ij} = \sigma_{ij} - p\delta_{ij} \qquad (9.2.16)$$

Thus equation (9.2.15) becomes

$$s_1 + s_3 = 0 \qquad (9.2.17)$$

and equation (9.2.14) becomes

$$s_1 - s_3 = 2p \sin \phi \qquad (9.2.18)$$

Hence,

$$s_1 = -s_3 = p \sin \phi \qquad (9.2.19)$$

and therefore,

$$s_1^2 + s_3^2 = 2p^2 \sin^2\phi \qquad (9.2.20)$$

Comparison of this form of the Coulomb yield criterion and the corresponding form of the Conical yield function, equation (9.2.13), shows considerable similarities and in particular, they take the same form whenever s_2 is equal to zero since under these conditions the mean stresses p and σ are the same. We therefore see that

$$M = \sin \phi \qquad (9.2.21)$$

provided the value of ϕ is obtained from measurements obtained in circumstances in which s_2 is zero. We will see later that in plane strain, s_2 closely approximates to zero and hence all three of the common shear cells, the Jenike, the Peschl and the annular cells, can be used to evaluate M from equation (9.2.21).

Thus we can write the Conical yield function in the form

$$s_1^2 + s_2^2 + s_3^2 = 2\sigma^2 \sin^2 \phi_p \qquad (9.2.22)$$

where ϕ_p is the angle of internal friction *measured in plane strain*.

We can determine which of these two criteria is appropriate for a particular material by conducting experiments in which s_2 is not equal to zero. A convenient apparatus for this purpose is the tri-axial tester described in §6.5. Here the material is subjected to an axial stress σ_a in the vertical direction and two equal horizontal stresses P. By symmetry these are all principal stresses. The tri-axial tester can be used in a compression mode in which $\sigma_a > P$ (and therefore $\sigma_1 = \sigma_a$ and $\sigma_2 = \sigma_3 = P$) and an extension mode in which $\sigma_a < P$ (and therefore $\sigma_1 = \sigma_2 = P$ and $\sigma_3 = \sigma_a$). We will denote these two values of σ_a as σ_c and σ_e. The apparent angle of friction in the compression test, ϕ_c is given by equation (9.2.2) as

$$\sin \phi_c = \frac{\sigma_c - P}{\sigma_c + P} \qquad (9.2.23)$$

and in extensive test

$$\sin \phi_e = \frac{P - \sigma_e}{P + \sigma_e} \qquad (9.2.24)$$

The corresponding values of M can be found from equation (9.2.13) as follows,

$$M_c = \sqrt{3} \, \frac{\sigma_c - P}{\sigma_c + 2P} \qquad (9.2.25)$$

and

$$M_e = \sqrt{3} \, \frac{P - \sigma_e}{\sigma_e + 2P} \qquad (9.2.26)$$

If the material obeys the Coulomb yield criterion all three values of ϕ (ϕ_c, ϕ_e and ϕ_p) will be the same and three different values of M will be obtained. A Conical material will give constant M and three different values of ϕ. Unfortunately, we often find that the precision of the test is insufficient to provide an unambiguous result.

It is instructive to compare equations (9.2.14) and (9.2.12) with the Tresca and von Mises criteria in common use for elastic solids. These take the forms

$$\sigma_1 - \sigma_3 = \sigma_y \qquad (9.2.27)$$

and

$$(\sigma_1 - \sigma_2)^2 + (\sigma_2 - \sigma_3)^2 + (\sigma_3 - \sigma_1)^2 = 2\sigma_y^2 \qquad (9.2.28)$$

respectively. Thus we can see that the Coulomb yield criterion can be regarded as the frictional analogue of Tresca's criterion, and the Conical yield criterion is the frictional analogue of von Mises' criterion. For this reason some authors prefer the phrase *Extended von Mises' Yield Criterion*.

Experience shows that Tresca's criterion is better than von Mises' for some elastic materials though the converse is true for others. However, the failure conditions predicted by these two criteria do not differ greatly. By analogy we must expect that some granular materials will obey the Coulomb model and others will obey the Conical model but that both models will predict similar stress distributions.

The Conical yield function is often used for axi-symmetric systems. Under these circumstances the circumferential stress (σ_{xx} in spherical co-ordinates) is a principal stress and we can therefore draw a Mohr's circle for rotation about the χ-axis as shown in figure 9.1. In this figure we have drawn Mohr's circle on deviatoric stress axes. This plot differs from a conventional Mohr's circle only in the position of the origin of co-ordinates. The deviatoric stresses (s_{rr}, $\tau_{r\theta}$) and ($s_{\theta\theta}$, $\tau_{\theta r}$), lie on this circle, and denoting the centre of the circle by a and its radius by R we have that

$$a = \tfrac{1}{2}(s_{rr} + s_{\theta\theta}) \tag{9.2.29}$$

and

$$R^2 = [\tfrac{1}{2}(s_{rr} - s_{\theta\theta})]^2 + \tau_{r\theta}^2 \tag{9.2.30}$$

The principal deviatoric stresses are s_{xx} and $a \pm R$, and hence their sum of the squares is given by

$$s_1^2 + s_2^2 + s_3^2 = (a - R)^2 + (a + R)^2 + s_{xx}^2$$
$$= 2a^2 + 2R^2 + s_{xx}^2 \tag{9.2.31}$$

Thus substituting from equations (9.2.29), (9.2.31) and (9.2.22) gives

$$s_{rr}^2 + s_{\theta\theta}^2 + s_{xx}^2 + 2\tau_{r\theta}^2 = 2\sigma^2 \sin^2\phi \tag{9.2.32}$$

In cylindrical co-ordinates (r,z,θ), this equation takes the form

$$s_{rr}^2 + s_{zz}^2 + s_{\theta\theta}^2 + 2\tau_{rz}^2 = 2\sigma^2 \sin^2\phi \tag{9.2.33}$$

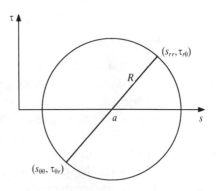

Figure 9.1 Mohr's circle for deviatoric stresses.

9.3 Levy's flow rule

The Conical yield function gives a relationship between all three principal stresses in contrast with the Coulomb yield function in which the intermediate principal stress does not appear. Thus an extra equation is required and in von Mises' original work this was provided by the Levy flow rule and as a result von Mises' analysis is often referred to as the Levy–von Mises model.

Levy's flow rule can be invoked also in the case of frictional materials and states that the deviatoric strain rates are proportional to the deviatoric stresses, i.e.

$$s_{ij} = \lambda \, \dot{e}_{ij} \qquad (9.3.1)$$

where the deviatoric strain rates are defined by

$$\dot{e}_{ij} = \dot{\varepsilon}_{ij} - \frac{\dot{\varepsilon}_V}{3} \delta_{ij} \qquad (9.3.2)$$

where $\dot{\varepsilon}_V$ is the volumetric strain rate. Equation (9.3.2) is seen to be analogous to the definition of the deviatoric stresses given by equation (9.2.7). The appearance of the factor of 3 in equation (9.3.2) results from the definition of the volumetric strain rate as the *sum* of the three principal strain rates, unlike the octahedral stress which was the *mean* of the principal stresses. The parameter λ appearing in equation (9.3.1) is a scalar whose value, like that of the λ of equation (8.3.7) need not be determined since it is always eliminated at an early stage in the analysis. It should be noted that Levy's flow rule imposes co-axiality since from equation (9.3.1) we see that the shear stresses are zero on the planes on which the shear strain rates are zero. We must, however, be careful to ensure that we have compatible definitions for the rates of strain. As can be seen from equations (A3.14) and (A3.15), the traditional definitions of direct and shear strain are incompatible by a factor of 2. Thus if we interpret $\dot{\varepsilon}_{ii}$ as the normal strain rate, $\dot{\varepsilon}_{ij}$ must be *half* the shear strain rate $\dot{\gamma}_{ij}$.

It must be appreciated that the Levy flow rule like the Coulomb and Conical failure criteria is a mathematical model, the accuracy of which must be checked by comparison with experiment.

In this book we will confine ourselves to the flow of incompressible materials so that $\dot{\varepsilon}_V$ is zero and the distinction between the deviatoric strain rates and the actual strain rates disappears. The reader is referred to Jenike's paper (1987) for details of the use of Levy's flow rule for compressible materials. In plane strain the intermediate strain rate $\dot{\varepsilon}_2$

is zero and hence s_2 is zero and the distinction between the Coulomb and Conical yield criteria disappears. The same result is almost true for compressible materials in plane strain. Under these circumstances $\dot{\varepsilon}_2$ is zero but $\dot{\varepsilon}_V$ and hence $\dot{e}_2$ are non-zero. However, the magnitude of $\dot{e}_2$ is inevitably less than that of the other principal strain rates and hence the magnitude of s_2 is less than the magnitudes of the other principal stresses. Thus s_2^2 is very much less than the sum $s_1^2 + s_3^2$ and there is little difference between the Conical and Coulomb yield criteria. It is for these reasons that we deduced in §9.2 that M is equal to the sine of the angle of friction *measured in plane strain*. This result is exact for incompressible materials and a good approximation for compressible materials.

The use of Levy's flow rule requires a knowledge of the kinematics of the system and in general this is not known *ab initio*. Thus it is usually necessary to evaluate the stress and velocity fields simultaneously, which makes the use of the Conical yield function much more difficult than the Coulomb yield function, for which we can find the stress field first and then proceed to evaluate the velocity field. However, in the case of radial flow in a conical hopper we know that the velocity must be given by

$$v_r = -\frac{f(\theta)}{r^2} \, , v_\theta = v_\chi = 0 \tag{9.3.3}$$

as in equation (8.4.1), and hence the strain rates are given by equations (8.4.2) to (8.4.4), i.e.

$$\dot{e}_{rr} = \dot{\varepsilon}_{rr} = -\frac{2f(\theta)}{r^3} \tag{9.3.4}$$

$$\dot{e}_{\theta\theta} = \dot{e}_{\chi\chi} = \frac{f(\theta)}{r^3} \tag{9.3.5}$$

$$\dot{e}_{r\theta} = \tfrac{1}{2}\dot{\gamma}_{r\theta} = -\frac{f'(\theta)}{2r^3} \tag{9.3.6}$$

Thus,

$$\dot{e}_{\theta\theta} = \dot{e}_{\chi\chi} = -\tfrac{1}{2}\dot{e}_{rr} = -2\frac{f(\theta)}{f'(\theta)}\dot{e}_{r\theta} \tag{9.3.7}$$

and hence from Levy's flow rule

$$s_{\theta\theta} = s_{\chi\chi} = -\tfrac{1}{2}s_{rr} = -2\frac{f(\theta)}{f'(\theta)}\tau_{r\theta} \tag{9.3.8}$$

Substituting into equation (9.2.32) gives

$$s_{\theta\theta}^2 \left[6 + \tfrac{1}{2} \left(\frac{f'(\theta)}{f(\theta)} \right)^2 \right] = 2\,\sigma^2 \sin^2\!\phi \qquad (9.3.9)$$

This equation is put in more convenient algebraic form if we define an angle ξ such that

$$\tan \xi = -\frac{1}{\sqrt{12}} \frac{f'(\theta)}{f(\theta)} \qquad (9.3.10)$$

Substituting into equation (9.3.9) gives

$$s_{\theta\theta} = \frac{1}{\sqrt{3}}\, \sigma \sin \phi \cos \xi \qquad (9.3.11)$$

or

$$\sigma_{\theta\theta} = \sigma \left(1 + \frac{1}{\sqrt{3}} \sin \phi \cos \xi \right) \qquad (9.3.12)$$

Similarly, from equation (9.3.8) we have that

$$\sigma_{rr} = \sigma \left(1 - \frac{2}{\sqrt{3}} \sin \phi \cos \xi \right) \qquad (9.3.13)$$

$$\sigma_{xx} = \sigma \left(1 + \frac{1}{\sqrt{3}} \sin \phi \cos \xi \right) \qquad (9.3.14)$$

and

$$\tau_{r\theta} = -\tau_{\theta r} = \sigma \sin \phi \sin \xi \qquad (9.3.15)$$

This set of equations can be used instead of the equivalent set for a Coulomb material, equations (7.2.14) to (7.2.16), to predict the stress distribution in a conical hopper. It will be noted that in the analysis immediately above, we have made certain choices of sign. Since the quantity $f'(\theta)/f(\theta)$ is inevitably negative we inserted a negative sign into equation (9.3.10) thereby confining ξ to the range $0 < \xi < 90$. Also, on proceeding from equation (9.3.9) to (9.3.11) we took a square root giving an ambiguity of sign in equations (9.3.11) and (9.3.12). In the passive case $\sigma_{\theta\theta} > \sigma$ and, with $\cos \xi$ being positive, this is achieved by the +ve sign in equation (9.3.12).

If we are concerned with the flow in a cylindrical bunker we can say by symmetry that

$$\dot{e}_{\theta\theta} = 0 \tag{9.3.16}$$

and hence

$$s_{\theta\theta} = 0 \tag{9.3.17}$$

and the Conical yield criterion reduces to the Coulomb yield criterion, and σ becomes equal to p^*. We see that the cohesionless form of equations (7.2.21) to (7.2.23) i.e.

$$\sigma_{rr} = \sigma(1 + \sin \phi \cos 2\psi) \tag{9.3.18}$$

$$\sigma_{zz} = \sigma(1 - \sin \phi \cos 2\psi) \tag{9.3.19}$$

$$\tau_{rz} = -\tau_{zr} = \sigma \sin \phi \sin 2\psi \tag{9.3.20}$$

satisfy equation (9.2.33). However, the stress distribution is not the same as that given by the Coulomb yield function and presented in §7.3 since the circumferential stress is given by equation (9.3.17) as

$$\sigma_{\theta\theta} = \sigma \tag{9.3.21}$$

which is not the same as its Coulomb counterpart, equation (7.2.24).

It is worth emphasising that in the analysis of Coulomb materials we invoke the Haar–von Karman hypothesis to evaluate the circumferential stress and the principle of co-axiality to provide the flow rule. By contrast, in the analysis of a Conical material, Levy's flow rule both determines the circumferential stress and imposes co-axiality.

9.4 The radial stress and velocity fields

Having established the relationship between the stresses in radial incompressible flow in a conical hopper, we can solve for the stress distribution by substituting equations (9.3.11) to (9.3.14) into the stress equilibrium equations (7.2.27) and (7.2.28). As with the radial stress field of a Coulomb material we can make the substitution

$$\sigma = \gamma r q(\theta) \tag{9.4.1}$$

and assume that ξ is a function of θ only. On doing so we find that

$$A \frac{dq}{d\theta} + B \frac{d\xi}{d\theta} + C = 0 \tag{9.4.2}$$

$$D \frac{dq}{d\theta} + E \frac{d\xi}{d\theta} + F = 0 \tag{9.4.3}$$

where,

$$A = -\sin\phi\sin\xi \qquad\qquad (9.4.4)$$

$$B = -q\sin\phi\cos\xi \qquad\qquad (9.4.5)$$

$$C = q\left(1 - \frac{8}{\sqrt{3}}\sin\phi\cos\xi - \sin\phi\sin\xi\cot\theta\right) + \cos\theta \quad (9.4.6)$$

$$D = 1 + \frac{1}{\sqrt{3}}\sin\phi\cos\xi \qquad\qquad (9.4.7)$$

$$E = -\frac{q}{\sqrt{3}}\sin\phi\sin\xi \qquad\qquad (9.4.8)$$

$$F = -4q\sin\phi\sin\xi - \sin\theta \qquad\qquad (9.4.9)$$

It is seen that these equations are similar to those of §7.4 and can be solved in the same manner. By analogy with equations (7.4.12) and (7.4.13) we have

$$\frac{dq}{d\theta} = \frac{CE - BF}{BD - AE} \qquad\qquad (9.4.10)$$

and

$$\frac{d\xi}{d\theta} = \frac{AF - CD}{BD - AE} \qquad\qquad (9.4.11)$$

These equations can be solved subject to the boundary conditions;

(i) on the centre line ($\theta = 0$), $\sigma_{\theta\theta}$ is the major principal stress i.e. from equation (9.3.11), $\xi = 0$, and

(ii) on the wall ($\theta = \alpha$), $\tau_{\theta r} = -\sigma_{\theta\theta}\tan\phi_w$, the minus sign appearing since we are considering a left-hand wall, and hence from equations (9.3.11) and (9.3.14)

$$\tan\phi_w = \frac{\sin\phi\sin\xi_w}{1 + \frac{1}{\sqrt{3}}\sin\phi\cos\xi_w} \qquad\qquad (9.4.12)$$

which is a quadratic in $\cos\xi_w$ for which we require the root of ξ_w in the range $0 < \xi_w < 90°$ since we have defined ξ to lie in this range.

Because of the split boundary conditions we must use a shooting technique as in §7.4. Again numerical problems occur on $\theta = 0$ due to indeterminacy of the $\sin \xi \cot \theta$ term in equation (9.4.6) and these are solved by the same techniques as in §7.4, giving

$$\frac{dq}{d\theta} = 0 \qquad (9.4.13)$$

and

$$\frac{d\xi}{d\theta} = \frac{1 + q(1 + (8/\sqrt{3}) \sin \phi)}{2q \sin \phi} \qquad (9.4.14)$$

Cleaver (1991) has compared the wall stresses determined by this method with those found from the radial stress field for a Coulomb material. He considered the cases of $\phi = 30°$ and $40°$, values of α covering the range $0 < \alpha < 40°$ and ϕ_w in the range $5° < \phi_w < \phi$. The difference between the two predictions was greatest for large values of α and ϕ_w but over the range of variables considered the difference was always less than 20% with the wall stresses predicted by Coulomb's model always being greater than those resulting from the Conical model. Thus it seems that Coulomb's model is to be recommended when predicting the wall stresses for the purpose of designing the strength of the walls as it gives a conservative, or safe, estimate.

Figure 9.2 shows typical plots of the variation of ξ with θ and we can predict the velocity profiles from these results since, from equation (9.3.10),

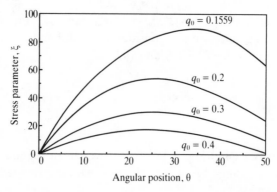

Figure 9.2 Variation of the stress parameter ξ with angular position θ.

$$f(\theta) = A \exp\left(- \sqrt{12} \int_0^{\theta} \tan \xi \, d\theta\right) \qquad (9.4.15)$$

Figure 9.3 shows typical results obtained by this method and figure 9.4 compares the velocity profiles with those predicted from the Coulomb yield criterion. It is seen that markedly different velocity profiles are obtained for a given combination of ϕ and ϕ_w. Experimental results by Cleaver (1991) suggest that the prediction of the Conical model are closer than those of the Coulomb model to the experimental results. This gives support to the contention that the Conical model is

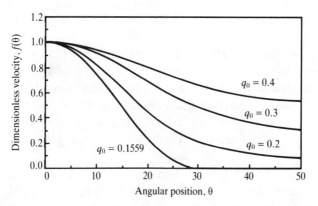

Figure 9.3 Variation of the dimensionless velocity $f(\theta)$ with angular position θ.

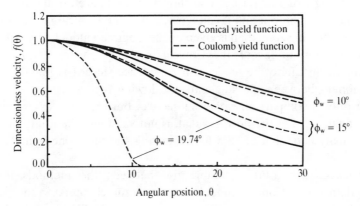

Figure 9.4 Comparison of the velocity predictions of the Coulomb yield criterion and the Conical yield criterion.

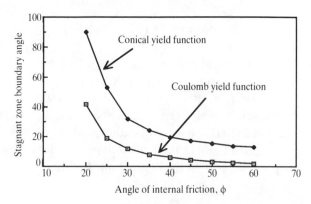

Figure 9.5 The dependence of the stagnant zone boundary angle with angle of wall friction as predicted by the Coulomb and Conical yield criteria.

the better of the two. However, the experiments have not yet been conducted over a sufficiently wide range of variable to determine whether this is universally correct or a particular feature of the materials used in Cleaver's experiments. Fortunately, the wall stress results presented above suggest that we need not worry which of the two models is best if we are only concerned with the prediction of wall stresses.

It can be seen from figure 9.2 that ξ can pass through a maximum within the range $0 < \theta < \alpha$ and can thus reach the value of $90°$, which from equation (9.4.15) is the critical value for the formation of a stagnant zone. From this we can obtain a prediction of the position of the stagnant zone boundary and this is shown in figure 9.5 as a function of the angle of friction ϕ. Also shown in the figure are the results for a Coulomb material and it can be seen that the Conical model predicts a wider flowing zone than is given by the Coulomb model. Experimental results seem to favour the Conical model (Cleaver, 1991) but the comparison cannot be made with confidence since it is found that the boundary between the stagnant and flowing regions is curved. Thus, not only is comparison rather subjective, but the basic assumption of the analysis, that the flow is radial, is not valid and the utility of either this analysis or that of §8.5 for predicting stagnant zone boundaries is questionable.

Cleaver has used a compromise between the methods for the Coulomb and Conical models presented in chapters 8 and 9. His method is based on the combination of the Coulomb yield criterion and Levy's flow rule. For the case of radial flow this has the sole effect of replacing equation (7.2.24), which for the passive failure of a cohesionless material takes the form

$$\sigma_{xx} = \sigma(1 + \sin \phi) \qquad (9.4.16)$$

with

$$\sigma_{xx} = \sigma_{\theta\theta} = \sigma(1 - \sin \phi \cos 2\psi) \qquad (9.4.17)$$

The velocity profiles predicted by this method are barely distinguishable from those of the Conical model.

As mentioned above the use of the Conical yield function necessitates the simultaneous solution of the stress and velocity fields except in a few special cases such as radial flow. The resulting calculations are therefore very much more difficult than those for the Coulomb model, for which the stress field can always be evaluated first and the velocity field later. Furthermore, the equations resulting from the Conical yield function are not hyperbolic and solution by the method of characteristics is not possible. Thus it does not seem likely that solutions for materials obeying the Conical yield function will be available, except for a few special cases, in the foreseeable future. Fortunately, as we noted in the previous section, the stress fields predicted by the two methods do not seem to differ significantly.

10

The prediction of mass flow rate

10.1 Introduction

In this final chapter we will consider the discharge rate of granular materials through orifices. In §10.2 we will review the correlations that have been proposed, paying particular attention to the Beverloo correlation and minor modifications to it. In §10.3 we will present the early theoretical attempts at the prediction of mass flow rate and discuss the validity of the 'free-fall arch', which forms the basis of most of these predictions. We will then consider in §10.4 and §10.5 some more recent theoretical approaches based on the so-called *Hour-Glass Theory*. The effect of interstitial pressure gradient on the flow rate is discussed in §10.6 and this leads on to consideration of the problems associated with the flow of fine powders in §10.7. Finally in §10.8 and §10.9 we will consider two notorious difficulties associated with discharge from hoppers, the problems of arching and flooding. The former can prevent discharge and the latter gives very large and often uncontrollable flow rates.

It may be noted that in the theoretical sections of this chapter, we are departing for the first time from the subject matter implied by the title of this book, since we are now considering dynamics, the effect of forces on the acceleration of the material.

10.2 Mass flow rate correlations

Figure 10.1 shows a schematic representation of both a cylindrical bunker and a conical hopper to define the terminology of this chapter. In particular we are concerned with bins with circular orifices of diameter D_0 discharging a free-flowing material of diameter d.

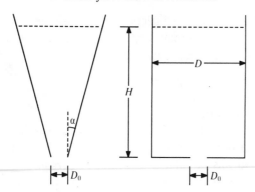

Figure 10.1 Definition of symbols.

Intuition suggests that the mass discharge rate W from a flat-bottomed bunker depends on the bulk density ρ_b, the height H and diameter D of the bunker, the orifice and particle diameters, the acceleration due to gravity and the coefficient of friction μ. In the case of a conical hopper the half-angle α becomes important. The influence of cohesion and orifice shape is considered later. Early experimentalists discovered that the influence of the bunker diameter and the height were negligible except when the bunker was nearly empty. Some authors have suggested that the mass flow rate is proportional to $H^{0.04}$ but it now seems generally agreed that the flow rate is independent of height provided $H \geqslant 2D_0$. This result must be interpreted with caution since in the later stages of discharge the top surface normally has a conical depression and we must take H to be the height to the centre of this depression. Similarly, the observation that the mass flow rate is independent of bunker diameter is only valid provided $D > 2D_0$; for narrower bunkers somewhat larger flow rates are observed and in the limit $W \to \infty$ as $D \to D_0$ since under these circumstances we cease to have an orifice and the material accelerates indefinitely under gravity.

There also exists a range of particle diameters over which the mass flow rate seems to be effectively independent of d. In fact as we will see later this represents a shallow maximum in the dependence of W upon d. However, within this range we can say that the mass flow rate depends on density, acceleration due to gravity, orifice diameter and coefficient of friction only, i.e.

$$W = f(\rho_b, g, D_0, \mu) \tag{10.2.1}$$

Noting that the coefficient of friction is dimensionless, we can apply the principles of dimensional analysis and deduce that the only permissible form of this relationship is

$$W = C \rho_b \sqrt{g} \, D_0^{5/2} \tag{10.2.2}$$

where C is a function of μ.

However, early experimentalists, plotting their results in the form $\ln W$ *vs* $\ln D_0$ did not obtain slopes of 5/2 but found values closer to 3. A value of 2.96 was popular with some workers as it seemed to tie in with the apparent dependence on $H^{0.04}$. Thus correlations of the type

$$W = C' \rho_b \sqrt{g} \, D_0^{2.96} \, H^{0.04} \tag{10.2.3}$$

were frequently proposed. Credit must be given to Beverloo, Leniger and Van de Velde (1961), who had confidence in the dependence of W on $D_0^{5/2}$. By plotting their experimental results in the form $W^{2/5}$ *vs* D_0, they obtained a straight line as expected but their results showed a non-zero intercept z on the D_0 axis as illustrated in figure 10.2. It was found that z was proportional to d for particles of the same shape and this gave rise to the correlation

$$W = C \rho_b \sqrt{g} \, (D_0 - kd)^{5/2} \tag{10.2.4}$$

where C seems to be almost independent of μ and takes a value close to 0.58. Values as large as 0.64 have, however, been reported for exceptionally smooth particles such as spherical glass beads. The parameter k is about 1.5 for spherical particles but takes somewhat larger values for angular particles, though in this case its precise value depends on what measure is taken for the equivalent particle diameter.

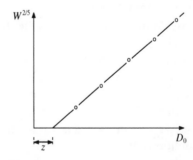

Figure 10.2 Experimental results plotted as recommended by Beverloo.

This correlation is perhaps the most reliable for general use though modifications have been proposed for particular situations. It is commonly called the Beverloo correlation though it should be pointed out that a very similar correlation had been proposed by Hagen in 1856.

The term kd appearing in the Beverloo correlation is compatible with the concept of the 'empty annulus' proposed by Brown and Richards (1970). They observed that there was an annular zone adjacent to the periphery of the orifice within which there were fewer particles than in the central region. The origin of this effect is unclear, but the term kd may simply represent the displacement thickness of the shear layer in the material approaching the orifice. The argument that no particle centre can approach within $d/2$ of the periphery of the orifice and therefore that all particle centres pass through a region of diameter $D_0 - d$ is weak; not only does it fail to predict the observation that $k > 1$ but it presumes incorrectly that the mass of the particle is concentrated at the centre. However, geometrical factors ensure that there is a greater void fraction in contact with a containing wall so that the outer regions of the orifice will undoubtedly contain a lower particle density. Whatever its origin, the concept of the empty annulus is a convenient mnemonic for the observation that the effective orifice diameter is less than D_0.

The bulk density ρ_b of the material must be determined with care. It was originally thought that this was the original packed density of the material in the bunker. However, it is easily demonstrated experimentally that the mass flow rate is independent of the original degree of compaction of the material. It seems that on initiation of flow the bulk density adjusts to a value appropriate for a flowing material, and this is consistent with the ideas of the critical state theory discussed in §6.5 and Reynolds' *Principle of Dilatancy*. However, since the parameter C must in principle be evaluated for each material, a standard density can be selected for that material and the appropriate value of C evaluated. By this means, the results of small-sized models can be scaled to industrial size. This approach, however, denigrates the value of the Beverloo correlation as a predictive method. The observation that C rarely differs much from 0.58 when ρ_b is taken to be thee bulk density measured by loosely filling a vessel will give a prediction of mass flow rate to within about 5% for coarse free-flowing materials, though some materials with particularly low coefficients of friction, such as glass beads, may discharge at somewhat greater flow rates.

The Beverloo correlation fails when the particle size is great enough for mechanical blocking of the orifice to occur and this happens when the particle diameter is about one-sixth of the orifice diameter. The Beverloo correlation also ceases to be valid for fine particles, for reasons which are discussed in detail in §10.7; a rough rule of thumb being that the flow rate is less than the Beverloo prediction when $d < 400$ μm. Thus we conclude that the Beverloo correlation is reliable in the range $D_0/6 > d > 400$ μm.

In the form of equation (10.2.4), the Beverloo correlation is applicable only for cylindrical bunkers and core flow hoppers. In mass flow conical hoppers, the effect of the half-angle α becomes important. A pre-Beverloo correlation by Rose and Tanaka (1956) states that the mass flow rate is proportional to $(\tan \alpha \tan \phi_d)^{-0.35}$ provided $\alpha < 90 - \phi_d$, where ϕ_d is the angle between the stagnant zone boundary and the horizontal. This can be incorporated into the Beverloo correlation in the form

$$W = W_B F(\alpha,\phi_d) \tag{10.2.5}$$

where W_B is the mass flow rate predicted from the Beverloo correlation, equation (10.2.4), and $F(\alpha,\phi_d)$ is given by

$$F(\alpha,\phi_d) = (\tan \alpha \tan \phi_d)^{-0.35} \qquad \text{for } \alpha < 90 - \phi_d$$

$$F = 1 \qquad\qquad\qquad \text{for } \alpha > 90 - \phi_d \tag{10.2.6}$$

As mentioned in chapters 8 and 9, there is no reliable method for predicting ϕ_d and this quantity must be determined experimentally. Values of 45° may be used in the absence of more reliable information.

The Beverloo correlation can be modifed for non-circular orifices by following a suggestion by Fowler and Glastonbury (1959) that the flow rate was proportional to the product of the orifice area and the square root of the hydraulic mean diameter. The latter quantity is defined as four times the cross-sectional area divided by the perimeter. Recalling the concept of the empty annulus, we can say that

$$W \propto A^* \sqrt{D_H^*} \tag{10.2.7}$$

where A^* and D_H^* are the area and hydraulic mean diameter of the orifice after the removal of the empty annulus. For a circular orifice, these quantities are given by

$$A^* = \frac{\pi}{4} (D_0 - kd)^2 \tag{10.2.8}$$

and

$$D_H^* = (D_0 - kd) \tag{10.2.9}$$

and we can therefore write equation (10.2.4) as

$$W = \frac{4C}{\pi} \rho_b A^* \sqrt{(gD_H^*)} \tag{10.2.10}$$

This form seems applicable to most orifice shapes. In particular for a $b \times l$ slot orifice, we have

$$A^* = (b - kd)(l - kd) \tag{10.2.11}$$

and

$$D_H^* = \frac{4(b - kd)(l - kd)}{2(l + b - 2kd)} \tag{10.2.12}$$

If $l \gg b$, $D_H^* \approx 2(b - kd)$ and equation (10.2.10) reduces to

$$W = \frac{4C\sqrt{2}}{\pi} \rho_b \sqrt{g}(l - kd)(b - kd)^{3/2} \tag{10.2.13}$$

Putting in the typical value of 0.58 for C gives

$$W = 1.03 \, \rho_b \sqrt{g}(l - kd)(b - kd)^{3/2} \tag{10.2.14}$$

a result which agrees to within 1% with the experimental results of Myers and Sellers (1977) who measured the flow rates of spherical particles from wedge-shaped hoppers.

It was noted at the start of this section that the mass flow rate is effectively independent of the quantity of material in the hopper, typified by the height H. Thus equation (10.2.4) can be compared with the corresponding result for an inviscid liquid

$$W = 0.64 \frac{\pi}{4} \rho D_0^2 \sqrt{(2gH)} \tag{10.2.15}$$

and it is seen that effectively D_0 replaces the H of equation (10.2.15). The reasons for this and the observation that C is a very weak function of μ form the subject matter of §10.3 to §10.5.

In the mean time we can note that predicting the flow rate of a cohesive material seems to be an unsolved problem. Rose and Tanaka (1956) include a multiplicative factor of $\exp(-c/\rho_b g d^3)$ in their correlation but not only is this dimensionally inconsistent but it is not made clear under what circumstances the cohesion c must be evaluated. An alternative approach by Johanson (1965) gives

$$W = W_0 \sqrt{\left(1 - \frac{ff}{ff_c}\right)} \qquad (10.2.16)$$

where W_0 is the mass flow rate of a cohesionless material of the same density, ff is the flow factor, defined in §10.8, and ff_c is the critical flow factor at which arching occurs.

The correlation presented by Al Din and Gunn (1984) is in effect a modification of the Beverloo correlation and is said to give somewhat more accurate predictions of the flow rate.

10.3 The free-fall arch and the minimum energy theory

All the early theoretical predictions of the mass flow rate from hoppers were based on the concept of the 'free-fall arch'. This is a surface spanning the orifice and represents the lower surface of the packed material. Above the free-fall arch the particles are assumed to be in contact with each other and inter-particle stresses occur. Below the free-fall arch, the particles lose contact and accelerate freely under gravity.

There are very considerable theoretical difficulties in the concept of a free-fall arch. As envisaged above it is the surface on which the normal stress falls to zero. Thus above the free-fall arch the material is subjected to forces due to its own weight plus those due to the stress gradient but below the arch the material is subjected only to its own weight. Therefore on passing through the arch there is a reduction in the accelerating force and this is incompatible with the increased acceleration required to achieve the dilation associated with the separation of the particles. This argument is sometimes known as Jackson's paradox and this paradox can be resolved by postulating one or a combination of the following three factors;

    (i) that a stress discontinuity within the material replaces the free-fall arch,

    (ii) that the stress gradient is zero at the free-fall arch, i.e. $d\sigma_{rr}/dr = 0$ on $r = r_0$, or,

    (iii) that there is a gradual dilation of the material as it passes through the orifice region.

Experimental evidence obtained from X- or γ-ray absorption shows that this last possibility is certainly a contributory factor, though probably not the whole explanation. Consequently, it is not possible to identify a free-fall arch unambiguously.

One of the first theoretical predictions of the flow rate from a discharging mass flow hopper was presented by Brown and Richards (1970). Their analysis is based on radial flow, so that

$$v_r = -\frac{f(\theta)}{r^2} \tag{10.3.1}$$

as in §8.4. The total energy content T per unit mass of material is given by analogy with Bernoulli's equation as

$$T = \frac{\sigma}{\rho_b} + \frac{v_r^2}{2} + gr \cos \theta \tag{10.3.2}$$

and, since the material is frictional, it is assumed that T must decrease in the direction of flow, i.e. dT/dr must be positive. It is further assumed that σ is constant in the vicinity of the orifice. This seems to have been based on Janssen's analysis, see §5.2, but cannot be true since σ must tend to zero near the orifice; the radial stress field, for example, predicting that the stress varies linearly with r.

Substituting equation (10.3.1) into (10.3.2) gives

$$\frac{dT}{dr} = -\frac{2f^2(\theta)}{r^5} + g \cos \theta \geq 0 \tag{10.3.3}$$

Brown and Richards have assumed that dT/dr is zero on the free-fall arch which is taken to be a spherical surface of radius r_0 spanning the orifice. For an orifice diameter D_0 and hopper half-angle α, we have that

$$r_0 = \frac{D_0}{2} \operatorname{cosec} \alpha \tag{10.3.4}$$

Thus from equation (10.3.3)

$$f(\theta) = \left(\frac{r_0^5 g \cos\theta}{2}\right)^{1/2} \tag{10.3.5}$$

but this prediction is quite at variance both with the observed velocity distribution and with that predicted by the radial velocity field. Despite this, the mass flow rate obtained by integrating equation (10.3.5)

$$W = 2\pi\rho_b \int_0^\alpha f(\theta) \sin\theta \, d\theta = \frac{\pi}{6} \rho_b \sqrt{g} \, D_0^{5/2} \left(\frac{1 - \cos^{3/2}\alpha}{\sin^{5/2}\alpha} \right) \quad (10.3.6)$$

is closer to that found experimentally than many of the more convincing predictions given below.

This theory is based on continuum arguments and like all such theories cannot give any information about the effect of particle size d. An empirical allowance can be made by invoking the concept of the empty annulus and replacing D_0 by $(D_0 - kd)$. In this case equation (10.3.6) can be written in the form

$$W = C_{ME}\rho_b \sqrt{g} \, (D_0 - kd)^{5/2} \quad (10.3.7)$$

where,

$$C_{ME} = \frac{\pi}{6} \left(\frac{1 - \cos^{3/2}\alpha}{\sin^{5/2}\alpha} \right) \quad (10.3.8)$$

Figure 10.3 compares C_{ME} with the prediction of the Beverloo/Rose and Tanaka correlations, $C = 0.58 \, F(\alpha, \phi_d)$ where $F(\alpha, \phi_d)$ is given by equation (10.2.6). It is seen that good agreement is obtained if ϕ_d is taken to be 45°. The success of the minimum energy theory in predicting the mass flow rate, coupled with its total failure to predict a reasonable velocity profile is one of the great puzzles of this subject and is dismissed by many authors as coincidental. However, some of the success of the minimum energy theory is inevitable. By putting $d\sigma/dr = 0$, we are dismissing both the height and the diameter of the bunker from the analysis and the correct dependence of the mass flow rate on the orifice diameter and the bulk density is assured.

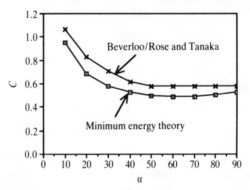

Figure 10.3 Comparison of the minimum energy theory with the Rose and Tanaka correlation.

10.4 The hour-glass theory

Savage (1967), Sullivan (1972) and Davidson and Nedderman (1973) have produced apparently independent theoretical predictions of the flow rate from a narrow-angled, smooth-walled hopper. Though differing in detail these analyses are basically similar and the version known as the *Hour-Glass Theory* is presented below.

Let us consider an incompressible cohesionless material of bulk density ρ_b and angle of internal friction ϕ discharging from a narrow-angled hopper of half-angle α. We will use a set of spherical co-ordinates (r,θ,χ) with origin at the virtual apex of the hopper. Since the walls are assumed to be smooth, the wall shear stress $(\tau_{\theta r})_w$ will be zero.

By symmetry $\tau_{\theta r}$ is also zero on the centre-line and hence it is reasonable to assume that $\tau_{\theta r}$ is zero throughout. Under these circumstances the stresses σ_{rr}, $\sigma_{\theta\theta}$ and $\sigma_{\chi\chi}$ are principal stresses and, since the flow is converging, σ_{rr} is the minor principal stress and $\sigma_{\theta\theta}$ is the major principal stress. From the Haar–von Karman hypothesis (§7.2), $\sigma_{\chi\chi}$ is also equal to the major principal stress and hence we can say from equation (7.2.10) that

$$\sigma_{\theta\theta} = \sigma_{\chi\chi} = K\,\sigma_{rr} \qquad (10.4.1)$$

where K is Rankine's coefficient of earth pressure, and is given by equation (3.4.11),

$$K = \frac{1 + \sin\phi}{1 - \sin\phi} \qquad (10.4.2)$$

For most materials $\phi \geq 30°$ and hence $K \geq 3$.

With the shear stress being zero throughout, we can assume that the shear strain rate is zero and hence the velocity is radial and independent of angular position. Thus we can say

$$v_r = -\frac{V}{r^2}, \; v_\theta = v_\chi = 0 \qquad (10.4.3)$$

where the parameter V can be related to the mass flow rate W by

$$W = \int_0^\alpha 2\pi r \sin\theta\, \rho_b\, v_r\, r\, d\theta = 2\pi\rho_b\, V(1 - \cos\alpha) \qquad (10.4.4)$$

We can now invoke Euler's equation which is derived in appendix 2 and is simply a statement of Newton's second law of motion, that the rate of change of momentum equals the applied force. In the absence of shear stresses this takes the form

$$\rho_b v_r \frac{dv_r}{dr} = -\frac{d\sigma_{rr}}{dr} - \frac{2\sigma_{rr} - \sigma_{\theta\theta} - \sigma_{xx}}{r} - \rho_b g \cos\theta \qquad (10.4.5)$$

Substituting from equations (10.4.1) and (10.4.3) and approximating $\cos\theta$ by 1 since the hopper is narrow, we have

$$\frac{2\rho_b V^2}{r^5} = \frac{d\sigma_{rr}}{dr} + \frac{2 - 2K}{r}\sigma_{rr} + \rho_b g \qquad (10.4.6)$$

The terms in σ_{rr} can be made into a perfect derivative by multiplying through by the integrating factor r^{2-2K} giving,

$$2\rho_b V^2 r^{-3-2K} - \rho_b g r^{2-2K} = r^{2-2K}\frac{d\sigma_{rr}}{dr}$$

$$+ (2 - 2K)\sigma_{rr} r^{1-2K} = \frac{d}{dr}(\sigma_{rr} r^{2-2K}) \qquad (10.4.7)$$

and on integration we find that

$$\sigma_{rr} r^{2-2K} = -\frac{2\rho_b V^2}{2 + 2K}r^{-2-2K}$$

$$-\frac{\rho_b g}{3 - 2K}r^{3-2K} + A \qquad (10.4.8)$$

where A is an arbitrary constant of integration.

We can evaluate the arbitrary constant and the parameter V by applying two boundary conditions. First, we will assume that there is a free-fall arch of radius r_0 and secondly, we will assume that the top surface is stress-free and part of a spherical surface of radius r_1. Thus our boundary conditions can be expressed as

$$\sigma_{rr} = 0 \text{ on } r = r_0 \text{ and } r = r_1 \qquad (10.4.9)$$

Inserting these boundary conditions and eliminating A by subtraction gives

$$\frac{\rho_b V^2}{1 + K}(r_0^{-2-2K} - r_1^{-2-2K}) + \frac{\rho_b g}{3 - 2K}(r_0^{3-2K} - r_1^{3-2K})$$

$$= 0 \qquad (10.4.10)$$

Making the substitution $y = r_1/r_0$, this yields

$$V^2 = gr_0^5 \left(\frac{1+K}{3-2K}\right)\left(\frac{1-y^{3-2K}}{1-y^{-2-2K}}\right) \tag{10.4.11}$$

Under all realistic circumstances the radius to the top surface r_1 is much greater than the radius of the free-fall arch r_0 and hence $y \gg 1$. We have seen above that K is typically greater than about 3, so that y appears in equation (10.4.11) only raised to large and negative indices. Thus all the terms in y are small compared to 1 and the last bracket of equation (10.4.11) is therefore close to unity. We can therefore say that

$$V = \sqrt{g}\, r_0^{5/2} \sqrt{\left(\frac{K+1}{2K-3}\right)} \tag{10.4.12}$$

From equation (10.4.4), noting from equation (10.3.4) that the orifice diameter D_0 is given by $2r_0 \sin \alpha$, we have

$$W = \frac{\pi\rho_b \sqrt{g}\, D_0^{5/2}}{2} \sqrt{\left[\frac{1+K}{2(2K-3)}\right]} \frac{1-\cos\alpha}{\sin^{5/2}\alpha} \tag{10.4.13}$$

or, since α is small,

$$W = \frac{\pi\rho_b \sqrt{g}\, D_0^{5/2}}{4\sin^{1/2}\alpha} \sqrt{\left[\frac{1+K}{2(2K-3)}\right]} \tag{10.4.14}$$

It is seen that this equation predicts the observed trends, namely that the mass flow rate is proportional to the density, the square root of the gravitational acceleration and the orifice diameter to the power of 5/2. As with all continuum analyses, this theory can give no information about the effect of particle diameter but we can always invoke the concept of the empty annulus (§10.2) and replace D_0 by $(D_0 - kd)$. Also, being based on the assumption of constant density, this theory cannot provide information about the effect of density changes near the orifice. Finally it should be noted that the hour-glass theory predicts that the flow rate is proportional to $\sin^{-1/2}\alpha$ in contrast to the experimental observation of Rose and Tanaka that W was proportional to $\tan^{-0.35}\alpha$. In fact, there is very little difference between these two quantities since it can readily be confirmed numerically that they are proportional to each other to within 5% over the range $15° < \alpha < 40°$.

However, the main theoretical importance of the hour-glass theory lies in the prediction that the mass flow rate is effectively independent

Table 10.1

y	2	3	4	5
W/W_∞	0.935	0.981	0.992	0.996

of the quantity of material in the hopper. This is in contrast to the minimum energy theory which *assumes* that W is independent of H. The actual dependence of the flow rate on r_1 and hence y can be seen from equation (10.4.11) which we can write in the form

$$W = W_\infty \sqrt{\left(\frac{1 - y^{3 - 2K}}{1 - y^{-2 - 2K}}\right)} \qquad (10.4.15)$$

where W_∞ is the flow rate from a hopper filled to infinite depth. The values of W/W_∞ are given as a function of y in table 10.1 for the case of $K = 3$, i.e. $\phi = 30°$ and it can be seen that W is indistinguishable from W_∞ for $r_1 \geqslant 4r_0$ as observed experimentally.

Similarly, replacing the boundary condition on the top surface to allow for a surcharge Q_0, shows that this has little effect on the flow rate, and both these results are clearly related to the observation, made in §5.8, that the stress distribution in the lower part of a hopper is effectively independent of the height of the material or the presence of a surcharge. Thus we can conclude that it is pointless to try to extrude a granular material through a converging passage by the application of a stress. This merely has the effect of increasing the wall stresses and hence the retarding frictional forces.

We can express equation (10.4.14) in a form similar to the Beverloo correlation

$$W = C(K) \frac{\rho_b \sqrt{g} D_0^{5/2}}{\sin^{1/2}\alpha} \qquad (10.4.16)$$

where

$$C(K) = \frac{\pi}{4} \sqrt{\left[\frac{1 + K}{2(2K - 3)}\right]} \qquad (10.4.17)$$

The dependence of $C(K)$ on K and hence the angle of friction ϕ is given in table 10.2 and it can be seen that $C(K)$ is a mild function of the angle of internal friction, just as observed by Beverloo.

Thus we see that the hour-glass theory gives an excellent qualitative

Table 10.2

ϕ (°)	30	35	40	45	50
K	3.0	3.69	4.60	5.83	7.55
$C(K)$	0.641	0.575	0.528	0.493	0.467

prediction of the observed dependence of the mass flow rate on density, gravitational acceleration, orifice diameter, wall slope and angle of internal friction. Unfortunately, it does not give an accurate quantitative prediction, the values calculated from equation (10.4.14) being typically about twice those observed in practice. One of the reasons for this is the assumption that the hopper walls are smooth. In reality, all walls are rough to some extent and the effect of wall roughness is considered in the following section.

In the mean time it is instructive to consider the stress distribution predicted by equation (10.4.8). Evaluating the arbitrary constant A from the boundary conditions (10.4.9) and back substituting yields

$$\sigma_{rr} = \frac{\rho_b g}{2K - 3}\left[r - \frac{r_0^5}{r^4} - \left(r_1 - \frac{r_0^5}{r_1^4}\right)\left(\frac{r}{r_1}\right)^{2K-2}\right] \quad (10.4.18)$$

It is seen that the dependence of σ_{rr} on r consists of three terms. First, there is a term in r which is analogous to the linear asymptotes of Walker's analysis (§5.8), Enstad's analysis (§5.10) and the radial stress field (§7.4). The term in r^{2K-2} gives the approach to the asymptote from the imposed boundary condition on the top surface. Since, as we have noted above, K is usually > 3, this term dies away rapidly with decreasing r just as in Walker's and Enstad's analyses. The sum of these two terms gives the stress distribution that would occur in the absence of motion and the term in r^{-4} gives the difference between the static and dynamic stress distribution. The fact that this difference is of the order of r^{-4} is inevitable since it reflects the gain in the momentum of the material as it approaches the orifice, the momentum being proportional to the square of the velocity which in turn is inversely proportional to the square of the radius.

Since most of the effects of importance occur in the lower part of the hopper, where the effect of the imposed surcharge can be neglected, we can approximate the stress distribution of equation (10.4.18) by

$$\sigma_{rr} = \frac{\rho_b g}{2K - 3}\left(r - \frac{r_0^5}{r^4}\right) \quad (10.4.19)$$

We see that the region in which the dynamic stress distribution differs significantly from the static stress distribution is confined to the immediate vicinity of the orifice. By $r = 2r_0$ the difference is only one part in 32 and by $3r_0$ it is one part in 243, i.e. the difference falls to about 1% by $r \approx 2.5\ r_0$.

By analogy with equation (10.4.19), we can obtain modifications to Walker's and Enstad's analyses for a flowing material simply by replacing co-ordinate h or r in the linear term by $h - h_0^5/h^4$ or $r - r_0^5/r^4$ respectively. Similarly, we can produce an approximation to the radial stress field in a flowing material by putting

$$\sigma = \gamma \left(r - \frac{r_0^5}{r^4} \right) q(\theta) \tag{10.4.20}$$

where $q(\theta)$ can be calculated by the methods of §7.4.

Finally, we must note that the analysis of this section has been based on the assumption that the stresses fall to zero on the free-fall arch, which we have taken to be a spherical surface of radius r_0. As pointed out in §10.3, the concept of a free-fall arch is suspect and this gives rise to unquantifiable uncertainties in these predictions.

10.5 Rough-walled hoppers

The effect of wall roughness on the discharge rate from a hopper can be analysed in various ways of which we will consider three in this section. First, we will present a modification to Walker's analysis for a conical hopper and then pay brief attention to a similar modification to Enstad's analysis. Finally, and perhaps most satisfactorily, we will consider improvements suggested by Kaza and Jackson (1984), and Thorpe (1984) to the hour-glass analysis of the previous section.

We can adapt Walker's analysis, which was presented in §5.8, for the case of a flowing material simply by including momentum terms in the force balance of equation (5.8.1). We will assume that the vertical component of the velocity is constant across any horizontal cross-section so that we can write

$$v_h = \frac{V}{h^2} \tag{10.5.1}$$

where h is the vertical co-ordinate measured upward from the virtual apex and V is a constant related to the mass flow rate W by,

$$W = \pi (h \tan \alpha)^2 \, \rho_b v_h = \pi \rho_b \, V \tan^2 \alpha \qquad (10.5.2)$$

The momentum flux at height h is given by

$$M = W v_h$$

and hence the rate of change of momentum flux with height is

$$\frac{dM}{dh} = W \frac{dv_h}{dh} = -\frac{2WV}{h^3} = -\frac{2W^2}{\pi \, h^3 \tan^2 \alpha} \qquad (10.5.3)$$

This term can be incorporated into the force balance of equation (5.8.1) giving

$$\pi (h \tan \alpha)^2 \frac{d\sigma_h}{dh} + \pi (h \tan \alpha)^2 \rho_b g - 2\pi (h \tan \alpha)(\tau_{rh})_w$$

$$= \frac{2W^2}{\pi \, h^3 \tan^2 \alpha} \qquad (10.5.4)$$

or

$$\frac{d\sigma_{rr}}{dr} + \rho_b g - \frac{m\sigma_{rr}}{h} = \frac{2W^2}{\pi^2 \rho_b h^5 \tan^4 \alpha} \qquad (10.5.5)$$

where m is defined by equation (5.8.10), or, if the inclusion of a distribution factor is thought desirable, by either equation (5.8.21) or equation (5.8.27).

The solution of equation (10.5.5) follows the same lines as that of equation (5.8.11). The integrating factor is again h^{-m}, giving

$$\frac{d}{dh}(h^{-m}\sigma_{hh}) + \rho_b g h^{-m} = \frac{2W^2}{\pi^2 \rho_b \tan^4 \alpha} h^{-5-m} \qquad (10.5.6)$$

and, on integration, we have

$$h^{-m}\sigma_{hh} + \frac{\rho_b g h^{1-m}}{1-m} = -\frac{2W^2}{\pi^2 \rho_b \tan^4 \alpha} \frac{h^{-4-m}}{4+m} + A \qquad (10.5.7)$$

Following the method of the previous section we can find the particular integral of this equation by saying that σ_{hh} varies linearly with h as $h \to \infty$, giving $A = 0$. We can then impose the boundary condition that the stress σ_{hh} falls to zero on a free-fall arch which we will take to be the horizontal plane at the orifice. This is at height h_0, where h_0 is related to the orifice diameter by

$$D_0 = 2h_0 \tan \alpha \qquad (10.5.8)$$

Substituting into equation (10.5.7) gives

$$W^2 = \frac{\pi^2 \rho_b^2 g h_0^5}{\tan^4 \alpha} \left(\frac{4 + m}{2(m - 1)} \right) \tag{10.5.9}$$

or

$$W = \frac{\pi \rho_b \sqrt{g} D_0^{5/2}}{8 \tan^{1/2} \alpha} \sqrt{\left(\frac{4 + m}{(m - 1)} \right)} \tag{10.5.10}$$

Comparison with equation (10.4.14) of the hour-glass theory shows that apart from the minor difference between $\sin^{1/2}\alpha$ and $\tan^{1/2}\alpha$, the only difference between the predictions is that m replaces $2K - 2$.

We can also obtain the stress distribution in the lower part of the hopper by back-substituting from equation (10.5.9) into (10.5.7), giving

$$\sigma_{hh} = \frac{\rho_b g h}{m - 1} \left[1 - \left(\frac{h_0}{h} \right)^5 \right] \tag{10.5.11}$$

as we deduced by inspection in §10.4. Thus once again we find that the effect of flow on the stress distribution is confined to a very small region immediately above the orifice.

These results suffer from all the deficiencies of Walker's analysis as discussed in §5.8 and §7.5 and furthermore can be criticised on the grounds that we have assumed that the free-fall arch is a horizontal plane at the orifice. In fact, such a free-fall arch avoids many of the difficulties associated with Jackson's paradox (§10.3) but there is no direct evidence that such an arch exists.

A very similar modification can be made to Enstad's analysis which was presented for the static case in §5.10. For the case of zero surcharge and a deeply filled hopper, equation (5.10.16) becomes

$$\sigma_w = \frac{\gamma Y r}{X - 1} \left[1 - \left(\frac{r_0}{r} \right)^5 \right] \tag{10.5.12}$$

and the mass flow rate is given by

$$W = \frac{\pi \rho_b \sqrt{g} D_0^{5/2}}{8} \left[\frac{(1 - \cos \alpha) f(\alpha, \beta)}{\sin^2 \alpha} \left(\frac{X + 4}{X - 1} \right) \right]^{1/2} \tag{10.5.13}$$

We see that in this case X replaces m and an extra geometric term appears due to the assumption that the radial velocity is constant across the surface of the Enstad element as opposed to constant across a spherical surface with centre at the origin in the hour-glass theory and constant across a horizontal plane in the modified Walker method.

A somewhat more convincing analysis of the effect of wall roughness on the flow rate has been presented by Kaza and Jackson (1984) and Thorpe (1984). This is a first order correction to the hour-glass theory presented in §10.4. In the basic theory we assumed that the shear stress $\tau_{\theta r}$ was zero throughout, so that the terms $1/r(\partial\tau_{\theta r}/\partial\theta)$ and $\tau_{\theta r}\cot\theta/r$ appearing in Euler's equation, equation (A2.17), could be omitted. In fact the shear stresses will not be zero but will be given by equation (7.2.16) which, for a cohesionless material takes the form

$$\tau_{\theta r} = p\sin\phi\sin 2\psi^* \qquad (10.5.14)$$

We will now make a rough estimate of these terms by assuming that the shear stress varies linearly with angular position from the value $(\tau_{\theta r})_w$ on $\theta = \alpha$ to zero on the centre-line, $\theta = 0$, so that

$$\tau_{\theta r} = B\theta \qquad (10.5.15)$$

where

$$B = \frac{\partial\tau_{\theta r}}{\partial\theta} \qquad (10.5.16)$$

Recalling that we are confining our attention to narrow-angled hoppers, we can also approximate $\cot\theta$ by $1/\theta$, so that

$$\frac{1}{r}\frac{\partial\tau_{\theta r}}{\partial\theta} + \frac{\tau_{\theta r}\cot\theta}{r} = \frac{B}{r} + \frac{B\theta}{\theta r} = \frac{2B}{r} = \frac{2}{r}\frac{\partial\tau_{\theta r}}{\partial r} \qquad (10.5.17)$$

Ignoring any variation in p with θ, we have from equation (10.5.14) that

$$\frac{\partial\tau_{\theta r}}{\partial\theta} = 2\,p\sin\phi\cos 2\psi^*\frac{\partial\psi^*}{\partial\theta} \approx -2\,p\sin\phi\,\frac{\psi_w^* - \psi_0^*}{\alpha} \qquad (10.5.18)$$

since $\psi^* \approx 90°$. From equation (7.2.14)

$$\sigma_{rr} \approx p(1 - \sin\phi) \qquad (10.5.19)$$

and from equations (7.4.15) and (7.4.16) we have

$$\psi_w^* - \psi_0^* = \tfrac{1}{2}(\omega + \phi_w) \qquad (10.5.20)$$

Thus from (10.5.17) we require an extra term in equation (10.4.6) of magnitude

$$\frac{2B}{r} \approx \frac{2\sigma_{rr} \sin \phi}{1 - \sin \phi} \frac{\omega + \phi_w}{\alpha} \qquad (10.5.21)$$

so that this equation becomes

$$\frac{2\rho_b V^2}{r^5} = \frac{\mathrm{d}\sigma_{rr}}{\mathrm{d}r} + \left(2 - 2K - \frac{2(\omega + \phi_w)\sin \phi}{\alpha(1 - \sin \phi)}\right) \qquad (10.5.22)$$

$$\frac{\sigma_{rr}}{r} + \rho_b g$$

If we define a quantity κ such that

$$\kappa = K + \frac{(\omega + \phi_w) \sin \phi}{\alpha(1 - \sin \phi)} \qquad (10.5.23)$$

equation (10.5.21) becomes

$$\frac{2\rho_b V^2}{r^5} = \frac{\mathrm{d}\sigma_{rr}}{\mathrm{d}r} + (2 - 2\kappa) \frac{\sigma_{rr}}{r} + \rho_b g \qquad (10.5.24)$$

which is identical to equation (10.4.6) except that κ replaces K.

Thus all the results of the hour-glass theory apply, provided κ is substituted for K. It is obvious from equation (10.5.23) that $\kappa > K$ and thus we see from table 10.2 that not only does this modification predict a lower flow rate than the basic hour-glass theory but also, since $\mathrm{d}C(K)/\mathrm{d}K$ is a decreasing function of K, the modified theory predicts that the flow rate is less dependent on the physical properties than the original version. Both of these effects are in accord with experimental observation.

10.6 Air-augmented flows

None of the variables appearing in the Beverloo correlation, equation (10.2.4), is convenient for controlling the discharge from a hopper. At the best the orifice area can be varied by means of a slide valve, but this causes problems due to the jamming of particles in the mechanism. People have therefore tried to control the flow rate by injecting air either above the material or through a porous section of the hopper walls. Interstitial pressure effects also occur when material is being discharged from one process vessel to another at a different pressure or when material is being discharged from an unventilated hopper.

The first major study of the effect of interstitial pressure on the flow rate was that of Bulsara *et al.* (1964) who found experimentally that

$$W \propto (\Delta p)^{0.5} \tag{10.6.1}$$

where Δp is the gauge pressure of the air above the material in the hopper. McDougal and Knowles (1969) modified this to the form

$$W \propto (\Delta p + \Delta p_0)^{0.5} \tag{10.6.2}$$

where Δp_0 was described as the adverse pressure gradient required to prevent flow. Resnick *et al.* (1966), however, pointed out that in the absence of air-augmentation, equation (10.6.2) must reduce to that for gravity flow and therefore the Δp_0 term must be closely related to the prediction of the Beverloo correlation.

Neither of the equations above was derived theoretically, though the analogy with the predictions of Bernoulli's equation was noted. The first attempt at a theoretical derivation was that of Crewdson *et al.* (1977) who proposed a modification to the Beverloo correlation on the following intuitive argument. In gravity flow the material is driven by its weight $\rho_b g$ but in the presence of an interstitial pressure gradient there will be an extra body force equal to the vertical pressure gradient dp/dz which, for a narrow-angled conical hopper, will be approximately equal to the radial pressure gradient dp/dr. Thus in air-augmented flow the term $\sqrt{g}$ appearing in the Beverloo correlation should be replaced by

$$\left(g + \frac{1}{\rho_b} \frac{dp}{dr} \right)^{1/2}$$

The pressure gradient dp/dr is a strong function of position but as we have seen in the previous section the flow rate is determined by conditions close to the orifice. Thus Crewdson argued that dp/dr should be evaluated at the orifice and that the Beverloo/Rose and Tanaka correlation should therefore be modified to the form

$$W = C\rho_b \left[g + \frac{1}{\rho_b} \left(\frac{dp}{dr} \right)_0 \right]^{1/2} (D_0 - kd)^{5/2} F(\alpha, \phi_d) \tag{10.6.3}$$

or

$$W = W_0 \left[1 + \frac{1}{\rho_b g} \left(\frac{dp}{dr} \right)_0 \right]^{1/2} \tag{10.6.4}$$

where $(dp/dr)_0$ is the pressure gradient evaluated at the free fall arch, $r = r_0$, and W_0 is the mass flow rate in the absence of interstitial pressure gradients.

We will now consider the simplest case of a narrow-angled conical hopper subject to a modest pressure difference so that the interstitial gas flow will be small and therefore Darcy's law, equation (6.4.1), will apply. Thus we can say that

$$\frac{dp}{dr} \propto v_R \qquad (10.6.5)$$

where v_R is the relative velocity between the gas and the solid. Since from continuity both the gas and the solid velocities are proportional to r^{-2} we can say that the relative velocity is also proportional to r^{-2} and we can rewrite equation (10.6.5) in the form

$$\frac{dp}{dr} = \frac{A}{r^2} \qquad (10.6.6)$$

where A is a constant. On integrating

$$\Delta p = \int_{r_0}^{r_1} \frac{A}{r^2} \, dr = A \left(\frac{1}{r_0} - \frac{1}{r_1} \right) \qquad (10.6.7)$$

where r_1 is the radius to the top surface of the material and r_0 is the radius of the free-fall arch. Normally $r_1 \gg r_0$ and hence

$$\Delta p \approx \frac{A}{r_0} \qquad (10.6.8)$$

and thus from equation (10.6.6)

$$\frac{dp}{dr} = \frac{r_0 \Delta p}{r^2} \qquad (10.6.9)$$

and

$$\left(\frac{dp}{dr} \right)_0 = \frac{\Delta p}{r_0} \qquad (10.6.10)$$

Equation (10.6.4) now becomes

$$W = W_0 \left(1 + \frac{\Delta p}{\rho_b g r_0} \right)^{1/2} \qquad (10.6.11)$$

which is seen to be of the same form as that suggested by McDougal and Knowles.

Crewdson's intuitive approach can be put on a firmer theoretical basis by modifying the hour-glass theory to make allowance for interstitial pressure gradients. Equation (10.4.6) then becomes

$$\frac{2\rho_b\,V^2}{r^5} = \frac{d\sigma_{rr}}{dr} + \frac{2-2K}{r}\sigma_{rr} + \frac{dp}{dr} + \rho_b g \tag{10.6.12}$$

and on substitution from equation (10.6.9), we have

$$\frac{2\rho_b\,V^2}{r^5} = \frac{d\sigma_{rr}}{dr} + \frac{2-2K}{r}\sigma_{rr} + \frac{r_0\Delta p}{r^2} + \rho_b g \tag{10.6.13}$$

This is seen to be similar in form to equation (10.4.6) and can be solved by the same method, giving

$$W = W_0\left(1 + \frac{2K-3}{2K-1}\frac{\Delta p}{\rho_b g r_0}\right)^{1/2} \tag{10.6.14}$$

and it is seen by comparison with equation (10.6.11) that a correction factor of $(2K-3)/(2K-1)$ has appeared. We have noted that K often has a value in excess of 3, so that the corection factor is in the range 0.6 to 1, telling us that instead of the pressure gradient at the orifice, we require the average pressure gradient over some small region immediately above the orifice. The same analysis can be performed using the Thorpe/Jackson assumption discussed in §10.5 in which case the correction factor becomes $(2\kappa-3)/(2\kappa-1)$ and, since $\kappa > K$ the correction factor is closer to unity.

In the above analyses we have assumed that Darcy's law can be applied, but this is only valid if the gas Reynolds number (defined by equation (6.4.10)) is less than about 10. At higher Reynolds numbers, inertial effects become important and we must use the Ergun equation, equation (6.4.5), which can be written as

$$\frac{dp}{dr} = \frac{150\mu_g v_R(1-\varepsilon)^2}{d^2\varepsilon^2} + \frac{1.75\rho_g v_R^2(1-\varepsilon)}{d^2\varepsilon} \tag{10.6.15}$$

Again v_R is proportional to r^{-2} and hence putting $v_R = B/r^2$ into equation (10.6.15) we have

$$\frac{dp}{dr} = \frac{150\mu_g B(1-\varepsilon)^2}{d^2\varepsilon^2 r^2} + \frac{1.75\rho_g B^2(1-\varepsilon)}{d^2\varepsilon r^4} \tag{10.6.16}$$

On integration, assuming that r_1 is large, we obtain

$$\Delta p = \frac{150\mu_g B(1-\varepsilon)^2}{d^2\varepsilon^2 r_0} + \frac{1.75\rho_g B^2(1-\varepsilon)}{3d^2\varepsilon r_0^3} \tag{10.6.17}$$

Equation (10.6.17) can be solved for B and hence $(dp/dr)_0$ can be evaluated from equation (10.6.16). It can be noted that at high

Reynolds numbers the second term in the Ergun equation dominates and comparing equation (10.6.16) and (10.6.17) shows that

$$\left(\frac{dp}{dr}\right)_0 = \frac{3\Delta p}{r_0} \tag{10.6.18}$$

giving from equation (10.6.11)

$$W = W_0 \left(1 + \frac{3\Delta p}{\rho_g g r_0}\right)^{1/2} \tag{10.6.19}$$

At intermediate Reynolds numbers, solution of equations (10.6.16) and (10.6.17) gives

$$\left(\frac{dp}{dr}\right)_0 = f(Re_0) \frac{\Delta p}{r_0} \tag{10.6.20}$$

where

$$f(Re_0) = \frac{150 + 5.25\, Re_0}{150 + 1.75\, Re_0} \tag{10.6.21}$$

and Re_0 is the Reynolds number, evaluated at the orifice. Hence, from equation (10.6.4),

$$W = W_0 \left(1 + f(Re_0) \frac{\Delta p}{\rho_b g r_0}\right)^{1/2} \tag{10.6.22}$$

Thus a plot of $(W/W_0)^2$ against $\Delta p/\rho_b g r_0$ shows linear asymptotes at low and high Reynolds numbers with slopes of 1 and 3, with a transition zone at intermediate Reynolds numbers as shown in figure 10.4.

In all these analyses we have assumed that we are concerned with a narrow-angled hopper and that both the gas and solid are incompressible. The effect of changing void fraction is considered in §10.7 and Thorpe (1984) has considered the effect of gas compressibility. Due to the complexity of this last analysis, it will not be considered here and the reader is referred to Thorpe's work. For flat-bottomed bunkers the analysis is much more complicated and the problem remains effectively unsolved. Various sweeping assumptions can be made, such as assuming that there is no air percolation through the stagnant zone, in which case the bunker can be treated as a hopper of angle equal to that of the flowing core. In the absence of any other information, this angle can be assumed to be about 45° giving $r_0 = D_0/\sqrt{2}$.

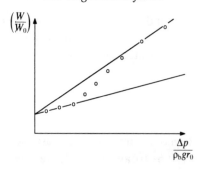

Figure 10.4 Dependence of normalised mass flow rate W/W_0 on normalised pressure drop $\Delta p/\rho_b \, gr_0$.

In order to evaluate Re_0 for use in equation (10.6.21), we must consider the flow of the interstitial air, and this is also necessary if we need to know the quantity of air that must be supplied to achieve a specified increase in the mass flow rate of solid.

The superficial velocity U appearing in the Ergun equation, equation (6.4.10), is related to the interstitial velocity v_i, by

$$U = \varepsilon v_i \qquad (10.6.23)$$

and it is this velocity that must be replaced by the relative velocity in the case of a moving bed.

Assuming that both the gas and solid velocities are independent of angular position, we can say that the relative velocity in a conical hopper is given by

$$v_R = \frac{1}{2\pi r^2(1 - \cos \alpha)} \left(\frac{W}{\rho_s(1 - \varepsilon)} - \frac{G}{\rho_g \varepsilon} \right) \qquad (10.6.24)$$

where G is the mass flow rate of the gas through the hopper. Hence, from equation (10.6.22)

$$U = \frac{1}{2\pi r^2(1 - \cos \alpha)} \left(\frac{W\varepsilon}{\rho_s(1 - \varepsilon)} - \frac{G}{\rho_g} \right) \qquad (10.6.25)$$

For low Reynolds number flow the evaluation of G is easy. The pressure gradient at the orifice is $\Delta p/r_0$ from equation (10.6.10) and hence, using the Carman–Kozeny equation, equation (6.4.3), we have

$$\frac{\Delta p}{r_0} = - \frac{180 \mu_g (1 - \varepsilon)^2}{2\pi r_0^2 d^2 \varepsilon^3 (1 - \cos \alpha)} \left(\frac{W\varepsilon}{(1 - \varepsilon)\rho_s} - \frac{G}{\rho_g} \right) \qquad (10.6.26)$$

from which G can be found directly. It must, however, be appreciated that this is the gas flow rate through the orifice. Gas must also be supplied to fill the space vacated by the solid. Thus the total gas supply rate G_T is given by

$$G_T = G + \frac{W\rho_g}{\rho_s} \qquad (10.6.27)$$

At intermediate Reynolds numbers, we must use the Ergun equation and from equations (10.6.20) and (6.4.12) we have

$$\left(\frac{150 + 5.25\ Re_0}{150 + 1.75\ Re_0}\right) \frac{\Delta p}{r_0} = \frac{\mu_g^2(1 - \varepsilon)^2}{\rho_s d^3 \varepsilon^3} (150\ Re_0 + 1.75\ Re_0^2) \qquad (10.6.28)$$

We can evaluate Re_0 from this equation, hence W from equation (10.6.22) and thus G from equations (10.6.25) and (6.4.10). G_T can then be found from equation (10.6.27).

It should be noted that in many cases the value of the void fraction is not known to great precision. In the case of low Reynolds number flow, this does not affect the accuracy of equation (10.6.11) since the value of the voidage is not required in the derivation of this equation, though it is necessary to assume that the voidage is constant. However, the value of the voidage is required when using equation (10.6.26) to evaluate G and what is more the voidage appears in this equation roughly as its fifth power. Thus we can obtain accurate prediction of the solid flow rate in the absence of an accurate determination of the voidage, but the gas flow rate is very sensitive to uncertainties in the value of ε.

As an alternative to injecting air into the space above the material in the hopper, it can be supplied through a porous section of the wall extending from radius r_a to radius r_b as shown in figure 10.5. Assuming that Darcy's law applies, we can integrate equation (10.6.6) to give

$$p = B - \frac{A}{r} \qquad (10.6.29)$$

and making the assumption that the pressure is independent of angular position we have the boundary conditions, $p = 0$ on $r = r_0$; $p = p_1$ for $r_a < r < r_b$ and $p \to 0$ as $r \to \infty$, where p_1 is the pressure at which the gas is supplied. From these boundary conditions we find that for $r < r_a$

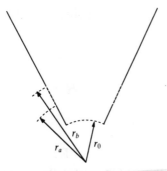

Figure 10.5 Definition of symbols.

$$p = p_1 \frac{r_a(r - r_0)}{r(r_a - r_0)} \qquad (10.6.30)$$

so that

$$\left(\frac{\mathrm{d}p}{\mathrm{d}r}\right)_0 = \frac{p_1 r_a}{r_0(r_a - r_0)} \qquad (10.6.31)$$

and the mass flow rate can be found from equation (10.6.4).

For $r > r_b$ the pressure profile is given by

$$p = \frac{p_1 r_b}{r} \qquad (10.6.32)$$

If the greatest upward pressure gradient exceeds the weight density γ, fluidisation will occur and an easy passage for gas upwards from the injection zone will be formed. This will occur when

$$\frac{p_1}{r_b} = \gamma \qquad (10.6.33)$$

and it is not possible to exceed this pressure by increasing the quantity of air injected.

The quantity of air injected can be found from the Carman–Kozeny equation, equation (6.4.3), using the pressure gradients derived above. Allowance must be made both for the downflow through the orifice and the upflow through the upper parts of the hopper.

The results derived in the earlier parts of this section can be used to analyse the case of an un-ventilated hopper, i.e. one for which the top is sealed so that a volume of air equal to that of the solid discharged must enter through the orifice, counter-current to the flow of solid.

From consideration of equality of volumetric flow rate, we can relate the gas and solid velocities at the orifice, v_g and v_s by

$$v_g \varepsilon = - v_s(1 - \varepsilon) \qquad (10.6.34)$$

and hence the relative velocity v_R is given by

$$v_R = v_g - v_s = - \frac{v_s}{\varepsilon} \qquad (10.6.35)$$

Thus assuming we are in the range of Reynolds number for which the Carman–Kozeny equation applies, we have from equation (6.4.3)

$$\frac{dp}{dl} = - \frac{180 \mu_g v_s (1 - \varepsilon)^2}{d^2 \varepsilon^3} \qquad (10.6.36)$$

Assuming that the solid velocity is independent of angular position, we have that

$$W = 2\pi r_0^2 (1 - \cos \alpha) \, \rho_b \, v_s \qquad (10.6.37)$$

and hence,

$$\left(\frac{dp}{dr}\right)_0 = - \frac{180 \mu_g \, W(1 - \varepsilon)^2}{2\pi r_0^2 \, \rho_s d^2 (1 - \cos \alpha) \, \varepsilon^3} = - CW \qquad (10.6.38)$$

where C is defined by

$$C = \frac{180 \mu_g (1 - \varepsilon)^2}{2\pi r_0^2 \, \rho_s d^2 (1 - \cos \alpha) \, \varepsilon^3} \qquad (10.6.39)$$

Thus from equation (10.6.4)

$$W^2 = W_0^2 \left(1 - \frac{CW}{\rho_b g r_0}\right) \qquad (10.6.40)$$

A correction factor of order $(2K-3)/(2K-1)$ may be included in this equation if great accuracy is required.

Equation (10.6.40) is a quadratic in W and therefore has two roots of which one is inevitably negative and can therefore be dismissed. It can be seen that this equation has two limiting solutions as follows;

(i) When C is small, as will be the case if the particle diameter d is large, W is approximately equal to W_0 and the material discharges at the same rate as from a ventilated hopper. This is because there is negligible resistance to gas flow through the coarse material.

(ii) When C is large, i.e. for fine particles, W will be small so that the terms in the brackets of equation (10.6.40) will be approximately zero, so that

$$W \approx \frac{\rho_b g r_0}{C} = \frac{2\pi r_0^3 \rho_b^2 d^2 g(1 - \cos \alpha)\varepsilon^3}{180\mu_g(1 - \varepsilon)^3} \tag{10.6.41}$$

In this latter case we should strictly have included the correction factor of $(2K-3)/(2K-1)$ and if we do so we see from equation (10.6.14) that

$$\left(\frac{dp}{dr}\right)_0 \approx \frac{2K - 1}{2K - 3}\rho_b g > \rho_b g \tag{10.6.42}$$

Thus fluidisation will occur in the region immediately above the orifice. Since bubbling occurs in gas-fluidised beds, the flow through the orifice will be subject to fluctuations, a phenomenon that is clearly visible in the jet of sand in any hour-glass or egg timer.

10.7 The flow of fine powders

As mentioned in §10.2, the flow rate of particles of diameter less than about 400 µm is less than that predicted by the Beverloo correlation. Although the flow rate still seems to be proportional to $\rho\sqrt{g}D_0^{5/2}$, the parameter C ceases to be about 0.58 and becomes a strong function of the particle diameter.

Carleton (1972) and later Arnold *et al.* (1980) attempted to explain this effect by assuming that the air between the particles is stationary and that there is a drag force on the particles which can be calculated from a drag coefficient and the particle velocity. We will show below that the relative velocity between the particles and the interstitial air is very much less than the particle velocity, since the particles carry the air with them as they move through the hopper. Thus any attempt to calculate the drag force on the particles using their absolute velocity is bound to fail.

If we consider a narrow-angled hopper we can evaluate the mean velocity v_s of the particles at radius r as

$$v_s = \frac{W}{2\pi r^2(1 - \cos \alpha)\rho_s(1 - \varepsilon)} \tag{10.7.1}$$

where ρ_s is the solid density and ε is the void fraction. Similarly, the gas velocity v_g is given by

$$v_g = \frac{G}{2\pi r^2 (1 - \cos \alpha) \rho_g \varepsilon}$$ (10.7.2)

where G is the mass flow rate of the air through the hopper. Hence the relative velocity v_R is given by

$$v_R = \frac{1}{2\pi r^2 (1 - \cos \alpha)} \left(\frac{W}{\rho_s(1 - \varepsilon)} - \frac{G}{\rho_g \varepsilon} \right)$$ (10.7.3)

Assuming for the moment that the void fraction ε is constant, v_R is proportional to r^{-2} and hence from Darcy's law, equation (6.4.1)

$$\frac{dp}{dr} = \frac{A}{r^2}$$ (10.7.4)

as in equation (10.6.6). This integrates to

$$p = B - \frac{A}{r}$$ (10.7.5)

If we consider the case of a material discharging from a hopper that is open to atmosphere at both top, $r = r_1$, and bottom $r = r_0$, we have, $p = p_a$ on $r = r_1$ and $r = r_0$ from which we deduce that $A = 0$ and $B = p_a$.

With $A = 0$, we have from equation (10.7.3) that

$$\frac{W}{\rho_s(1 - \varepsilon)} = \frac{G}{\rho_g \varepsilon}$$ (10.7.6)

and hence from equations (10.7.1) and (10.7.2) that

$$v_s = v_g$$ (10.7.7)

Thus in the absence of voidage changes, the gas and solid flow through the hopper with the same velocities, the interstitial pressure is constant at atmospheric pressure and there is no drag on the particles. Although this result has been derived using Darcy's law, it is equally valid within the Ergun regime.

The observation that the flow rate of fine particles is less than that predicted from the Beverloo correlation shows that some drag force does occur and the analysis above proves that this can only be due to relative velocities induced by the voidage changes.

As a material flows through a hopper, it is subjected to varying stresses as predicted by the analysis of §5.8. The stress is zero at the top surface, passes through a maximum and falls to zero on the free-

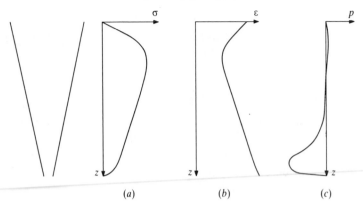

Figure 10.6 (a) Stress, (b) voidage and (c) pressure profiles during the flow of a fine powder.

fall arch as illustrated in figure 10.6(a). Since the material is shearing, the density is free to adjust to that in equilibrium with the local stresses and therefore a voidage profile will occur, which mirrors the stress profile as shown in figure 10.6(b). Thus in the upper part of the hopper the material is being compressed and the interstitital air must be expelled, whereas in the lower part of the hopper the material is dilating and air is being drawn in. As a result, elevated pressures occur in the upper part of the hopper and sub-atmospheric pressures occur in the lower part. However, because the velocities behave roughly as r^{-2}, the magnitude of the elevated pressure peak is reduced to a barely detectable value whereas the pressure deficiency in the lower part of the hopper is magnified, figure 10.6(c).

If we assume for the moment that all materials have similar compressibilities, the quantities of air that have to be expelled or drawn in, and hence the relative velocities, will be roughly the same for all materials. However, with the permeability being proportional to the square of the particle diameter, the magnitudes of the resulting pressure gradients will be inversely proportional to d^2. Thus the pressure gradients due to dilation are negligible for coarse materials but become significant for fine materials.

The ideas presented above have been confirmed experimentally by Verghese (1991) who measured the interstitial pressure profiles. The pressure gradient at the orifice $(dp/dr)_0$ was found to correlate well with the measured flow rate according to equation (10.6.4) and furthermore, the magnitude of the measured pressure gradient increased rapidly with decreasing particle size. He also discovered that for a particular material the pressure profiles measured in hoppers with different orifice diameters could be brought into a common form by plotting $\Delta p/r_0$ vs r/r_0 though there is no obvious reason why this should be so. Accepting this experimental fact we see that the pressure gradient at the orifice is independent of the orifice size so that from equation (10.6.3) the dependence of the flow rate on D_0 remains in the form of the Beverloo correlation. However, we expect that the orifice pressure gradient will be roughly proportional to d^{-2} and Verghese has proposed the correlation

$$W = W_0 \left(1 - \frac{\lambda}{\rho_b g d^2}\right)^{1/2} \qquad (10.7.8)$$

where W_0 is the flow rate predicted from the Beverloo/Rose and Tanaka correlation, equation (10.2.5) and λ is a function of the compressibility of the material and the gas properties. The results for a series of fine sands each with a narrow size distribution were well correlated by $\lambda/\rho_b g = 1.48 \times 10^{-8}$ m^2. Though Verghese's correlation seems promising it has not yet been developed sufficiently to give reliable predictions of the flow rate. There is, for example, no reason to expect that the same value of λ will correlate the behaviour of materials of different compressibility. For materials with a wide size distribution, the appropriate value of d would seem to be that evaluated from the measured permeability using equation (6.4.4) but this has yet to be confirmed experimentally.

10.8 Cohesive arching

The discharge of cohesive materials from hoppers is notoriously unreliable. Often a material will discharge freely, but if the flow is stopped for a while it may be very difficult to restart the discharge. A stable arch of material is formed across the orifice which has to be broken before the material will flow again. Devices as crude as poles and sledge hammers are frequently used and 'hammer rash' is a common sight around the orifices of hoppers storing difficult materials.

Jenike (1961) analysed this situation and the following account is based on his work. As originally presented this took the form of a design calculation, with the objective of determining the minimum orifice diameter for trouble-free flow. The complementary performance calculation, determining whether an existing hopper will arch, is perhaps easier to understand and will be presented first.

The analysis begins by considering the stress distribution in the material when it is freely discharging. Since it is in steady flow the material will be effectively cohesionless as discussed in §6.6 and the effective angle of internal friction must be used when evaluating the stress distribution. Jenike assumed that the stress distribution was given by the radial stress field as derived in §7.4. In fact, the stresses will fall to zero at the orifice, $r = r_0$, as explained in §10.4 and during flow the stresses will differ from the radial stress field by a factor of $1 - (r/r_0)^{-5}$. Jenike's analysis implicitly assumes that the stresses revert to those of the radial stress field when the flow is stopped, though there seems to be no experimental confirmation of this.

As we will see below, the force supporting the arch is that due to the unconfined yield stress acting on the wall, and hence we will assume that it is conditions at the wall that determine the behaviour. We can therefore calculate the major principal stress at the wall from the equations of §7.4, i.e.

$$(\sigma_1)_w = r_0 \gamma q(\alpha)(1 + \sin \phi) \qquad (10.8.1)$$

and it is postulated that the major principal stress remains at this value until flow restarts.

While the material is stationary, consolidation takes place under this principal stress and an unconfined yield stress f_c develops. The value of f_c can be determined from the incipient yield locus of the material and is most conveniently presented in the form of the *Flow Function*, which is a plot of f_c vs σ_1, as described in §6.6. On attempting to restart the flow we must treat the material as being cohesive with an unconfined yield stress determined as above. It is seen that the analysis hinges on the distinction between the effective yield locus which determined the stress distribution during flow and the incipient yield locus which determines whether flow can be restarted.

Compared to the sophistication of the stress analysis during flow, Jenike presents a relatively crude analysis of the forces required to restart flow. He considers an arch such as that shown in figure 10.7 and investigates the stability of a zone of a constant vertical height δz immediately above the arch. Since the normal stress on the lower surface of the arch is clearly zero, the maximum tangential stress is the unconfined yield stress f_c. The force supporting the arch is that due to the unconfined yield stress acting on a plane normal to the arch

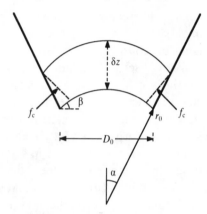

Figure 10.7 Jenike's model for the formation of an arch.

and the forces tending to break the arch are its own weight and the force exerted by the material above the arch. Since we are trying to predict the widest possible span of a stable arch, we will consider the case when this last force is zero.

Denoting the angle between the horizontal and the tangent to the arch at the wall by β, we see that the width of the arch at the wall is $\delta z \cos \beta$ and hence the vertical force due to the unconfined yield stress is, $2\pi(r_0 \sin \alpha) \, \delta z \cos \beta \, f_c \sin \beta$ which can be rearranged to, $\pi r_0 \sin \alpha \, f_c \sin 2\beta \, \delta z$. The largest value this can take is when $\sin 2\beta = 1$, i.e. the value $\pi r_0 \sin \alpha \, f_c \delta z$. The weight of the arch is $\pi(r_0 \sin \alpha)^2 \, \gamma \delta z$ and we can say that if the weight is greater than the supporting force the arch will collapse. Thus the maximum diameter D_m across which a stable arch can form is given by

$$D_m = 2r_0 \sin \alpha = 2\frac{f_c}{\gamma} \qquad (10.8.2)$$

Later semi-empirical modifications of this analysis gave

$$D_m = (2 + 0.0137 \, \alpha)\frac{f_c}{\gamma} \qquad (10.8.3)$$

where α is measured in degrees. A similar analysis for a wedge-shaped hopper gave

$$D_m = \frac{f_c}{\gamma} \qquad (10.8.4)$$

which was later modified to

$$D_m = (1 + 0.004\,67\,\alpha)\frac{f_c}{\gamma} \tag{10.8.5}$$

It must be realised that there are many assumptions in this analysis which gives a highly conservative estimate of the maximum diameter for the formation of a stable arch. Not only have we assumed that the material above the arch exerts no force but we have also assumed that β has its optimum value of $45°$. Neither of these seems to have been confirmed experimentally. It is hard to envisage any mechanism that would give a stable arch over a wider orifice but there is no reason to believe that arching will necessarily occur if $D_0 < D_m$.

The performance calculation is straight-forward. The major principal stress during flow is calculated from the radial stress field, the unconfined yield stress is determined from the flow function and the minimum diameter for trouble-free flow determined from equation (10.8.3). If the actual orifice diameter is greater than this value, no arching will occur.

For design purposes, it is convenient to eliminate the product $r_0\gamma$ from equations (10.8.1 and 10.8.2) giving

$$(\sigma_1)_w = \frac{q(\alpha)(1 + \sin\phi)}{\sin\alpha}f_c = ff\,f_c \tag{10.8.6}$$

where ff is defined by

$$ff = \frac{q(\alpha)(1 + \sin\phi)}{\sin\alpha} \tag{10.8.7}$$

and is, confusingly, known as the flow factor. Jenike (1961), Arnold et al. (1980) and the various codes of practice for silo design give graphs of ff as a function of ϕ, ϕ_w and α to avoid the necessity of the designer having to solve the radial stress field equations himself.

Equation (10.8.6) can be plotted on the flow function graph as a straight line of slope $(ff)^{-1}$ as shown in figure 10.8. It intersects the flow function at some value of f_c and hence the minimum orifice diameter for trouble-free flow can be found from equation (10.8.2).

We noted in §6.6 that for simple materials the flow function is a straight line passing through the origin. Under these circumstances equation (10.8.6) cannot intersect the flow function and this analysis predicts that arching will never occur. Thus this analysis is only appropriate for materials which deviate from the concept of a 'simple'

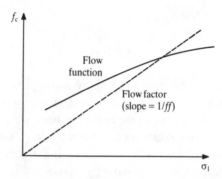

Figure 10.8 Jenike's design method.

material and it is therefore necessary to be very careful indeed when measuring the failure properties of a material. There is an additional philosophic problem. If the material is 'simple' the analysis breaks down and if it is not 'simple' the effective yield locus will not be cohesionless and the radial stress field analysis will not apply. However, materials that undergo time-consolidation, have flow functions which are not straight lines passing through the origin and the present analysis seems much more appropriate for such materials. Caution must still be exercised when determining the flow function to ensure that it is representative of the longest time the material is likely to be held stationary in the bunker. Agricultural products often have to be stored for a year, or even longer and for these materials time-consolidation is difficult to quantify. It is therefore good practice to circulate a small proportion of the material from time to time so that none of it is held stationary for long periods.

10.9 Flooding

The flow of fine powders is often found to be irregular and intermittent. Under normal operation the flow rate may be somewhat slow for the reasons explained in §10.7 but, suddenly, very much larger flows can occur causing severe operating difficulties. The reasons for this are not fully understood, but the argument below, which is based on the work of Rathbone and Nedderman (1987) may provide a partial explanation.

Flooding seems to occur with materials that are prone to rat-holing. This is an extreme case of core flow in which the material in a narrow flow channel discharges completely, leaving most of the material in the hopper as shown in figure 10.9. Rathbone and Nedderman postulate

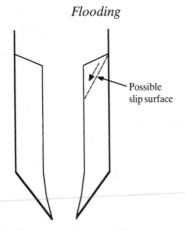

Figure 10.9 Sloughing of material into an empty rat-hole.

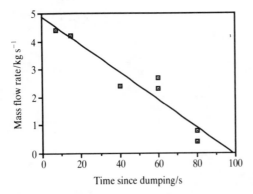

Figure 10.10 Dependence of the mass flow rate and the time elapsed since the dumping of a fine material.

that flooding occurs when the upper part of the rat-hole becomes unstable and material falls down the rat-hole.

Rathbone and Nedderman have demonstrated experimentally that when a mass of material is dumped, i.e. falls under gravity and hits the base of a bunker, considerable dilation and aeration of the material can occur. The material de-aerates at a rate which is determined by the permeability and compressibility of the material. For coarse materials the de-aeration is almost instantaneous but for finer material the rate of de-aeration is slow as a result of the low permeability. Their analysis shows that the de-aeration time is roughly proportional to H^2/k where H is the depth of the material and k is the permeability.

Before de-aeration is complete, the interstitial air is under an elevated pressure with the result that if the orifice is opened air-augmented flow will take place. Rathbone and Nedderman measured the solid flow rate as a function of the time elapsed since the material was dumped. Figure 10.10 shows the results for a 20 μm powder and it can be seen that the material discharges freely for a period shortly after dumping but that the discharge rate decreases rapidly with time eventually falling to zero. The initial rate, immediately after dumping was found to be close to that for an inviscid fluid of the same density.

It may be that the sudden increases in flow rate found when discharging fine powders is due to spontaneous aeration by a mechanism similar to that described by Rathbone and Nedderman but it has not been confirmed experimentally that collapse of a rat-hole inevitably precedes flooding.

Problems

Chapter 2

2.1 A sample of material is subjected to the following stresses; $\sigma_{xx} = 50$ kN m^{-2}, $\sigma_{yy} = 120$ kN m^{-2}, $\tau_{xy} = -\tau_{yx} = 60$ kN m^{-2}. Find the magnitude and direction of the principal stresses and evaluate the greatest shear stress.
   Ans. 154.46 kN m^{-2}, 15.54 kN m^{-2}, the major principal stress being 29.87° anticlockwise from the y-direction. 69.46 kN m^{-2}.

2.2 The velocities in a two-dimensional flow pattern are given by

$$u = 0.1(x^2 - y^2), \quad v = -0.2xy$$

Find the strain rates $\dot{\epsilon}_{xx}$, $\dot{\epsilon}_{yy}$ and $\dot{\gamma}_{xy}$ at the point $x = 0.5$ m, $y = 0.2$ m. Show that the flow is incompressible and evaluate the largest shear strain rate at this point.
   Ans. -0.1 s^{-1}, $+0.1$ s^{-1}, -0.08 s^{-1}, 0.215 s^{-1}.

2.3 A velocity distribution is given in spherical co-ordinates by

$$v_r = -\frac{0.06}{r^2}\cos 2\theta, \quad v_\theta = v_\chi = 0.$$

Evaluate the strain rates $\dot{\epsilon}_{rr}$, $\dot{\epsilon}_{\theta\theta}$, $\dot{\epsilon}_{\chi\chi}$ and $\dot{\gamma}_{r\theta}$ at the point $r = 0.5$ m, $\theta = 20°$. Find also the magnitude and direction of the principal strain rates.
   Ans. -0.735 s^{-1}, 0.368 s^{-1}, 0.368 s^{-1}, 0.617 s^{-1}, 0.448 s^{-1}, 0.368 s^{-1}, -0.816 s^{-1} 12.5° anticlockwise from the θ-direction.

Chapter 3

3.1 The major principal stress at a point within a cohesionless material with angle of friction 25° is 1.5 kN m^{-2}. Evaluate the minor principal stress and the angles between the minor principal plane and the slip planes.
   Ans. 0.609 kN m^{-2}, $\pm 32.5°$.

3.2 A material with $c = 1.8$ kN m^{-2} and $\phi = 35°$ slips when subjected to a shear stress of 5.0 kN m^{-2} along the slip plane. Determine the normal stress on the slip plane and the principal stresses.
   Ans. 4.57 kN m^{-2}, 14.18 kN m^{-2}, 1.97 kN m^{-2}.

3.3 A material has $\rho = 1500$ kg m^{-3}, $\phi = 15°$ and $c = 5$ kN m^{-2}. What are the stresses 1.0 m below a horizontal surface in (a) the active state and (b) the passive state? What is the greatest depth to which tension cracks can develop in the active state?
Ans. (a) 14.7 kN m^{-2}, 0.99 kN m^{-2}; (b) 14.7 kN m^{-2}, 38.0 kN m^{-2}; 89 cm.

3.4 Determine the maximum stable height of a surface inclined at 70° to the horizontal if the material has $c = 1.5$ kN m^{-2}, $\phi = 35°$ and $\gamma = 17$ kN m^{-3}. At what angle to the horizontal will the slip plane form?
Ans. 1.50 m, 52.5°.

3.5 A vertical trench is dug in a material with weight density 15 kN m^{-3}. When the trench reaches a depth of 50 cm, slip planes are formed at an angle of 62° to the horizontal. Determine the angle of friction and the cohesion.
Ans. 34°, 1.0 kN m^{-2}.

3.6 The stresses on the x-plane in a material with $c = 3.0$ kN m^{-2} and $\phi = 35°$ are $\sigma_{xx} = 7.5$ kN m^{-2} and $\tau_{xy} = -1.3$ kN m^{-2}. Evaluate the stress parameters p^* and ψ.
Ans. Either $p^* = 7.62$ kN m^{-2}, $\psi = 8.66°$ or $p^* = 27.5$ kN m^{-2}, $\psi = 87.64°$.

3.7 A cohesionless material with $\phi = 35°$ is in passive failure adjacent to a right-hand vertical wall with $\phi_w = 20°$. Determine the angle measured anticlockwise from the wall plane to (a) the major principal plane and (b) the slip planes.
Ans. (a) $-28.30°$; (b) 34.2°, 89.2°.

Chapter 4

4.1 A fully rough vertical wall supports a horizontal fill of a cohesionless material of weight density 12.0 kN m^{-3} and angle of internal friction 32°. Find the normal and shear forces on the top 2 m of wall and the stresses at a depth of 3 m.
Ans. 5.64 kN m^{-1}, 3.52 kN m^{-1}, 8.46 kN m^{-2}, 5.29 kN m^{-2}.

4.2 A vertical wall of height 2.5 m supports a horizontal fill of a cohesionless material of weight density 8.0 kN m^{-3} and angle of internal friction 35°. If the angle of wall friction is 20°, find the inclination of the slip plane through the base of the wall and the components of the resulting force on the wall. Find also the stresses at depth h.
Ans. 59.39°, 5.76 kN m^{-1}, 2.10 kN m^{-1}, 1.84 h kN m^{-2}, 0.67 h kN m^{-2}.

4.3 Repeat problem 4.2 using Reimbert's rule-of-thumb.
Ans. 56.67°, 5.71 kN m^{-1}, 2.08 kN m^{-1}, 1.83 h kN m^{-2}, 0.66 h kN m^{-2}.

4.4 A cohesionless material of bulk density 1300 kg m^{-3} and internal angle of friction 30° is stored behind a wall inclined at 15° to the vertical, with the material being on the underside of the wall so that in the terminology of figure 4.6, $\eta = -15°$. Find the inclination of the slip plane and the normal and shear forces on the wall down to a (vertical) depth of 3 m. The angle of wall friction is 25°.
Ans. 50°, 10.3 kN m^{-1}, 4.8 kN m^{-1}.

4.5 A vertical wall with angle of wall friction 22° supports a cohesionless material of angle of internal friction 30° and bulk density 800 kg m^{-3}.

The top surface of the fill is inclined upwards from the top of the wall at 10° to the horizontal. Evaluate the inclination of the slip planes and the wall stresses at a depth of 2.0 m.
Ans. 52.7°, 4.94 kN m^{-2}, 2.00 kN m^{-2}.

4.6 A cohesionless material is supported by a vertical fully rough wall. If the top surface of the material slopes upwards at the angle of repose, show that the slip planes are inclined at ϕ to the horizontal and that the normal stress on the wall at depth h is $\gamma h \cos^2 \phi$.

4.7 A cohesionless material of bulk density 1 300 kg m^{-3} and angle of internal friction 30° is supported by a vertical wall of height 2 m and angle of wall friction 20°. The top surface of the material rises from the top of the wall at the angle of repose through a vertical height of 1 m and is then horizontal. Confirm that the inclination of the slip plane through the base of the wall is at about 51° to the horizontal and hence find the normal and shear forces on the wall.
Ans. 12.0 kN m^{-1}, 4.36 kN m^{-1}.

4.8 A cohesionless granular material of weight density γ and angle of internal friction 33° is supported by a vertical wall of angle of friction 25°. The horizontal top surface of the fill is subjected to a surcharge γx where x is distance from the top of the wall. Show that the normal stress on the wall at depth h is 0.414 γh.

4.9 A vertical wall of height 2.5 m and angle of wall friction 20° supports a horizontal fill of a cohesionless material of weight density 8.0 kN m^{-3} and angle of internal friction 35°. The top surface is subjected to a concentrated load of (a) 1.0 kN m^{-1} and (b) 5.0 kN m^{-1} at a distance of 2.0 m from the top of the wall. Using the results of question 4.2, find the normal force on the wall.
Ans. (a) 5.76 kN m^{-1}; (b) 6.62 kN m^{-1}.

4.10 Parallel vertical walls with a spacing of 1.0 m and height 2.5 m support a cohesionless material of weight density 8.0 kN m^{-3} and angle of internal friction 35°. The angle of wall friction is 20°. Using the results of question 4.2, show that the slip plane from the base of one of the walls meets the other wall at a depth of about 0.67 m from the top and hence evaluate the normal force on each of the walls.
Ans. 5.61 kN m^{-1}.

Chapter 5

5.1 A cylindrical bunker of diameter 1.5 m and height 10 m is filled with a cohesionless material of weight density 8.0 kN m^{-3} and angle of internal friction 35°. The angle of wall friction is 20°. Using the simplest form of Janssen's analysis, find the wall stresses at a depth of (a) 1.0 m and (b) at the bottom. Find also the force on the base of the hopper and the vertical force in the wall at its base.
Ans. (a) 1.91 kN m^{-2}, 0.69 kN m^{-2}; (b) 7.65 kN m^{-2}, 2.78 kN m^{-2}; 49.9 kN, 91.5 kN.

5.2 Revise the answers to problem 5.1 using Walker's method of evaluating the stress ratio K_w but assuming a distribution factor of 1.
Ans. (a) 2.03 kN m^{-2}, 0.74 kN m^{-2}; (b) 7.75 kN m^{-2}, 2.82 kN m^{-2}; 47.1 kN, 94.3 kN.

5.3 Evaluate the distribution factor for the situation of problem 5.2 and hence produce improved answers.

Ans. 0.964; (a) 1.96 kN m^{-2}, 0.71 kN m^{-2}; (b) 7.70 kN m^{-2}, 2.80 kN m^{-2}; 48.6 kN, 92.8 kN.

5.4 Wheat is stored in a parallel-sided bunker of hexagonal cross-section of side 2 m and height 15 m. The wheat has a bulk density of 850 kg m^{-3} and angle of internal friction 22°. The walls may be considered fully rough. Using Walker's modification of Janssen's analysis but with a distribution factor of 1, find the stress on the base of the bunker and the normal and shear forces on each of the side walls.
Ans. 23.6 kN m^{-2}, 435 kN, 176 kN.

5.5 A material of weight density 13.0 kN m^{-3} and angle of internal friction 30° is stored in a tall bunker of diameter 1.6 m and angle of wall friction 20°. Find the great depth values of the radial, shear and axial stresses at the wall. Assuming that the radial stress is constant across the bunker evaluate the axial stress and the shear stress τ_{rz} at the centre-line and at a radius of 0.4 m. Hence find the inclinations of the slip planes to the horizontal at the wall, at $r = 0.4$ m and on the centre-line.
Ans. 14.3 kN m^{-2}, 5.20 kN m^{-2}, 38.6 kN m^{-2}; 42.9 kN m^{-2}, 0; 41.9 kN m^{-2}, 2.60 kN m^{-2}; 71.6°, 48.4°; 65.3°, 54.7°; 60°, 60°.

5.6 Parallel vertical walls with a spacing of 1.0 m and height 2.5 m support a cohesionless material of weight density 8.0 kN m^{-3} and angle of internal friction 35°. The angle of wall friction is 20°. Using Walker's modification of the Janssen analysis but with a distribution factor of 1, find the normal and shear stresses at the base of the walls and hence the total normal force on each of the walls. Active failure may be assumed. Compare the answer with that of question 4.10.
Ans. 4.51 kN m^{-2}, 1.64 kN m^{-2}, 6.14 kN.

5.7 The walls in question 5.6 are pushed together until passive failure occurs and the material is about to be expelled upwards. Calculate the stresses at the base of the walls.
Ans. 353 kN m^{-2}, 128 kN m^{-2}.

5.8 A cohesionless material of bulk density 1800 kg m^{-3} and angle of internal friction 40° is stored in a conical hopper of half-angle 15° and angle of wall friction 20°. Assuming a distribution factor of 1, evaluate the stresses on the wall at a height of 0.6 m above the apex if the hopper is filled to a depth of (a) 4 m and (b) 0.8 m.
Ans. 3.85 kN m^{-2}, 2.99 kN m^{-2}.

5.9 The base of a bin is in the form of a square pyramid, the sides of which are inclined at 10° to the vertical and have an angle of wall friction of 18°. The bin is loaded to great depth with a cohesionless material of weight density 12.0 kN m^{-3} and angle of internal friction 35°. Estimate the wall stresses at a height of 0.5 m above the apex.
Ans. 1.54 kN m^{-2}, 0.50 kN m^{-2}.

5.10 A cohesionless material of weight density 11.0 kN m^{-3} and angle of internal friction 33° is stored in a cylindrical bunker of diameter 2.0 m. The angle of wall friction is 22°. A switch plane is formed at a depth of 2.5 m. Using Walker's method for predicting wall stresses, but with a distribution factor of 1, and using Walters' switch stress analysis, evaluate the wall stresses immediately above the switch plane, immediately below the switch plane and 0.5 m below the switch plane.
Ans. 6.58 kN m^{-2}, 2.66 kN m^{-2}; 31.9 kN m^{-2}, 12.9 kN m^{-2}; 22.3 kN m^{-2}, 8.99 kN m^{-2}.

5.11 A cohesionless material of weight density 11.0 kN m^{-3} and angle of internal friction of 33° is stored in a bin with angle of wall friction 22°. The bin consists of a cylindrical section of diameter 2.0 m surmounting a conical hopper of half-angle 10°. The bin is filled to a height of 2.5 m above the hopper/bunker transition plane. Assuming a horizontal switch plane at the height of the hopper/bunker transition and using the results of question 5.10 to give the stresses above the switch, predict the wall stresses immediately below the switch plane. *Ans.* 25.8 kN m^{-2}, 10.4 kN m^{-2}.

Chapter 6

6.1 A material has the following sieve analysis,

Mesh size (μm)	1000	900	700	600	500	420	250
Wt% retained	2.0	5.0	39.0	20.6	22.6	6.9	3.7

Confirm that this conforms approximately to the log-normal distribution with $D_{gm} \approx 655$ μm and $\sigma \approx 0.228$.

Using these values find the geometric mean diameter, the surface mean diameter and the volumetric mean diameter.
Ans. 560 μm, 590 μm, 606 μm.

6.2 Analyse the material of question 6.2 in terms of the Rosin–Rammler distribution and evaluate k and D_R.
Ans. 5.5, 743 μm.

6.3 Determine the permeability of the material of question 6.1 for a fluid of viscosity 0.89 × 10^{-3} N s m^{-2} when compacted to a voidage of 0.36.
Ans. 2.90 × 10^{-7} m^4 N^{-1} s^{-1}.

6.4 Air with a viscosity 1.7 ×10^{-5} N s m^{-2} and density 1.29 kg m^{-3} flows at a superficial velocity of 0.05 m s^{-1} through a bed of spherical particles of diameter 450 μm and voidage 0.39. Confirm that the flow is laminar and predict the pressure gradient using (a) the Carman–Kozeny equation and (b) the Ergun equation.
Ans. (a) 4740 N m^{-3}; (b) 4080 N m^{-3}.

6.5 Air with viscosity 1.7 × 10^{-5} N s m^{-2} and density 1.29 kg m^{-3} percolates through a bed of spherical particles of diameter 700 μm contained in a tube of diameter 0.12 m. The height of the bed is 0.5 m and the voidage is 0.38. Determine the volumetric flow rate when the overall pressure drop is 0.1 bar.
Ans. 4.45 × 10^{-3} m^3 s^{-1}.

6.6 A granular material of diameter 600 μm and voidage 0.4 is contained in a conical hopper of half-angle 15°. The top and bottom surfaces of the material can be considered to be parts of spherical surfaces centred on the virtual apex and with radii 0.3 m and 0.03 m. Evaluate the flow rate of air (viscosity 1.7 × 10^{-5} N s m^{-2} and density 1.29 kg m^{-3}) through the bed under the influence of an overall pressure drop of 0.1 bar. Find also the pressure gradient at the lower surface of the bed.
Ans. 7.73 × 10^{-4} m^3 s^{-1}, 7.27 × 10^5 N m^{-3}.

6.7 Tests on a material consolidated under a normal stress of 8.7 kN m^{-2} gave the following results for the incipient yield locus,

Normal stress σ (kN m^{-2})	0.4	0.7	1.3	3.4	4.7	6.3	8.7
Shear stress τ (kN m^{-2})	1.0	1.1	1.4	2.1	2.6	3.0	3.6

Separate tests at the same degree of consolidation gave an ultimate tensile stress of 1.2 kN m^{-2}. Evaluate the parameters n, C and T in the Warren Spring Equation.

Find also the unconfined yield stress, the termination locus angle, the effective angle of internal friction and the principal consolidating stresses.

Ans. 1.4, 0.81 kN m^{-2}, 1.2 kN m^{-2}; 2.55 kN m^{-2}, 22.5°, 22.8°, 13.4 kN m^{-2}, 5.9 kN m^{-2}.

6.8 The material of question 6.7 is stored in a cylindrical bunker of diameter 1.0 m. The wall is adhesionless and has an angle of wall friction of 25°. Assuming that the density of the material is 750 kg m^{-3} and that it is consolidated as in question 6.7, find the magnitudes of the principal stresses at great depth and the inclinations of the slip planes at the wall.

Ans. 9.0 kN m^{-2}, 3.3 kN m^{-2}, 50° anticlockwise and 10° clockwise from the wall plane.

6.9 The following results were obtained for a material in an annular shear cell.

When consolidated under a normal stress of 1620 N m^{-2}

Normal Stress (N m^{-2})	175	495	790	1095	1620
Shear Stress (N m^{-2})	480	725	920	1125	1500

When consolidated under a normal stress of 3100 N m^{-2}

Normal Stress (N m^{-2})	175	790	1620	2345	3100
Shear Stress (N m^{-2})	820	1310	1800	2300	2930

Show that the material is adequately described as a 'simple' Coulomb material and estimate the cohesion and angle of friction when consolidated under a normal stress of 2500 N m^{-2}. Find also the end point locus angle and the effective angle of friction during sustained yield.

Ans. 566 N m^{-2}, 35°, 42.8°, 43.4°.

Chapter 7

7.1 The stress parameters and co-ordinates of two points A and B in a cohesionless material of angle of internal friction 35° and weight density 13.0 kN m^{-3} are

A $x = 0.02$ m $y = 0.10$ m $p = 1.5$ kN m^{-2} $\psi = 90°$
B $x = 0.01$ m $y = 0.09$ m $p = 1.4$ kN m^{-2} $\psi = 95°$

Find the co-ordinates and stress parameters at the point of intersection of the β-characteristic through A with the α-characteristic through B and also at the point of intersection of the β-characteristic through B with a vertical wall with an angle of wall friction of 25° at $x = 0$. Assume active failure.

Ans. 0.0169 m, 0.1057 m, 1.454 kN m^{-2}, 92.52°; 0, 0.1040 m, 1.286 kN m^{-2}, 101.23°.

7.2 A vertical wall of angle of friction 20° supports a cohesionless material of angle of internal friction 35° in active failure. The horizontal top surface is subjected to a surcharge of 3.0 kN m^{-2}. Evaluate the width of the fan, the values of p on the top surface and at the top of the wall and the wall stresses at the top of the wall.

Ans. 8.302°, 1.906 kN m^{-2}, 1.556 kN m^{-2}, 0.701 kN m^{-2}, 0.255 kN m^{-2}.

7.3 If the wall of question 7.2 is inclined at 8.303° to the vertical so that there is no fan, evaluate the wall stresses at the top of the wall.
Ans. 0.858 kN m^{-2}, 0.312 kN m^{-2}.

7.4 If the wall of question 7.2 is inclined at 16° to the vertical, determine the direction of the plane of the discontinuity, the value of p below the discontinuity and the wall stresses at the top of the wall.
Ans. 31.53° clockwise from the x-plane, 2.299 kN m^{-2}, 1.035 kN m^{-2}, 0.377 kN m^{-2}.

7.5 A cohesionless material of weight density 11.0 kN m^{-3} and internal angle of friction 33° is supported by a wall with angle of friction 22°. At some point, the stress state switches from active to passive. On the active side of the switch the normal stress on the wall is 6.58 kN m^{-2}. Find the wall stress on the passive side of the switch assuming that this takes the form of a stress discontinuity. Compare your answer with that of question 5.10.
Ans. 40.69 kN m^{-2}.

7.6 Reconsider problem 7.5 assuming that the switch takes the form of a fan of characteristics.
Ans. 47.60 kN m^{-2}.

7.7 A cohesionless material of weight density 12.0 kN m^{-3} and angle of internal friction 35° is stored in a bin consisting of a cylindrical bunker of diameter 2.0 m surmounting a conical hopper of half-angle 20°. The angle of wall friction for both parts of the bin is 22°. The bin is filled to great depth and active failure may be assumed in the bunker and passive failure in the hopper. Assuming a stress discontinuity occurs at the bunker/hopper junction, evaluate the wall stresses at (a) the base of the bunker and at (b) the top of the hopper.
Ans. (a) 14.85 kN m^{-2}, 6.00 kN m^{-2}; (b) 80.26 kN m^{-2}, 32.43 kN m^{-2}.

7.8 Reconsider problem 7.7, assuming that a fan occurs.
Ans. (a) 14.85 kN m^{-2}, 6.00 kN m^{-2}; (b) 83.81 kN m^{-2}, 33.86 kN m^{-2}.

Chapter 8

8.1 In a region in which $\psi = 90°$, the velocities at and co-ordinates of two points A and B are

A	$x = 0.01$ m	$y = 0.800$ m	$u = 0.01$ m s^{-1}	$v = 0.75$ m s^{-1}
B	$x = 0.02$ m	$y = 0.802$ m	$u = 0.02$ m s^{-1}	$v = 0.77$ m s^{-1}

Find the co-ordinates and velocity (a) at the point of intersection of the α-characteristic through A with the β-characteristic through B and (b) at the point of intersection of the β-characteristic through A with a smooth vertical wall at $x = 0$.
Ans. (a) 0.016 m, 0.806 m, 0.005 m s^{-1}, 0.755 m s^{-1}; (b) 0, 0.810 m, 0, 0.740 m s^{-1}.

8.2 Re-evaluate problem 8.1 if the value of ψ varies with position according to $\psi = 101.68 - 400x$.
Ans. (a) 0.0130 m, 0.8039 m, 0.0059 m s^{-1}, 0.7532 m s^{-1}; (b) 0, 0.8071 m, 0, 0.7359 m s^{-1}.

Chapter 9

9.1 A cohesionless material with angle of internal friction 30° flows radially through a conical hopper of half-angle 15° and angle of wall friction 20°. At some point on the wall the normal stress $\sigma_{\theta\theta}$ is 2.0 kN m^{-2}. Assuming that the material obeys the Conical yield function, evaluate $\tau_{r\theta}$, σ_{rr}, σ_{xx} and the principal stresses at this point.
Ans. 0.728 kN m^{-2}, 1.19 kN m^{-2}, 2.0 kN m^{-2}, 2.428 kN m^{-2}, 2.0 kN m^{-2}, 0.762 kN m^{-2}.

Chapter 10

10.1 Using the Beverloo correlation, find the discharge rate of a material of bulk density 800 kg m^{-3} consisting of spherical particles of diameter 400 μm through an orifice of diameter 25 mm in the base of a wide cylindrical bunker.
Ans. 0.135 kg s^{-1}.

10.2 The material of question 10.1 discharges through a 25 mm diameter orifice in the base of a conical hopper of half-angle 20°. Using the Rose and Tanaka correlation with $\phi_d = 45°$, find the flow rate.
Ans. 0.192 kg s^{-1}.

10.3 The material of question 10.1 discharges through a 40 mm × 15 mm rectangular orifice in the base of a wide cylindrical bin. Evaluate the flow rate.
Ans. 0.152 kg s^{-1}.

10.4 The air above the material in the hopper of question 10.2 is pressurised to a gauge pressure of 500 N m^{-2}. Assuming laminar flow and neglecting the Mushin and Harrison correction, predict the mass flow rate of the material. Find the Reynolds number at the orifice if the voidage is 0.4 and hence confirm that the flow is laminar. Find also the volumetric outflow of air and the volumetric flow rate of air that must be supplied to the top of the hopper. For air the viscosity is 1.7×10^{-5} N s m^{-2} and the density is 1.29 kg m^{-3}.
Ans. 0.320 kg s^{-1}, 6.43, 2.24×10^{-4} m^3 s^{-1}, 4.64×10^{-4} m^3 s^{-1}.

10.5 Glass ballotini of diameter 1.5 mm, solid density 2940 kg m^{-3} and void fraction 0.39, discharge from a wide wedge-shaped hopper with walls inclined at 15° to the vertical. The mass flow rate per unit width W is found to depend on the spacing b of the walls at their base as follows,

b (mm)	10	15	20	30	40
W (kg m^{-1} s^{-1})	4.92	10.6	17.6	34.7	55.2

Correlate these data and hence predict the mass flow rate when $b = 17$ mm.
Ans. $C = 1.36$, $k = 1.67$, 13.3 kg s^{-1}.

10.6 The hopper of question 10.5 with $b = 17$ mm is filled to a depth of 0.5 m and the air above the fill is pressurised to 300 N m^{-2} gauge. Evaluate the mass flow ratte. For air the viscosity is 1.7×10^{-5} N s m^{-2} and the density 1.29 kg m^{-3}.
Ans. 15.0 kg s^{-1}.

10.7 When discharging freely from a conical hopper of half-angle 25° with orifice diameter of 30 mm, the flow rate of 0.3 mm ballotini of bulk density 1800 kg m^{-3} and void fraction 0.4 was found to be

0.619 kg s^{-1}. Find the flow rate that would occur if the top of the hopper was sealed. Assume laminar flow of the interstitial air, which has a viscosity of $1.7 \times 10^{-5} \text{ N s m}^{-2}$.

Ans. 0.119 kg s^{-1}.

10.8 A 200 μm sand of bulk density 1800 kg m^{-3} discharges through a 50 mm orifice on the base of a conical hopper of half-angle 20°. Using the Beverloo and Rose and Tanaka correlations with $\phi_d = 45°$, predict the mass flow rate. Compare this with the flow rate predicted by Verghese's correlation.

Ans. 2.56 kg s^{-1}; 1.61 kg s^{-1}.

10.9 A material with bulk density 800 kg m^{-3} discharges through a 90 mm orifice in the base of a conical hopper of half-angle 20°. Assuming that the particle size is small compared with the orifice diameter predict the mass flow rate. The mass flow rate is increased by injecting air through nozzles at a radius of 0.4 m from the virtual apex. The air injection rate is such that the pressure at this radius is 0.02 bar gauge. Assuming laminar flow of the air, determine the resulting flow rate.

Ans. 5.03 kg s^{-1}; $9.96 \text{ kg s}^{-1.}$

Appendix 1

Resolution of forces and stresses by matrix or tensorial methods

The components of a force $\mathbf{F}$ in one set of Cartesian co-ordinates (x,y,z) can be related to those in another set (u,v,w) by the cosine law of vector resolution, which can be written in the matrix form

$$\begin{pmatrix} F_u \\ F_v \\ F_w \end{pmatrix} = \begin{pmatrix} \cos\theta_{xu} & \cos\theta_{yu} & \cos\theta_{zu} \\ \cos\theta_{xv} & \cos\theta_{yv} & \cos\theta_{zv} \\ \cos\theta_{xw} & \cos\theta_{yw} & \cos\theta_{zw} \end{pmatrix} \begin{pmatrix} F_x \\ F_y \\ F_z \end{pmatrix} \tag{A1.1}$$

which on multiplying out gives the familiar expressions

$$F_u = F_x \cos\theta_{xu} + F_y \cos\theta_{yu} + F_z \cos\theta_{zu} \tag{A1.2}$$

etc.

In these expressions θ denotes the angle between the appropriate axes, i.e. θ_{xu} is the angle between the x- and u-axes. It is convenient to denote these direction cosines by the symbol a so that

$$a_{xu} = \cos\theta_{xu} \text{ etc.} \tag{A1.3}$$

We can therefore define a rotation matrix $\mathbf{L}$ such that

$$\mathbf{L} = \begin{pmatrix} a_{xu} & a_{yu} & a_{zu} \\ a_{xv} & a_{yv} & a_{zv} \\ a_{xw} & a_{yw} & a_{zw} \end{pmatrix} \tag{A1.4}$$

and write equation (A1.1) in the form

$$\mathbf{F}' = \mathbf{L}\,\mathbf{F} \tag{A1.5}$$

where $\mathbf{F}$ is the force expressed in (x,y,z) co-ordinates and $\mathbf{F}'$ is the force expressed in (u,v,w) co-ordinates.

Equation (A1.2) can be written in the form

$$F_u = F_x a_{xu} + F_y a_{yu} + F_z a_{zu} \tag{A1.6}$$

338

It may be noted that we have the sum of three terms, in each of which one of the subscripts x, y or z appears twice. Einstein noted that summation under these circumstances is more common than the absence of summation and in tensorial notation summation is assumed whenever a subscript is repeated. Thus using Einstein's summation convention we can write equation (A1.6) as

$$F_u = F_x a_{xu} \tag{A1.7}$$

or

$$F_l = F_i a_{il} \tag{A1.8}$$

where the subscript i permutes to all of x, y and z and l permutes to all of u, v and w.

A stress is defined so that the product of stress with area gives a force. Thus we can write

$$\mathbf{F} = \boldsymbol{\sigma} \mathbf{A} \tag{A1.9}$$

or

$$\begin{pmatrix} F_x \\ F_y \\ F_z \end{pmatrix} = \begin{pmatrix} \sigma_{xx} & \sigma_{xy} & \sigma_{xz} \\ \sigma_{yx} & \sigma_{yy} & \sigma_{yz} \\ \sigma_{zx} & \sigma_{zy} & \sigma_{zz} \end{pmatrix} \begin{pmatrix} A_x \\ A_y \\ A_z \end{pmatrix} \tag{A1.10}$$

where we have used the symbol σ for both normal and shear stresses. This equation requires the sign convention that the stress acting on an area facing greater values of the co-ordinate is positive if the associated force acts in the direction of the co-ordinate increasing. In this convention normal stresses are positive when tensile and complementary shear stresses occur in equal pairs. This differs from the sign convention used throughout the rest of the book and in two dimensions the sign conventions impose different signs on all the stress components except σ_{xy} ($=\tau_{xy}$) which has the same sign in the two conventions.

We can multiply equation (A1.9) by the rotation matrix $\mathbf{L}$ and insert the quantity $\mathbf{L}^{-1}\mathbf{L} = \mathbf{I}$ between $\boldsymbol{\sigma}$ and $\mathbf{A}$ giving

$$\mathbf{LF} = \mathbf{L\sigma L}^{-1}\mathbf{LA} \tag{A1.11}$$

But from (A1.5), $\mathbf{LF} = \mathbf{F}'$ and similarly $\mathbf{LA} = \mathbf{A}'$. Thus equation (A1.11) becomes

$$\mathbf{F}' = \mathbf{L\sigma L}^{-1}\mathbf{A}' \tag{A1.12}$$

which can be compared with the definition of stress in the new co-ordinate system

$$\mathbf{F}' = \boldsymbol{\sigma}'\mathbf{A}' \tag{A1.13}$$

showing that

$$\boldsymbol{\sigma}' = \mathbf{L\sigma L}^{-1} \tag{A1.14}$$

We have thereby devised the rule for the resolution of stresses analogous to that for forces given by equation (A1.5).

Since the determinant of the matrix $\mathbf{L}$ is inevitably equal to unity, the transpose and inverse of $\mathbf{L}$ are equal, so that equation (A1.14) is more conveniently written as

$$\boldsymbol{\sigma}' = \mathbf{L}\boldsymbol{\sigma}\mathbf{L}^T \qquad (A1.15)$$

or

$$\begin{pmatrix} \sigma_{uu} & \sigma_{uv} & \sigma_{uw} \\ \sigma_{vu} & \sigma_{vv} & \sigma_{vw} \\ \sigma_{wu} & \sigma_{wv} & \sigma_{ww} \end{pmatrix} = \begin{pmatrix} a_{xu} & a_{yu} & a_{zu} \\ a_{xv} & a_{yv} & a_{zv} \\ a_{xw} & a_{yw} & a_{zw} \end{pmatrix} \begin{pmatrix} \sigma_{xx} & \sigma_{xy} & \sigma_{xz} \\ \sigma_{yx} & \sigma_{yy} & \sigma_{yz} \\ \sigma_{zx} & \sigma_{zy} & \sigma_{zz} \end{pmatrix} \begin{pmatrix} a_{xu} & a_{xv} & a_{xw} \\ a_{yu} & a_{yv} & a_{yw} \\ a_{zu} & a_{zv} & a_{zw} \end{pmatrix} \qquad (A1.16)$$

Multiplying out this equation gives

$$\sigma_{lm} = a_{li}a_{mj}\sigma_{ij} \qquad (A1.17)$$

where i and j permute to all of x, y and z, l and m permute to all of u, v and w and summation with respect to both i and j is assumed.

In the two-dimensional system shown in figure 2.3, θ_{xu} is denoted by θ, θ_{xv} is given by $90 + \theta$, θ_{yu} is $90 - \theta$ and θ_{yv} is θ. Thus

$$\mathbf{L} = \begin{pmatrix} \cos\theta & \cos(90-\theta) \\ \cos(90+\theta) & \cos\theta \end{pmatrix} = \begin{pmatrix} \cos\theta & \sin\theta \\ -\sin\theta & \cos\theta \end{pmatrix} \qquad (A1.18)$$

From equation (A1.5) we see that

$$\begin{pmatrix} F_u \\ F_v \end{pmatrix} = \begin{pmatrix} \cos\theta & \sin\theta \\ -\sin\theta & \cos\theta \end{pmatrix} \begin{pmatrix} F_x \\ F_y \end{pmatrix} \qquad (A1.19)$$

i.e.

$$F_u = F_x \cos\theta + F_y \sin\theta \qquad (A1.20)$$

$$F_v = -F_x \sin\theta + F_y \cos\theta \qquad (A1.21)$$

which is the two-dimensional equivalent of equation (A1.2).

Similarly, from equation (A1.16),

$$\begin{pmatrix} \sigma_{uu} & \sigma_{uv} \\ \sigma_{vu} & \sigma_{vv} \end{pmatrix} = \begin{pmatrix} \cos\theta & \sin\theta \\ -\sin\theta & \cos\theta \end{pmatrix} \begin{pmatrix} \sigma_{xx} & \sigma_{xy} \\ \sigma_{yx} & \sigma_{yy} \end{pmatrix} \begin{pmatrix} \cos\theta & -\sin\theta \\ \sin\theta & \cos\theta \end{pmatrix} \qquad (A1.22)$$

or, on multiplying out,

$$\sigma_{uu} = \sigma_{xx} \cos^2\theta + \sigma_{xy} \sin\theta \cos\theta + \sigma_{yx} \sin\theta \cos\theta + \sigma_{yy} \sin^2\theta \qquad (A1.23)$$

$$\sigma_{uv} = -\sigma_{xx} \sin\theta \cos\theta + \sigma_{xy} \cos^2\theta - \sigma_{yx} \sin^2\theta + \sigma_{yy} \sin\theta \cos\theta \qquad (A1.24)$$

Apart from the differences in sign resulting from the use of two incompatible sign conventions, equations (A1.23) and (A1.24) are seen to be identical to equations (2.3.2) and (2.3.3).

It can also be shown that the principal stresses are the Eigen-values of the stress matrix and that the principal stress directions are given by the corresponding Eigen-vectors. Furthermore, it can be shown that all combinations of σ and τ, i.e. σ_{uu} and $\sqrt{(\sigma_{uv}^2 + \sigma_{uw}^2)}$ lie within the shaded region ABC of figure 2.7.

Appendix 2

Euler's equation and rates of strain

In §2.4, we derived Euler's equation for a two-dimensional set of Cartesian co-ordinates and in this appendix we list this result together with the corresponding results for other co-ordinate systems. The axes and the directions in which the velocities are taken to be positive are shown in figure A2. No attempt is made to derive these results since the derivations can be found in many standard texts. The reader is, however, cautioned to expect certain differences in sign since the sign convention for stresses used

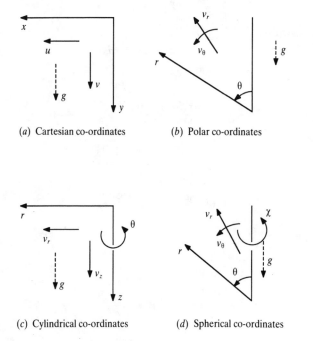

(a) Cartesian co-ordinates   (b) Polar co-ordinates

(c) Cylindrical co-ordinates   (d) Spherical co-ordinates

Figure A.2 Definition of co-ordinate directions and associated velocities.

in this book (defined in §2.3) differs from that normally used in books on elasticity or fluid mechanics. In these equations, normal stresses are taken to be positive when compressive and shear stresses in the first quadrant are taken to be positive when acting anticlockwise on the element.

The rates of strain in the various co-ordinate systems are also given in terms of the appropriate velocity gradients. The sign convention for strain rates is chosen so that positive stresses cause positive rates of strain.

Two-dimensional Cartesian coordinates (x,y) with the y-axis being directed vertically downwards

$$\rho_b\left(\frac{\partial u}{\partial t} + u\frac{\partial u}{\partial x} + v\frac{\partial u}{\partial y}\right) + \frac{\partial \sigma_{xx}}{\partial x} + \frac{\partial \tau_{yx}}{\partial y} = 0 \tag{A2.1}$$

$$\rho_b\left(\frac{\partial v}{\partial t} + u\frac{\partial v}{\partial x} + v\frac{\partial v}{\partial y}\right) - \frac{\partial \tau_{xy}}{\partial x} + \frac{\partial \sigma_{yy}}{\partial y} - \rho_b g = 0 \tag{A2.2}$$

$$\dot{\varepsilon}_{xx} = -\frac{\partial u}{\partial x} \tag{A2.3}$$

$$\dot{\varepsilon}_{yy} = -\frac{\partial v}{\partial y} \tag{A2.4}$$

$$\dot{\gamma}_{xy} = -\dot{\gamma}_{yx} = \frac{\partial u}{\partial y} + \frac{\partial v}{\partial x} \tag{A2.5}$$

Two-dimensional polar co-ordinates (r, θ) with the line θ = 0 directed vertically upwards

$$\rho_b\left(\frac{\partial v_r}{\partial t} + v_r\frac{\partial v_r}{\partial r} + \frac{v_\theta}{r}\frac{\partial v_r}{\partial \theta} - \frac{v_\theta^2}{r}\right)w9$$

$$+ \frac{\partial \sigma_{rr}}{\partial r} + \frac{1}{r}\frac{\partial \tau_{\theta r}}{\partial \theta} + \frac{\sigma_{rr} - \sigma_{\theta\theta}}{r} + \rho_b g \cos\theta = 0 \tag{A2.6}$$

$$\rho_b\left(\frac{\partial v_\theta}{\partial t} + v_r\frac{\partial v_\theta}{\partial r} + \frac{v_\theta}{r}\frac{\partial v_\theta}{\partial \theta} + \frac{v_r v_\theta}{r}\right)$$

$$- \frac{\partial \tau_{r\theta}}{\partial r} + \frac{1}{r}\frac{\partial \sigma_{\theta\theta}}{\partial \theta} - \frac{2\tau_{r\theta}}{r} + \rho_b g \sin\theta = 0 \tag{A2.7}$$

$$\dot{\varepsilon}_{rr} = -\frac{\partial v_r}{\partial r} \tag{A2.8}$$

$$\dot{\varepsilon}_{\theta\theta} = -\frac{1}{r}\frac{\partial v_\theta}{\partial \theta} - \frac{v_r}{r} \tag{A2.9}$$

$$\dot{\gamma}_{r\theta} = -\dot{\gamma}_{\theta r} = \frac{\partial v_\theta}{\partial r} - \frac{v_\theta}{r} + \frac{1}{r}\frac{\partial v_r}{\partial \theta} \tag{A2.10}$$

Axi-symmetric cylindrical co-ordinates (r,z,θ) with the z-axis directed
vertically downwards

$$\rho_b\left(\frac{\partial v_r}{\partial t} + v_r\frac{\partial v_r}{\partial r} + v_z\frac{\partial v_r}{\partial z}\right) + \frac{\partial \sigma_{rr}}{\partial r} + \frac{\sigma_{rr}-\sigma_{\theta\theta}}{r} + \frac{\partial \tau_{zr}}{\partial z} = 0 \qquad \text{(A2.11)}$$

$$\rho_b\left(\frac{\partial v_z}{\partial t} + v_r\frac{\partial v_z}{\partial r} + v_z\frac{\partial v_z}{\partial r}\right) - \frac{\partial \tau_{rz}}{\partial r} - \frac{\tau_{rz}}{r} + \frac{\partial \sigma_{zz}}{\partial z} - \rho_b g = 0 \qquad \text{(A2.12)}$$

$$\dot{\varepsilon}_{rr} = -\frac{\partial v_r}{\partial r} \qquad \text{(A2.13)}$$

$$\dot{\varepsilon}_{zz} = -\frac{\partial v_z}{\partial z} \qquad \text{(A2.14)}$$

$$\dot{\varepsilon}_{\theta\theta} = -\frac{v_r}{r} \qquad \text{(A2.15)}$$

$$\dot{\gamma}_{rz} = -\dot{\gamma}_{zr} = \frac{\partial v_r}{\partial z} + \frac{\partial v_z}{\partial r} \qquad \text{(A2.16)}$$

Axi-symmetric spherical co-ordinates (r,θ,χ) with the line $\theta = 0$
vertically upwards

$$\rho_b\left(\frac{\partial v_r}{\partial t} + v_r\frac{\partial v_r}{\partial r} + \frac{v_\theta}{r}\frac{\partial v_r}{\partial \theta} - \frac{v_\theta^2}{r}\right) + \frac{\partial \sigma_{rr}}{\partial r}$$

$$+\frac{1}{r}\frac{\partial \tau_{\theta r}}{\partial \theta} + \frac{2\sigma_{rr}-\sigma_{\theta\theta}-\sigma_{\chi\chi}}{r} + \frac{\tau_{\theta r}\cot\theta}{r} + \rho_b g\cos\theta = 0 \qquad \text{(A2.17)}$$

$$\rho_b\left(\frac{\partial v_\theta}{\partial t} + v_r\frac{\partial v_\theta}{\partial r} + \frac{v_\theta}{r}\frac{\partial v_\theta}{\partial \theta} + \frac{v_r v_\theta}{r}\right) - \frac{\partial \tau_{r\theta}}{\partial r}$$

$$+\frac{1}{r}\frac{\partial \sigma_{\theta\theta}}{\partial \theta} - \frac{3\tau_{r\theta}}{r} + \frac{(\sigma_{\theta\theta}-\sigma_{\chi\chi})\cot\theta}{r} + \rho_b g\sin\theta = 0 \qquad \text{(A2.18)}$$

$$\dot{\varepsilon}_{rr} = -\frac{\partial v_r}{\partial r} \qquad \text{(A2.19)}$$

$$\dot{\varepsilon}_{\theta\theta} = -\frac{1}{r}\frac{\partial v_\theta}{\partial \theta} - \frac{v_r}{r} \qquad \text{(A2.20)}$$

$$\dot{\varepsilon}_{\chi\chi} = -\frac{v_r}{r} - \frac{v_\theta\cot\theta}{r} \qquad \text{(A2.21)}$$

$$\dot{\gamma}_{r\theta} = -\dot{\gamma}_{\theta r} = \frac{\partial v_\theta}{\partial r} - \frac{v_\theta}{r} + \frac{1}{r}\frac{\partial v_r}{\partial \theta} \qquad \text{(A2.22)}$$

Appendix 3

Mohr's circle for rate of strain

Let us take a set of Cartesian axes (x,y) and a second set (p,q) where the p-axis is inclined at angle θ anticlockwise from the x-axis as in figure A3. This is entirely equivalent to the axes (u,v) shown in figure 2.3 but p and q are used in this appendix to avoid confusion with the velocity components. We will denote the velocities in the (x,y) plane by (u,v) and those in the (p,q) plane by (U,V).

Since velocities are vectors

$$U = u \cos \theta + v \sin \theta \qquad (A3.1)$$

$$V = -u \sin \theta + v \cos \theta \qquad (A3.2)$$

Similarly, gradients are also vectors so that

$$\frac{\partial \alpha}{\partial p} = \frac{\partial \alpha}{\partial x} \cos \theta + \frac{\partial \alpha}{\partial y} \sin \theta \qquad (A3.3)$$

$$\frac{\partial \alpha}{\partial q} = -\frac{\partial \alpha}{\partial x} \sin \theta + \frac{\partial \alpha}{\partial y} \cos \theta \qquad (A3.4)$$

where α is any variable.

By analogy with equation (2.5.2), the strain rate $\dot{\varepsilon}_{pp}$ is defined by

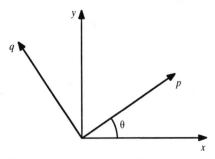

Figure A.3 Relationship between the (x,y) and (p,q) axes.

$$\dot{\varepsilon}_{pp} = -\frac{\partial U}{\partial p} = -\frac{\partial u}{\partial p}\cos\theta - \frac{\partial v}{\partial p}\sin\theta \tag{A3.5}$$

and substituting from equation (A3.3) gives

$$\dot{\varepsilon}_{pp} = -\frac{\partial u}{\partial x}\cos^2\theta - \frac{\partial u}{\partial y}\sin\theta\cos\theta - \frac{\partial v}{\partial x}\sin\theta\cos\theta - \frac{\partial v}{\partial y}\sin^2\theta \tag{A3.6}$$

Thus from equations (2.5.2) to (2.5.4), we have

$$\dot{\varepsilon}_{pp} = \dot{\varepsilon}_{xx}\cos^2\theta - \dot{\gamma}_{xy}\sin\theta\cos\theta + \dot{\varepsilon}_{yy}\sin^2\theta \tag{A3.7}$$

or

$$\dot{\varepsilon}_{pp} = \tfrac{1}{2}(\dot{\varepsilon}_{xx}+\dot{\varepsilon}_{yy}) + \tfrac{1}{2}(\dot{\varepsilon}_{xx}-\dot{\varepsilon}_{yy})\cos 2\theta - \frac{\dot{\gamma}_{xy}}{2}\sin 2\theta \tag{A3.8}$$

i.e.

$$\dot{\varepsilon}_{pp} = e + R\cos(2\theta+\Lambda) \tag{A3.9}$$

where

$$e = \tfrac{1}{2}(\dot{\varepsilon}_{xx}+\dot{\varepsilon}_{yy}) \tag{A3.10}$$

$$R^2 = \tfrac{1}{4}[(\dot{\varepsilon}_{xx}-\dot{\varepsilon}_{yy})^2+\dot{\gamma}_{xy}^2] \tag{A3.11}$$

and

$$\tan\Lambda = \frac{\dot{\gamma}_{xy}}{(\dot{\varepsilon}_{xx}-\dot{\varepsilon}_{yy})} \tag{A3.12}$$

We can manipulate the shear strain rate $\dot{\gamma}_{pq}$ in the same way, giving

$$\frac{\dot{\gamma}_{pq}}{2} = R\sin(2\theta+\Lambda) \tag{A3.13}$$

Thus it can be seen that the locus of $(\dot{\varepsilon}_{pp}, \dot{\gamma}_{pq}/2)$ is a circle with centre $(e,0)$ and radius R as shown in figure 2.9. This is exactly analogous to Mohr's stress circle derived in §2.3. It should, however, be noted that the axes of the strain rate circle are the direct strain rate, $\dot{\varepsilon}$, and *one-half* of the shear strain rate, i.e. $\dot{\gamma}/2$. This results from the rather unfortunate definition of shear strain rate normally used. With shear strain rate being defined by

$$\dot{\gamma}_{yx} = -\left(\frac{\partial u}{\partial y} + \frac{\partial v}{\partial x}\right) \tag{A3.14}$$

the compatible definition of the direct strain rate $\dot{\varepsilon}_{yy}$ should be obtained by replacing x by y and u by v giving

$$\dot{\varepsilon}_{yy} = -2\frac{\partial v}{\partial y} \tag{A3.15}$$

This differs from the conventional definition, given by equation (A2.4), by a factor of two, which accounts for the factor of one-half that appears in the ordinate of Mohr's strain rate circle.

References and bibliography

Abbott, M. B. (1966) *An Introduction to the Method of Characteristics*, Thames and Hudson.

Abramowitz, M. and Stegun, I. A. (1965) *Handbook of Mathematical Functions*, Dover.

Airy, W. (1897) *Proc. Instn. Civ. Eng.*, **131**, 347.

Al Din, N. and Gunn, D. J. (1984) *Chem. Engng. Sci.* **39**, 121.

American Concrete Institute (1983) *Recommended Practice for Design of and Construction of Concrete Bins, Silos and Bunkers for Storing Granular Materials*, Detroit.

Arnold, P. C., McLean, A. G. and Roberts, A. W. (1980) *Bulk Solids; Storage, Flow and Handling*, TUNRA Ltd., University of Newcastle, New South Wales.

Atkinson, J. H. and Bransby, P. L. (1978) *The Mechanics of Soils*, McGraw Hill.

Bagster, D. F. and Nedderman, R. M. (1985) *Powder Technol.* **42**, 193.

Baldwin, A. J. and Hampson, H. A. (1988) Tripos Project Report, Department of Chemical Engineering, University of Cambridge.

Benink, E. J. (1989) *Flow and Stress Analysis of Cohesionless Bulk Materials in Silos Related to Codes*, Enschede.

Beverloo, W. A., Leniger, H. A. and Van de Velde, J. (1961) *Chem. Engng. Sci.* **15**, 260.

Birks, A. H. and Williams, J. C. (1967) *Powder Technol.* **1**, 189.

BMHB (1987) *Silos; Draft Design Code for Silos, Bins, Bunkers and Hoppers*.

Bransby, P. L. and Blair-Fish, P. M. (1974) *Chem. Engng. Sci.* **29**, 1061.

Briscoe, B. J. and Adams, M. J. (1987) *Tribology in Particulate Technology*, Adam Hilger.

Brown, R. L. and Hawksley, P. W. G. (1947) *Fuel* **26**, 159.

Brown, R. C. and Richards, J. C. (1970) *Principles of Powder Mechanics*, Pergamon.

Bulsara, P. U., Zeng, F. A. and Eckert, R. S. (1964) *Ind. Engng. Chem.* **3**, 348.

Capper, L. P. and Cassie, W. F. (1963) *The Mechanics of Engineering Soils*, Fourth Edition, Spon.

Carleton, A. J. (1972) *Powder Technol.* **6**, 91.

Carman, P. C. (1937) *Trans. Inst. Chem. Eng.* **15**, 150.

Cleaver, J. A. S. (1991) Ph.D. Thesis, University of Cambridge.
Coulomb, C. A. (1776) *Mem. de Math. de l'Acad. Royale des Science* **7**, 343.
Cousens, T. W. (1980) Ph.D. Thesis, University of Cambridge.
Crank, J. (1956) *Mathematics of Diffusion*, Oxford.
Crewdson, B. J., Ormond, A. L. and Nedderman, R. M. (1977) *Powder Technol.* **16**, 197.
Darcy, H. P. G. (1856) *Les Fontaines Publiques de la Ville de Dijon*, Victor Dalamont.
Davidson, J. F. and Nedderman, R. M. (1973) *Trans. Inst. Chem. Eng.* **51**, 29.
Davies, S. T., Horton, D. J. and Nedderman, R. M. (1980) *Powder Technol.* **25**, 215.
DIN 1055, (1981) Deutches Institut für Normung.
Drescher, A., Cousens, T. W. and Bransby, P. L. (1978) *Geotechnique* **28**, 27.
EFCE Working Party (1989) *Standard Shear Testing Techniques for Particulate Solids using the Jenike Shear Cell*, I. Chem. E.
Enstad, G. (1975) *Chem. Engng. Sci.* **30**, 1273.
Ergun, S. (1952) *Chem. Engng. Prog.* **48**, 89.
Fayed, M. E. and Otten, L. (1984) *Handbook of Powder Science and Technology*, Van Nostrand Reinhold.
Fowler, R. T. and Glastonbury, J. G. (1959) *Chem. Engng. Sci.* **10**, 150.
Hancock, A. W. (1970) Ph.D. Thesis, University of Cambridge.
Hancock, A. W. and Nedderman, R. M. (1974) *Trans. Inst. Chem. Eng.* **52**, 29.
Horne, R. M. (1977) Ph.D. Thesis, University of Cambridge.
Horne, R. M. and Nedderman, R. M. (1976) *Powder Technol.* **14**, 93.
Horne, R. M. and Nedderman, R. M. (1978a) *Powder Technol.* **19**, 235.
Horne, R. M. and Nedderman, R. M. (1978b) *Powder Technol.* **19**, 243.
Janssen, H. A. (1895) *Z. Ver. Dt. Ing.* **39**, 1045.
Jenike, A. W. (1961) *Gravity Flow of Bulk Solids*, Bulletin 108, Utah Engineering Station.
Jenike, A. W. (1967) *Flow and Storage of Solids*, Bulletin 123, Utah Engineering Station.
Jenike, A. W. (1987) *Powder Technol.* **50**, 229.
Jenike, A. W. and Johanson, J. R. (1962) *Stress and Velocity Fields in Gravity Flow of Bulk Solids*, Bulletin 116, Utah Engineering Station.
Johanson, J. R. (1965) *J. Appl. Mech.* **32**, 842.
Kay, J. M. and Nedderman, R. M. (1985) *Fluid Mechanics and Transfer Processes*, Cambridge University Press.
Kaza, K. R. (1982) Ph.D. Thesis, University of Houston.
Kaza, K. R. and Jackson, R. (1984) *Powder Technol.* **39**, 915.
Kocova, S. and Pilpel, N. (1971–2) *Powder Technol.* **5**, 329.
Kozeny, J. S. B. (1927) *Akad. Wiss. Wein* **136**, 271.
Kvapil, R. (1964) *Aufber. Techn.* **5**, 139.
Lindley, D. V. and Miller, J. C. P. (1953) Cambridge Elementary Statistical Tables, Cambridge University Press.
Litwiniszyn, J. (1971) Symposium Franco-Polonais, Warsaw.
McDougal, I. R. and Knowles, G. H. (1969) *Trans. Inst. Chem. Eng.* **47**, 73.
Molerus, O. (1982) *Powder Technol.* **33**, 81.

Moreea, S. (1990) CPGS Report, University of Cambridge.
Mróz, Z. and Szymanski, C. Z. (1971) *Arch. Mech.* **23**, 897.
Mullins, W. W. J. (1972) *Appl. Phys.* **43**, 665.
Myers, M. E. and Sellers, M. (1977) Final Year Project Report, Department of Chemical Engineering, Cambridge.
Nedderman, R. M. (1978) *Powder Technol.* **19**, 287.
Nedderman, R. M. and Tüzün, U. (1979*a*) *Powder Technol.* **22**, 243.
Nedderman, R. M. and Tüzün, U. (1979*b*) *Powder Technol.* **24**, 256.
Pitman, E. B. (1986) *Powder Technol.* **47**, 219.
Prakesh, J. R. and Rao, K. K. (1988) *Chem. Engng. Sci.* **43**, 479.
Prakesh, J. R. and Rao, K. K. (1991) *J. Fluid Mech.* **225**, 21.
Rankine, W. J. M. (1857) *Phil. Trans. Royal Soc.*
Rathbone, T. and Nedderman, R. M. (1987) *Powder Technol.* **51**, 115.
Reimbert, M. and Reimbert, A. (1976) *Silos—Theory and Practice*, Trans. Tech. Publications.
Reisner, W. and Rothe, M. von E. (1971) *Bins and Bunkers for Handling Bulk Materials*, Trans. Tech. Publications.
Resnick, W., Heled, Y., Klein, A. and Palm, E. (1966) *Ind. Engng. Chem.* **5**, 392.
Richards, J. C. (1966) *The Storage and Recovery of Particulate Solids*, I. Chem. E.
Rose, H. F. and Tanaka, T. (1956) *The Engineer (London)* **208**, Oct. 23.
Savage, S. B. (1967) *Br. J. Mech. Sci.* **16**, 1885.
Schofield, A. and Wroth, P. (1968) *Critical State Soil Mechanics*, McGraw Hill.
Schwedes, J. (1968) *Fliessverhalten von Schuttgetern in Bunkern*, Verlag Chemie.
Shamlou, P. A. (1988) *Handling of Bulk Solids: Theory and Practice*, Butterworths.
Sokolovskii, V. V. (1965) *Statics of Granular Materials*, Pergamon Press, Oxford.
Sullivan, W. N. (1972) Ph.D. Thesis, Californian Institute of Technology.
Svarovski, L. (1978) *Powder Testing Guide: Methods of Measuring the Physical Properties of Bulk Powders*, Elsevier.
Terzaghi, K. (1943) *Theoretical Soil Mechanics*, Wiley.
Thorpe, R. B. (1984) Ph.D. Thesis, University of Cambridge.
Tüzün, U. (1979) Ph.D. Thesis, University of Cambridge.
Verghese, T. M. (1991) Ph.D. Thesis, University of Cambridge (in preparation).
Walker, D. M. (1966) *Chem. Engng. Sci.* **21**, 975.
Walters, K. J. (1973) *Chem. Engng. Sci.* **28**, 779.
Walters, K. J. and Nedderman, R. M. (1973) *Chem. Engng. Sci.* **28**, 1907.
Wilms, H. (1984) Dr. Ing. Thesis, University of Braunschweig.

Index

Note: **bold page numbers indicate main mention**